Methods in Molecular Biology

Series Editor
John M. Walker
School of Life and Medical Sciences
University of Hertfordshire
Hatfield, Hertfordshire, UK

For further volumes:
http://www.springer.com/series/7651

For over 35 years, biological scientists have come to rely on the research protocols and methodologies in the critically acclaimed *Methods in Molecular Biology* series. The series was the first to introduce the step-by-step protocols approach that has become the standard in all biomedical protocol publishing. Each protocol is provided in readily-reproducible step-by-step fashion, opening with an introductory overview, a list of the materials and reagents needed to complete the experiment, and followed by a detailed procedure that is supported with a helpful notes section offering tips and tricks of the trade as well as troubleshooting advice. These hallmark features were introduced by series editor Dr. John Walker and constitute the key ingredient in each and every volume of the *Methods in Molecular Biology* series. Tested and trusted, comprehensive and reliable, all protocols from the series are indexed in PubMed.

Bioconjugation

Methods and Protocols

Edited by

Sam Massa

Cellular and Molecular Immunology Laboratory, Vrije Universiteit Brussel, Brussels, Belgium

Nick Devoogdt

In Vivo Cellular and Molecular Imaging Laboratory (ICMI), Vrije Universiteit Brussel, Brussels, Belgium

Humana Press

Editors
Sam Massa
Cellular and Molecular Immunology Laboratory
Vrije Universiteit Brussel
Brussels, Belgium

Nick Devoogdt
In Vivo Cellular and Molecular Imaging Laboratory (ICMI)
Vrije Universiteit Brussel
Brussels, Belgium

ISSN 1064-3745 ISSN 1940-6029 (electronic)
Methods in Molecular Biology
ISBN 978-1-4939-9653-7 ISBN 978-1-4939-9654-4 (eBook)
https://doi.org/10.1007/978-1-4939-9654-4

This Humana imprint is published by the registered company Springer Science+Business Media, LLC, part of Springer Nature.
The registered company address is: 233 Spring Street, New York, NY 10013, U.S.A.

Preface

Precision medicine is an expanding field in drug discovery research. The growing knowledge of the molecular basis of diseases allows us to increasingly develop smart diagnostic and therapeutic compounds that help in the clinical management of patients. Current drugs are most frequently *targeted*, i.e., are directed to diseased cells, proteins, and pathways, and thus only active at sites where the disease is manifested. This *personalized* approach allows us to bring the right drug to the right patient at the right time, thereby maximally inducing curative effects while avoiding toxicities. In the current precision medicine we target the disease with *magic bullets*, very frequently of proteinaceous nature (e.g., antibodies, antibody fragments, engineered scaffold proteins, peptides). By covalently conjugating the targeting biological to a diagnostic or therapeutic effectuator (e.g., dye, label, toxin, enzyme, cytokine) through a so-called bioconjugation reaction, we generate new compounds that can bring a new functionality to the site of action. In the past decade, the number of new and exciting bioconjugation methods has grown enormously, as described in some excellent recent reviews on this topic [1–3]. New methods tend to focus on methodologies that generate more homogenous products that are more stable in vivo, more *site-specific*, and more efficacious.

This book attempts to provide a comprehensive overview of well-established and emerging conjugation strategies that are relevant for proteins used in the field of precision medicine. All the contributions appear under the form of detailed protocols, which are directly applicable for the interested researcher, each including the typical notes section that provides detailed "tips and tricks" to help the newcomers in the field to maximize their success rate. A focus is to include techniques that have proven, among others, to be suitable for antibodies (Chapters 1, 2, 4, 5, 7, 10, 15, and 18); antibody fragments, such as Fabs, scFvs, or nanobodies (Chapters 6, 8, 9, and 16); scaffold proteins such as FN3 or DARPin (Chapters 14, 17, and 20); or peptides (Chapters 13 and 19) that are used as targeting agents in molecular diagnosis and therapy. Also emerging techniques are included that are demonstrated on model proteins such as albumin, calmodulin, ubiquitin, annexin V, and GFP (Chapters 3, 11, 12, and 16).

Several protocols (Chapters 2, 3, 15, 18, and 20) describe conjugation methods that are applicable to native proteins (i.e., no modification of the protein is required); others require some protein engineering, through the introduction or mutation of single residues (Chapters 1, 4, 5, 16–18) or the fusion of the targeting protein with a genetic tag or polypeptide (Chapters 6–14, and 19).

Traditional bioconjugation reactions deploy surface-exposed lysine and cysteine side chains, and these anchor sites, either naturally or genetically introduced, are also used in this book: Chapters 3, 4, 5, and 20 target the amine groups of lysines, while Chapters 1, 2, and 17 utilize thiol groups of cysteines. One protocol targets the canonical glycosylation of antibodies (Chapter 15), and three protocols (Chapters 16–18) conjugate to an unnatural amino acid that is introduced at a desired location. While cysteine and lysine conjugation are still the standards of industry, enzymatic techniques keep on revolutionizing the field. Our book follows this trend with more than ten protocols describing enzymatic conjugation strategies (Chapters 5–8, 10–15, and 18).

Although the focus of this book is on the development of bioconjugates rather than their application, still, most protocols also show, as a proof of concept, the conjugation of the targeting vehicle to a diagnostic or therapeutic entity. The end product of the majority of protocols is most often an antibody drug conjugate (Chapters 1, 2, 4, 5, 10, and 15), an optical probe (Chapters 7, 8, 11, 12, 14, and 16), a nanomedicine (Chapters 6 and 11), or a radiopharmaceutical (Chapters 13 and 20).

This book consists of 20 chapters involving expert contributions from all over the globe. Although most authors are from academic labs, four protocols were delivered by researchers from company RnD departments (Chapters 1, 5, 10, and 13), showing the importance of bioconjugation methodologies in the advancements of drug development pipelines at an industrial scale. Finally and most importantly, we would like to thank all the authors for their timely contributions and their active support towards the bioconjugation community.

References

1. Agarwal P, Bertozzi CR (2015) Site-specific antibody-drug conjugates: the nexus of bioorthogonal chemistry, protein engineering, and drug development. Bioconjug Chem 26(2):176–192
2. Schumacher D, Helma J, Schneider AFL et al (2018) Nanobodies: Chemical Functionalization Strategies and Intracellular Applications. Angew Chem Int Ed Engl 57(9):2314–2333
3. Massa S, Xavier C, Muyldermans S et al (2016) Emerging site-specific bioconjugation strategies for radioimmunotracer development. Expert Opin Drug Deliv 13(8):1149–1163

Brussels, Belgium

Sam Massa
Nick Devoogdt

Contents

Contributors

LOTFI ABOU-ELKACEM • *Department of Radiology, James H. Clark Center, Stanford University, Stanford, CA, USA*
PARTHA SARATHI ADDY • *Department of Chemistry, Boston College, Chestnut Hill, MA, USA*
PETER ADRIAENSENS • *Biomolecule Design Group, Institute for Materials Research (IMO), Hasselt University, Diepenbeek, Belgium; Applied and Analytical Chemistry, Institute for Materials Research (IMO), Hasselt University, Diepenbeek, Belgium*
EARL F. ALBONE • *EPAT, Eisai Inc., Exton, PA, USA*
KAREN ALT • *NanoBiotechnology Laboratory, Australian Centre for Blood Diseases, Central Clinical School, Monash University, Melbourne, VIC, Australia*
SHELBY AUGER • *Department of Chemistry, University of Minnesota, Minneapolis, MN, USA*
JAMES R. BAKER • *Department of Chemistry, University College London, London, UK*
ROBYN M. BARFIELD • *Catalent Pharma Solutions, Richmond, CA, USA*
GONÇALO J. L. BERNARDES • *Department of Chemistry, University of Cambridge, Cambridge, UK; Faculdade de Medicina, Instituto de Medicina Molecular, Universidade de Lisboa, Lisbon, Portugal*
BRECHT BILLEN • *Biomolecule Design Group, Institute for Materials Research (IMO), Hasselt University, Diepenbeek, Belgium*
FRANK BORDUSA • *Charles-Tanford-Protein Center, Institute of Biochemistry/Biotechnology, Martin-Luther-University Halle-Wittenberg, Halle, Germany*
FABIAN BRANDL • *Department of Biochemistry, University of Zurich, Zurich, Switzerland; Institute of Pharmacology, University of Bern, Bern, Switzerland*
JACOB S. BRENNER • *Department of Pharmacology, The Perelman School of Medicine, University of Pennsylvania, Philadelphia, PA, USA*
ABHISHEK CHATTERJEE • *Department of Chemistry, Boston College, Chestnut Hill, MA, USA*
VIJAY CHUDASAMA • *Department of Chemistry, University College London, London, UK*
NICK DEVOOGDT • *In Vivo Cellular and Molecular Imaging Laboratory (ICMI), Vrije Universiteit Brussel, Brussels, Belgium*
STEPHAN DICKGIESSER • *ADCs & Targeted NBE Therapeutics, Merck KGaA, Darmstadt, Germany*
MARK D. DISTEFANO • *Department of Chemistry, University of Minnesota, Minneapolis, MN, USA*
CHRISTOPHER R. DRAKE • *Sofie Co., Culver City, CA, USA*
KARIN F. K. EJENDAL • *Weldon School of Biomedical Engineering, Purdue University, West Lafayette, IN, USA*
SARAH B. ERICKSON • *Department of Chemistry, Boston College, Chestnut Hill, MA, USA*
MICHAEL J. EVANS • *Department of Radiology and Biomedical Imaging, University of California San Francisco, San Francisco, CA, USA; Department of Pharmaceutical Chemistry, University of California San Francisco, San Francisco, CA, USA*
NAFSIKA FORTE • *Department of Chemistry, University College London, London, UK*
JULIA G. FRASEUR • *Weldon School of Biomedical Engineering, Purdue University, West Lafayette, IN, USA*
GEERT-JAN GRAULUS • *Biomolecule Design Group, Institute for Materials Research (IMO), Hasselt University, Diepenbeek, Belgium*

WANDA GUEDENS • *Biomolecule Design Group, Institute for Materials Research (IMO), Hasselt University, Diepenbeek, Belgium*
CHRISTIAN P. R. HACKENBERGER • *Department of Chemical-Biology, Leibniz-Institut für Molekulare Pharmakologie (FMP), Berlin, Germany; Department of Chemistry, Humboldt Universität zu Berlin, Berlin, Germany*
CHRISTOPH E. HAGEMEYER • *NanoBiotechnology Laboratory, Australian Centre for Blood Diseases, Central Clinical School, Monash University, Melbourne, VIC, Australia*
REBEKKA HANSEN • *Biomolecule Design Group, Institute for Materials Research (IMO), Hasselt University, Diepenbeek, Belgium*
RANIA A. HASHAD • *NanoBiotechnology Laboratory, Australian Centre for Blood Diseases, Central Clinical School, Monash University, Melbourne, VIC, Australia*
JONAS HELMA • *Department of Biology II, Ludwig Maximilians Universität München and Center for Integrated Protein Science Munich, Martinsried, Germany*
WEI HUANG • *CAS Key Laboratory of Receptor Research, CAS Center for Excellence in Molecular Cell Science, Shanghai Institute of Materia Medica, Chinese Academy of Sciences, Pudong, Shanghai, China; University of Chinese Academy of Sciences, Beijing, People's Republic of China*
JAMES S. ITALIA • *Department of Chemistry, Boston College, Chestnut Hill, MA, USA*
GONZALO JIMÉNEZ-OSÉS • *Departamento de Química, Centro de Investigación en Síntesis Química, Universidad de La Rioja, Logroño, Spain; CIC bioGUNE, Bizkaia Technology Park, Derio, Spain*
ROLAND KELLNER • *ADCs & Targeted NBE Therapeutics, Merck KGaA, Darmstadt, Germany*
MAKAN KHOSHNEJAD • *Department of Pharmacology, The Perelman School of Medicine, University of Pennsylvania, Philadelphia, PA, USA*
TAMARA L. KINZER-URSEM • *Weldon School of Biomedical Engineering, Purdue University, West Lafayette, IN, USA*
HARALD KOLMAR • *Institute for Organic Chemistry and Biochemistry, Technische Universität Darmstadt, Darmstadt, Germany*
JACLYN L. LANGE • *NanoBiotechnology Laboratory, Australian Centre for Blood Diseases, Central Clinical School, Monash University, Melbourne, VIC, Australia*
HEINRICH LEONHARDT • *Department of Biology II, Ludwig Maximilians Universität München and Center for Integrated Protein Science Munich, Martinsried, Germany*
SANDRA LIEBSCHER • *Charles-Tanford-Protein Center, Institute of Biochemistry/Biotechnology, Martin-Luther-University Halle-Wittenberg, Halle, Germany*
JUNJIE LIU • *Catalent Pharma Solutions, Richmond, CA, USA*
SAM MASSA • *Cellular and Molecular Immunology Laboratory, Vrije Universiteit Brussel, Brussels, Belgium*
MARIA J. MATOS • *Department of Chemistry, University of Cambridge, Cambridge, UK*
HANNES MERTEN • *Department of Biochemistry, University of Zurich, Zurich, Switzerland*
MAURÍCIO MORAIS • *Department of Chemistry, University College London, London, UK*
SERGE MUYLDERMANS • *Cellular and Molecular Immunology Laboratory, Vrije Universiteit Brussel, Brussels, Belgium*
VLADIMIR R. MUZYKANTOV • *Department of Pharmacology, The Perelman School of Medicine, University of Pennsylvania, Philadelphia, PA, USA*
ALEX R. NANNA • *Department of Immunology and Microbiology, The Scripps Research Institute, Jupiter, FL, USA; Department of Chemistry, The Scripps Research Institute, Jupiter, FL, USA*
ARUTSELVAN NATARAJAN • *Department of Radiology, James H. Clark Center, Stanford University, Stanford, CA, USA*

JEAN-PAUL NOBEN • *Biomedical Research Institute (Biomed), Hasselt University, Diepenbeek, Belgium*
HAMIDEH PARHIZ • *Department of Pharmacology, The Perelman School of Medicine, University of Pennsylvania, Philadelphia, PA, USA*
ANDREAS PLÜCKTHUN • *Department of Biochemistry, University of Zurich, Zurich, Switzerland*
DAVID RABUKA • *Catalent Pharma Solutions, Richmond, CA, USA*
CHRISTOPH RADER • *Department of Immunology and Microbiology, The Scripps Research Institute, Jupiter, FL, USA*
NICOLAS RASCHE • *ADCs & Targeted NBE Therapeutics, Merck KGaA, Darmstadt, Germany*
ERIK ROYACKERS • *Biomedical Research Institute (Biomed), Hasselt University, Diepenbeek, Belgium*
JONAS V. SCHAEFER • *Department of Biochemistry, University of Zurich, Zurich, Switzerland*
DOMINIK SCHUMACHER • *Department of Biology II, Ludwig Maximilians Universität München and Center for Integrated Protein Science Munich, Martinsried, Germany*
WEI SHI • *CAS Key Laboratory of Receptor Research, CAS Center for Excellence in Molecular Cell Science, Shanghai Institute of Materia Medica, Chinese Academy of Sciences, Pudong, Shanghai, China; University of Chinese Academy of Sciences, Beijing, People's Republic of China*
JARED L. SPIDEL • *EPAT, Eisai Inc., Exton, PA, USA*
DUY TIEN TA • *Biomolecule Design Group, Institute for Materials Research (IMO), Hasselt University, Diepenbeek, Belgium; Downstream Processing Group, Bioprocessing Technology Institute, Agency for Science, Technology and Research (A*STAR), Singapore, Singapore*
NATASHA C. W. TAN • *NanoBiotechnology Laboratory, Australian Centre for Blood Diseases, Central Clinical School, Monash University, Melbourne, VIC, Australia*
FENG TANG • *CAS Key Laboratory of Receptor Research, CAS Center for Excellence in Molecular Cell Science, Shanghai Institute of Materia Medica, Chinese Academy of Sciences, Pudong, Shanghai, China; University of Chinese Academy of Sciences, Beijing, People's Republic of China*
HUONG TRAN • *Biomolecule Design Group, Institute for Materials Research (IMO), Hasselt University, Diepenbeek, Belgium*
ANDREW TSOURKAS • *Department of Bioengineering, University of Pennsylvania, Philadelphia, PA, USA*
HENRY F. VANBROCKLIN • *Department of Radiology and Biomedical Imaging, University of California San Francisco, San Francisco, CA, USA*
XIAO-WEI WANG • *Beijing National Laboratory for Molecular Sciences, Key Laboratory of Polymer Chemistry and Physics of Ministry of Education, Center for Soft Matter Science and Engineering, College of Chemistry and Molecular Engineering, Peking University, Beijing, People's Republic of China*
UWE ZANGEMEISTER-WITTKE • *Department of Biochemistry, University of Zurich, Zurich, Switzerland; Institute of Pharmacology, University of Bern, Bern, Switzerland*
FABIANA ZAPPALA • *Department of Bioengineering, University of Pennsylvania, Philadelphia, PA, USA*
WEN-BIN ZHANG • *Beijing National Laboratory for Molecular Sciences, Key Laboratory of Polymer Chemistry and Physics of Ministry of Education, Center for Soft Matter Science and Engineering, College of Chemistry and Molecular Engineering, Peking University, Beijing, People's Republic of China*
YI ZHANG • *Department of Chemistry, University of Minnesota, Minneapolis, MN, USA*

Chapter 1

Site-Specific Conjugation of Thiol-Reactive Cytotoxic Agents to Nonnative Cysteines of Engineered Monoclonal Antibodies

Stephan Dickgiesser, Roland Kellner, Harald Kolmar, and Nicolas Rasche

Abstract

Antibody–drug conjugates (ADCs) have been proven to be a successful therapeutic concept, allowing targeted delivery of highly potent active pharmaceutical ingredients (HPAPIs) selectively to tumor tissue. So far, HPAPIs have been mainly attached to the antibody via a chemical reaction of the payload with lysine or cysteine side chains of the antibody backbone. However, these conventional conjugation technologies result in formation of rather heterogeneous products with undesired properties. To overcome the limitations of heterogeneous ADC mixtures, several site-specific conjugation technologies have been developed over the last years. Originally pioneered by scientist from Genentech with their work on THIOMABs, several engineered cysteine mAb ADCs (ECM-ADCs) are now investigated in clinical trials. Here, we describe in detail how to engineer additional cysteines into antibodies and efficiently use them as highly site-specific conjugation sites for HPAPIs.

Key words Antibody–drug conjugate (ADC), Engineered cysteine mAb (ECM), THIOMAB, Site-specific conjugation, Targeted tumor therapy

1 Introduction

Over the last decade, antibody–drug conjugates (ADCs) have attracted strong interest in the area of biotherapeutics. With four molecules approved by the authorities and more than 60 ADCs currently investigated in clinical trials, the field is not only rapidly growing but the concept of ADCs in targeted therapy has been proven successful. Combining the high selectivity of monoclonal antibodies (mAbs) with the potency of HPAPIs such as microtubulin-disrupting agents (e.g., auristatins and maytansines) or DNA-damaging agents (e.g., calicheamicin and duocarmycins) has been shown to be very effective against various tumor types [1, 2]. However, a remaining problem is the attachment of a small-molecule drug to the large antibody. The first conjugation approaches addressed amino acid side chains, mainly lysine or

Sam Massa and Nick Devoogdt (eds.), *Bioconjugation: Methods and Protocols*, Methods in Molecular Biology, vol. 2033, https://doi.org/10.1007/978-1-4939-9654-4_1, © Springer Science+Business Media, LLC, part of Springer Nature 2019

cysteine, occurring naturally in the antibody backbone [3]. There are about 40 reactive lysine residues present on the surface of typical IgG1 molecules making it almost impossible to address certain positions specifically [4]. As a consequence, conjugation to lysine residues results in rather heterogeneous ADC products which are problematic to manufacture and are accompanied by undesirable properties like increased toxicity or suboptimal pharmacokinetics. To overcome these issues, several site-specific conjugation technologies have been developed over the last years. One of the most popular ones is the placing of engineered cysteines that can be site-specifically addressed by maleimide functionalized cytotoxic agents into the antibody backbone [5, 6]. These ECM-ADCs, also known as THIOMABs, have shown superior characteristics concerning homogeneity, pharmacokinetics, and safety aspects [7]. Additionally, it was demonstrated that careful choice of conjugation site heavily impacts the properties of the ADC and can therefore be used to further improve the final ADC product. Not only do cysteines placed at different positions in the mAb show varying reactivity toward maleimides, but the resulting ADCs also display differences in stability and hydrophobicity—two parameters strongly affecting the conjugate's in vivo characteristics [8–11]. In this protocol, we exemplarily introduced cysteines at two positions of IgG1 antibodies, one located in the light (LC) and the other one in the heavy chain (HC) (LC-V205C and HC-S442C, respectively; EU numbering) [5, 12].

However, introduction of additional cysteines does not enable a simple one-step conjugation of ECMs recombinantly expressed in mammalian cell culture. The incorporated cysteine is usually blocked by a disulfide linkage formed with thiol containing molecules present in the cell culture medium (e.g., glutathione or cysteine) [13]. To remove these thiol-caps, the engineered mAb needs to be treated under reducing conditions which not only leads to opening of the disulfide bonds between the engineered cysteines and its caps but also breaks the antibody's native interchain disulfide bridges. Reoxidation is then required to reclose these disulfide bridges which contribute heavily to the exceptional stability of antibodies. Only after this reduction–oxidation cycle is completed, the engineered cysteines can efficiently be targeted for conjugation [5, 6].

In this chapter we explain in detail how to introduce cysteine residues into the antibody backbone via conventional molecular biology tools. Additionally, we describe the expression and purification procedure of ECMs and share a typical conjugation protocol with a thiol reactive cytotoxic agent. The last part of this chapter includes tools for analytical and in vitro ADC characterization and some supportive notes.

2 Materials

2.1 Site-Directed Mutagenesis

1. Mammalian expression vector encoding for the light or heavy chain of an antibody, depending on the desired site of cysteine substitution.
2. Forward and reverse primers for site-directed mutagenesis (*see* **Note 1**).
3. QuikChange® Site-Directed Mutagenesis Kit (Agilent Technologies) including PfuTurbo® DNA polymerase, 10× reaction buffer, DpnI restriction enzyme, and dNTP mix.
4. PCR thermocycler apparatus.
5. One Shot™ TOP10 Chemically Competent *Escherichia coli* (*E. coli*) cells (S.O.C. medium included; ThermoFisher Scientific) or equivalent.
6. LB-Agar plates with appropriate antibiotics.
7. Plasmid DNA mini prep kit.
8. Plasmid DNA midi or maxi prep kit.

2.2 ECM Expression

1. Expi293F™ Suspension Cells (Gibco).
2. Expi293™ Expression Medium (Gibco).
3. ExpiFectamine™ 293 Transfection Kit (Gibco).
4. Opti-MEM™ I Reduced Serum Medium (Gibco).
5. CO_2 incubator with orbital shaker.
6. Mammalian expression vectors encoding for antibody with cysteine substitution (light and heavy chain).

2.3 ECM Purification

1. Filtered cell culture supernatant from Subheading 3.2.
2. Binding buffer: phosphate buffered saline (PBS), pH 7.4.
3. Elution buffer: 50 mM acetic acid in H_2O.
4. Neutralization buffer: 1 M Tris–HCl in H_2O, pH 8.0.
5. HiTrap™ Protein A HP column (GE Healthcare Life Sciences).
6. Äkta chromatography system (GE Healthcare Life Sciences).
7. Electrophoresis chamber and SDS-PAGE gels.
8. HiTrap desalting columns (GE Healthcare Life Sciences).
9. Amicon Ultra-15 Centrifugal Filter Units, 100 kDa (Merck KGaA).

2.4 ECM Reduction, Oxidation and Conjugation

1. PBS, pH 7.4.
2. 10 mg/mL ECM from Subheading 3.3.
3. 0.5 M EDTA, pH 8.0.

4. 500 mM Tris-(2-carboxyethyl)-phosphine (TCEP) in H_2O.
5. Dimethyl sulfoxide (DMSO).
6. 100 mM *N*-acetylcysteine (NAC) in H_2O.
7. PD10 column (GE Healthcare).
8. 50 mM DHAA (Dehydroascorbic acid) in DMSO.
9. Superdex 200 10/300 GL increase size exclusion chromatography (SEC) column (GE Healthcare).
10. SEC running buffer: PBS, pH 7.0.
11. 20 mM maleimido-caproyl-val-Cit-PAB-MMAE (MC-vcMMAE) in DMSO (Levena Biopharma).

2.5 DAR Determination via Hydrophobic Interaction Chromatography (HIC)

1. MAbPac HIC-Butyl column, 5 μm, 4.6 × 100 mm (Thermo Scientific).
2. 1260 Infinity LC system (Agilent).
3. OpenLAB CDS, ChemStation Edition (Agilent).
4. Mobile phase A: 2 M ammonium sulfate, 25 mM Tris–HCl pH 7.5.
5. Mobile phase B: 25 mM Tris–HCl, pH 7.5.
6. Mobile phase C: 100% isopropanol.
7. Storage buffer: 0.1% sodium azide in ddH_2O.

2.6 DAR Determination via Reduced Reversed-Phase (RP) HPLC

1. Polymeric reversed phase column (PLRP-S column), 5 μm, 2.1 × 50 mm (Agilent).
2. 1260 Infinity LC system (Agilent).
3. OpenLAB CDS, ChemStation Edition operating system (Agilent).
4. Mobile phase A: 0.1% trifluoroacetic acid (TFA)in H_2O.
5. Mobile phase B: 0.1% TFA in 100% acetonitrile.
6. 500 mM TCEP (Tris-(2-carboxyethyl)-phosphine) in H_2O.
7. 0.5 M iodoacetamide stock solution.
8. Storage buffer: 50% acetonitrile–50% ultrapure (LiChrosolv) water.

2.7 DAR Determination by Mass Spectrometry

1. Reversed Phase (RP) column Proteomix RP-100, 3.5 μm, 4.6 × 100 mm (Sepax).
2. U3000 HPLC system (Dionex).
3. Mobile phase A: 0.1% TFA in H_2O.
4. Mobile phase B: 0.1% TFA in 85% acetonitrile.
5. Synapt G2 High-definition Mass Spectrometry (HDMS) System (Waters).
6. Solvent A: 0.1% formic acid (FA) in H_2O.

7. Solvent B: 0.1% FA in acetonitrile.
8. 500 mM TCEP in H_2O.

2.8 In Vitro Cytotoxicity Assessment of ADCs

1. Target-positive cell lines (e.g., SKBR-3 and HCC-1954 for trastuzumab ADCs).
2. Target-negative cell line (e.g., MDA-MB-468 cells for trastuzumab ADCs).
3. Cell culture media (cell line specific, e.g., RPMI-Medium supplemented with 10% fetal bovine serum, 2 mM L-Gln, and 1 mM sodium pyruvate).
4. Flat bottom 96-well cell culture plates.
5. Microplate reader with luminescence detection.
6. Antigen-specific ECM-ADC of interest.
7. Nonbinding ECM-ADC (isotype control).
8. Antigen-specific unconjugated antibody (ADC progenitor, e.g., trastuzumab).
9. CellTiter-Glo® Luminescent Cell Viability Assay (Promega).

3 Methods

3.1 Site-Directed Mutagenesis

1. Prepare a 50 μL PCR reaction mixture with 25 ng light or heavy chain-encoding plasmid, 125 ng of each primer (forward and reverse), 1 μL dNTP mix, and 5 μL reaction buffer.
2. Run PCR in a thermocycler machine with 30 s preincubation at 95 °C and 16 cycles with 30 s at 95 °C, 1 min at 55 °C and 1 min/kb of plasmid length at 68 °C (*see* **Note 2**).
3. Add 1 μL of DpnI restriction enzyme and incubate for 1 h at 37 °C.
4. Thaw chemically competent *E. coli* on ice and add 1.5 μL of PCR product. Mix by tapping only and incubate on ice for 30 min.
5. Incubate for 30 s in a 42 °C water bath and place immediately back on ice.
6. Add 200 μL prewarmed S.O.C. medium and incubate for 1 h at 37 °C and 250 rpm.
7. Spread the whole suspension on a prewarmed agar plate containing respective antibiotics and incubate overnight in a 37 °C incubator.
8. Pick about three colonies and inoculate small-scale overnight cultures in LB medium with appropriate antibiotics.

9. Isolate plasmids by using a commercially available plasmid DNA mini prep kit and verify introduced mutations by DNA sequencing.
10. Prepare transfection grade plasmid preparation by using a suitable plasmid DNA midi- or maxi prep kit.

3.2 ECM Expression

1. Thaw Expi293F cells and subculture at least three times in shaker flasks at 80 rpm, 37 °C, and 8% CO_2 in a humidified atmosphere as per the manufacturer's instructions (*see* **Note 3**).
2. Count cells and dilute to a density of 2.9×10^6 viable cells per mL in a final volume of 50 mL with fresh Expi293™ medium. Cell viability should be higher than 95% (*see* **Note 4**).
3. Dilute 60 μg plasmid DNA (30 μg of light and heavy chain encoding plasmids, respectively) in a final volume of 3 mL Opti-MEM™ I medium.
4. Dilute 160 μL of ExpiFectamine™ 293 reagent in 3 mL Opti-MEM™ 1 medium and incubate for 5 min at room temperature.
5. Add the diluted DNA to the diluted ExpiFectamine™ 293 Reagent and incubate for 20 min at room temperature.
6. Add the DNA-ExpiFectamine™ mixture to the prediluted cells and incubate cells as described.
7. After 16–18 h add 300 μL of ExpiFectamine™ 293 Transfection Enhancer 1 and 3 mL of Transfection Enhancer 2 and incubate the culture for another 4 days (*see* **Note 5**).
8. Remove the cells by centrifugation for 30 min at 4000 × *g* and filter the supernatant using a bottle top filter with 0.45 μm pore size.

3.3 ECM Purification

1. Equilibrate a HiTrap™ Protein A HP column with 10 column volumes (CV) PBS using an Äkta chromatography system. Use 1 mL/min or 5 mL/min flow rates for 1 mL or 5 mL columns, respectively (*see* **Note 6**).
2. Prepare fraction tubes by prefilling each with 200 μL neutralization buffer (*see* **Note 7**).
3. Apply the filtered cell culture supernatant onto the column.
4. Wash with 10 CV binding buffer.
5. Apply 100% elution buffer and collect 1 mL fractions in tubes prefilled with neutralization buffer.
6. Evaluate fractions by SDS-PAGE and pool those containing purified protein.
7. Equilibrate HiTrap desalting columns with 25 mL PBS at a flow rate of 5 mL/min (*see* **Note 8**).

8. Apply up to 1.5 mL sample per desalting column, elute with 10 mL buffer, and collect fractions (*see* **Note 9**).
9. Concentrate purified ECM to 10 mg/mL using Amicon Centrifugal Filter Units at 4000 × *g*.

3.4 ECM Reduction, Oxidation, and Conjugation

1. Mix 500 μL of ECM (5 mg) with 500 μL reduction buffer containing 495.3 μL PBS, pH 7.4, 2 μL EDTA, and 2.7 μL TCEP (40 molar equivalents compared to the ECM) and incubate at 37 °C for 2 h.
2. Perform buffer exchange to PBS, pH 7.4, 1 mM EDTA using a PD10 column according to the manufacturer's instructions.
3. To reoxidize, mix 1 mL of reduced ECM (at a concentration of 5 mg/mL) with 13.3 μL of DHAA (20 molar equivalents compared to the ECM) and incubate at 25 °C for 1.5 h.
4. Add 10 μL of MC-vcMMAE (6 molar equivalents compared to the ECM) and incubate at 25 °C for 2 h (*see* **Note 10**).
5. Add 8.4 μL of NAC (25 molar equivalents compared to the ECM) to quench the reaction. Incubate at 37 °C for 15 min.
6. Inject 500 μL of the quenched reaction mix onto a Superdex 200 10/300 GL increase column preequilibrated with PBS, pH 7.0. Run SEC at 0.5 mL/min and collect and pool the ADC-containing fractions. An example of a typical SEC chromatogram is show in Fig. 1; *see* **Note 11**.
7. Filter-sterilize pooled ADC fractions and store at −80 °C to 4 °C, depending on the stability of the construct.

3.5 DAR Determination via Hydrophobic Interaction Chromatography (HIC)

1. Adjust the ADC concentration to 0.5 mg/mL by addition of H_2O and ammonium sulfate to a final concentration of 1.5 M ammonium sulfate (*see* **Note 12**).
2. Run the HPLC system at 75% A/25% B, a flow rate of 1 mL/min, and a column temperature of 30 °C.
3. Inject 40 μg of mAb/ADC onto the MAbPac HIC-Butyl column.
4. Elute the sample by applying a linear gradient from 75% A/25% B to 0% A/80% B/20% C over 20 min and monitor the mAb/ADC signal using UV absorption at 280 nm. Examples of typical HIC chromatograms are shown in Fig. 2.
5. Clean the HIC column by injections of 0.2 M NaOH.
6. Run the column with storage buffer at a flow rate of 1 mL/min for at least 30 min.
7. Remove the column, seal it properly, and store it at room temperature.

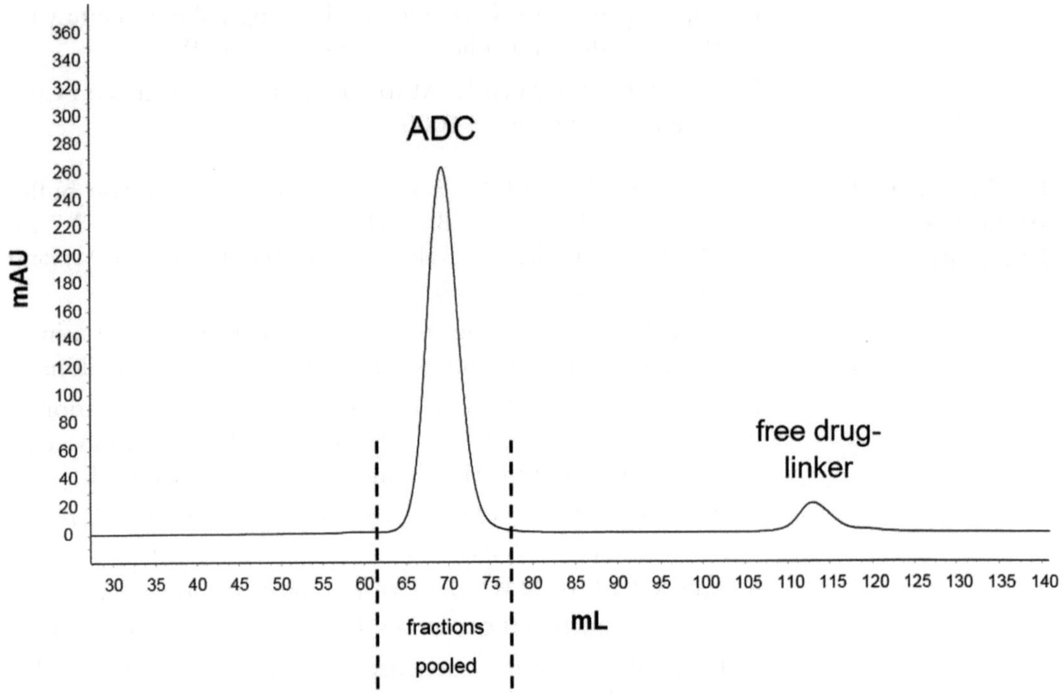

Fig. 1 Preparative SEC of ECM ADCs. ECM ADCs have been separated from free drug-linker via preparative SEC. Fractions containing purified ECM-ADC have been pooled as indicated

8. Integrate peak areas for individual species with drug load n and calculate the average drug-to-antibody ratio (DAR) according to the following formula:

$$\mathrm{DAR\ (HIC)} = \frac{\sum_{n=0}^{\infty}(n \times An)}{\sum_{n=0}^{\infty}An}$$

n = # of conjugated payloads.
An = peak area of species with drug load n.

3.6 DAR Determination via Reduced Reversed-Phase HPLC

1. Mix 20 μL of ADC solution (80 μg) with 16 μL PBS, pH 7.0 and 4 μL 0.5 M TCEP and incubate at 37 °C for 5 min.
2. To 40 μL reduction mixture add 40 μL 0.5 M iodoacetamide solution and incubate at 37 °C for 15 min, to alkylate.
3. Spin down with a benchtop centrifuge at 10,000 × g for 1 min.
4. Run the HPLC system in 70% A/30% B at a flow rate of 1 mL/min and a column temperature of 65 °C.
5. Inject 10 μL (10 μg) onto the PLRP-S column.
6. Elute the sample by applying a linear gradient from 70% A/30% B to 55% A/45% B over 7.5 min and monitor the mAb/ADC

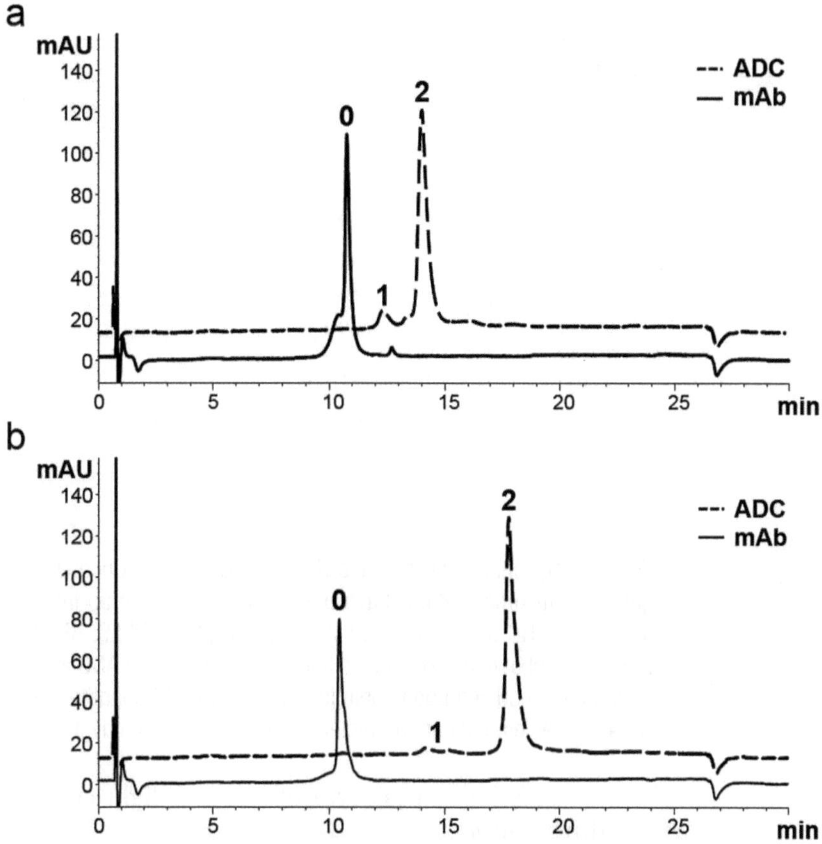

Fig. 2 Analytical HIC of ECM-ADCs. DAR of ADCs was determined via analytical HIC. HIC chromatogram of the mAb is shown as continuous line, of the respective ADC as dashed line. Numbers indicate the drug load of the mAb/ADC. (**a**) Trastuzumab LC-V205C ADC and (**b**) Trastuzumab HC-S442C ADC

signal using UV absorption at 214 nm. Examples of typical PLRP chromatograms are shown in Fig. 3.

7. Run the column with storage buffer at a flow rate of 1 mL/min for at least 30 min.
8. Remove the column, seal it properly and store at room temperature.
9. Integrate peak areas for individual species with drug load n and calculate the average drug-to-antibody ratio (DAR) according to the following formula for reduced samples:

$$\text{DAR (reduced)} = 2 \times \frac{\sum_{n=0}^{\infty} (n \times A(\text{LC})n)}{\sum_{n=0}^{\infty} A(\text{LC})n} + 2 \times \frac{\sum_{n=0}^{\infty} (n \times A(\text{HC})n)}{\sum_{n=0}^{\infty} A(\text{HC})n}$$

$n = \#$ of conjugated payloads.

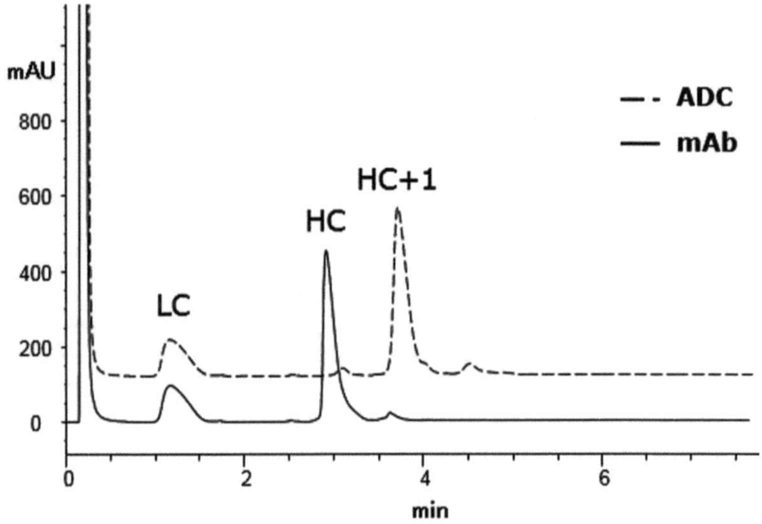

Fig. 3 Analytical reversed-phase HPLC of ECM-ADCs. Unconjugated and conjugated trastuzumab S442C have been reduced to open up interchain disulfides and LC and HC were separated via reversed-phase HPLC. RP chromatogram of the mAb is shown as continuous line, of the respective ADC as dashed line. The LC was not modified upon conjugation while the HC quantitatively shifted to the HC + 1 species, indicating efficient conjugation of exactly 1 payload to the HC

$A(XC)n$ = peak area of heavy or light chain species with drug load n.

3.7 DAR Determination via Mass Spectrometry

1. Dilute the sample in 0.1% TFA to final volume of 100 μL (0.1 mg/mL).
2. To reduce, add 2 μL 500 mM TCEP to 10 μg ADC and incubate for 5 min at room temperature.
3. Inject 10 μg sample onto the RP-column and use a splitter to reduce the flow rate directed to the spray inlet by 1:5.
4. For Electrospray ionization Mass Spectrometry (ESI-MS), apply standard instrument parameters with an acquisition range of 400–6000 Da.
5. For instrument control and data interpretation, use methods for LC and MS run under MassLynx (Waters), deconvolute with MaxEnt1.
6. Light and heavy chain need to be separated according to their number of conjugated payloads ($n = 0, 1, 2, \ldots$) either by chromatography or by mass ion extraction [14]. The resulting peak area is used for DAR calculation as described for reduced samples in Subheading 3.6 (*see* **Note 13**). Examples of typical mass spectra are shown in Fig. 4.

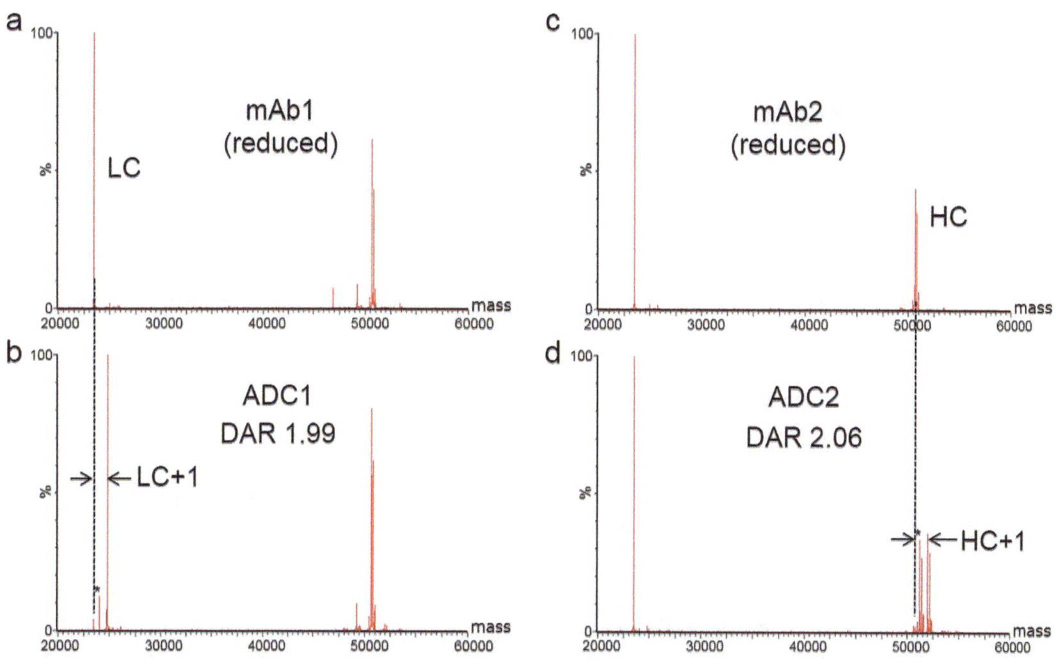

Fig. 4 Intact mass analysis of two reduced mAbs and corresponding ADCs. (**a**) Detection of unconjugated trastuzumab-ECM mAb1 with LC and HC carrying the LC-mutation V205C and (**b**) the corresponding ADC1 with conjugated LC + 1 and unchanged HC; (**c**) detection of unconjugated trastuzumab-ECM mAb2 with LC and HC carrying the HC-mutation S442C and (**d**) the corresponding ADC2 with conjugated HC + 1 and unchanged LC. Note that a mass spec-artefact (∗) occurs after fragmentation of the linker-toxin bond, both at LC + 1 as well as the HC + 1

3.8 In Vitro Cytotoxicity Assessment of ADCs

1. Thaw cells and subculture at least three times.
2. Seed 5000–10,000 cells in 80 μL growth medium in 96-well plates and incubate overnight (*see* **Note 14**).
3. Prepare dilution series with ADCs of interest, isotype control ADC, and unconjugated antibody (*see* **Note 15**).
4. Apply 20 μL of each dilution step to each well and incubate for another 3 days (*see* **Note 16**).
5. Prepare CellTiter-Glo reagents as per the manufacturer's instructions and equilibrate plates for 30 min at room temperature.
6. Add 100 μL of CellTiter-Glo reagent, mix, incubate for 10 min at room temperature and read luminescence using a microplate reader.
7. Normalize data using wells with cells but without test compound as 100% and those without cells as 0%. Fit data using a four-parameter regression model to determine IC_{50} values by using appropriate software (e.g., GraphPad Prism). Examples of typical cytotoxicity assays are shown in Fig. 5.

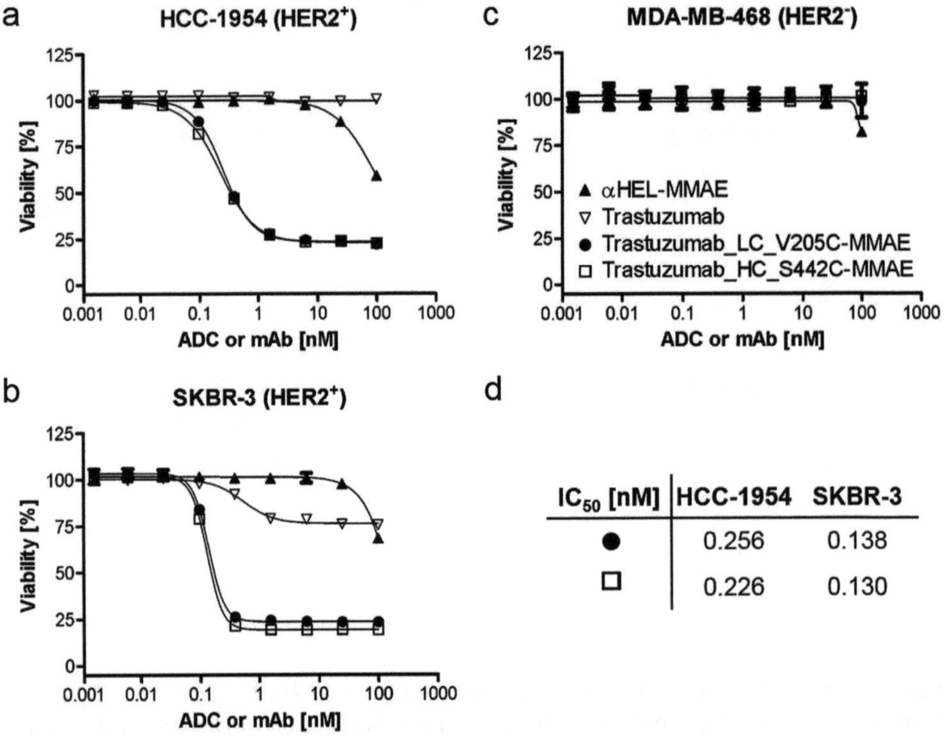

IC_{50} [nM]	HCC-1954	SKBR-3
●	0.256	0.138
□	0.226	0.130

Fig. 5 Evaluation of in vitro cytotoxicity of trastuzumab-based ECM-ADCs. Trastuzumab-ECMs conjugated with MC-vcMMAE at light chain position V205C (circles) or heavy chain position S442C (squares) were tested for cytotoxicity on HER2$^+$ cell lines: (**a**) HCC-1954, (**b**) SKBR-3, and (**c**) HER2$^-$ MDA-MB-468 cells. Anti-hen egg lysozyme (aHEL)-specific ADC (filled triangles) as well as unconjugated trastuzumab (empty triangles) served as controls. Data was fitted using a four-parameter regression model with GraphPad Prism 5.04 software. Trastuzumab-based ECM-ADCs showed effective and specific cell killing on both HER2$^+$ cell lines with (**d**) IC_{50} values in the subnanomolar range while no toxicity was observed to HER2$^-$ cells

4 Notes

1. Design primers that only partially overlap (about four nucleotides in 5′ and about eight nucleotides in 3′ direction of the mutation site), a predicted melting temperature of 73–85 °C, an overall length of around 22 nucleotides and G or C at their termini [15]. http://bioinformatics.org/primerx/ can be a useful help.
2. These conditions work well in most cases. In case of problems, consult the instructions given by the QuikChange® Mutagenesis Kit's manual.
3. Optimal shaking speed depends on the amplitude of your orbital shaker.
4. Transfection volume can be easily scaled up or down. Use 125 mL or 250 mL flasks for up to 40 mL or 80 mL final

transfection volume, respectively. Consider that the final transfection volume includes also the DNA transfection mixture and the feeding solution. More details and troubleshooting are given in the manufacturer's instructions.

5. Optimal expression time depends on your antibody and several other factors. However, five days are a good start.
6. For increased binding capacity, up to three HiTrap columns can be connected in series.
7. Check if addition of 1 mL elution buffer results in neutral pH. If not, change volume of neutralization buffer accordingly.
8. For volumes larger than 1.5 mL, up to five columns can be connected in series. Alternatively, HiPrep 26/10 columns or dialysis can be used for desalting.
9. Start collection of fractions simultaneously with sample application in order to prevent loss of protein.
10. Use of very hydrophobic payloads might result precipitation or aggregation during the conjugation reaction. In this case, increasing concentrations of certain excipients like DMSO or propylene glycol might help [16].
11. Alternative purification strategies are also feasible (tangential flow filtration, protein A affinity chromatography, preparative HIC, etc.).
12. If more hydrophobic payloads and/or a higher DAR is being used, adjustment of the sample to an ammonium sulfate concentration of 1.5 M may result in heavy precipitation of the sample. In this case simply reduce the ammonium sulfate concentration to a level where no precipitation is observed. Too low ammonium sulfate concentration might result in loss of binding of the residual free mAb.
13. Consider potential mass spec-artefact due to fragmentation of the linker-toxin bond.
14. Run assays with target-positive and target-negative cell lines to distinguish between target-specific and unspecific effects. Include wells without cells (medium only) as control. Optimal number of cells depends on the cell line and anticipated incubation time, but the given numbers are a good starting point.
15. Start with, for example, 500 nM concentration and use 1:4 dilution steps. Nonspecific ADCs and unconjugated antibody are controls to evaluate target specificity and the cytotoxic effect of conjugated payloads. Include steps with cells but without sample (medium only) as controls.
16. Optimal incubation time depends on cell lines, ADCs, and used cytotoxic payload.

References

1. Drake PM, Rabuka D (2017) Recent developments in ADC technology: preclinical studies signal future clinical trends. BioDrugs 31:1–11. https://doi.org/10.1007/s40259-017-0254-1
2. Beck A, Goetsch L, Dumontet C, Corvaïa N (2017) Strategies and challenges for the next generation of antibody–drug conjugates. Nat Rev Drug Discov 10:345–352. https://doi.org/10.1038/nrd.2016.268
3. Perez HL, Cardarelli PM, Deshpande S et al (2014) Antibody–drug conjugates: current status and future directions. Drug Discov Today 19:869–881. https://doi.org/10.1016/j.drudis.2013.11.004
4. Wang L, Amphlett G, Bla WA et al (2005) Structural characterization of the maytansinoid—monoclonal antibody immunoconjugate, huN901—DM1, by mass spectrometry. Protein Sci 4(9):2436–2446. https://doi.org/10.1110/ps.051478705.therapeutic
5. Junutula JR, Raab H, Clark S et al (2008) Site-specific conjugation of a cytotoxic drug to an antibody improves the therapeutic index. Nat Biotechnol 26:925–932. https://doi.org/10.1038/nbt.1480
6. Junutula JR, Bhakta S, Raab H et al (2008) Rapid identification of reactive cysteine residues for site-specific labeling of antibody-Fabs. J Immunol Methods 332:41–52. https://doi.org/10.1016/j.jim.2007.12.011
7. Panowski S, Bhakta S, Raab H et al (2014) Site-specific antibody drug conjugates for cancer therapy. MAbs 6:34–45
8. Tumey LN, Li F, Rago B et al (2017) Site selection: a case study in the identification of optimal cysteine engineered antibody drug conjugates. AAPS J. https://doi.org/10.1208/s12248-017-0083-7
9. Sussman D, Westendorf L, Meyer DW et al (2018) Engineered cysteine antibodies: an improved antibody-drug conjugate platform with a novel mechanism of drug-linker stability. Protein Eng Des Sel:1–8. https://doi.org/10.1093/protein/gzx067
10. Su D, Kozak KR, Sadowsky J et al (2018) Modulating antibody–drug conjugate payload metabolism by conjugation site and linker modification. Bioconjug Chem 29:1155–1167. https://doi.org/10.1021/acs.bioconjchem.7b00785
11. Ohri R, Bhakta S, Fourie-O'Donohue A et al (2018) High-throughput cysteine scanning to identify stable antibody conjugation sites for maleimide- and disulfide-based linkers. Bioconjug Chem 29(2):473–485. https://doi.org/10.1021/acs.bioconjchem.7b00791
12. Thompson P, Fleming R, Bezabeh B et al (2016) Rational design, biophysical and biological characterization of site-specific antibody-tubulysin conjugates with improved stability, efficacy and pharmacokinetics. J Control Release. https://doi.org/10.1016/j.jconrel.2016.06.025
13. Zhong X, He T, Prashad AS et al (2017) Mechanistic understanding of the cysteine capping modifications of antibodies enables selective chemical engineering in live mammalian cells. J Biotechnol 248:48–58. https://doi.org/10.1016/j.jbiotec.2017.03.006
14. Basa L (2013) Drug-to-antibody ratio (dar) and drug load distribution by LC-ESI-MS. In: Ducry L (ed) Antibody-drug conjugates, Methods in molecular biology (methods and protocols), pp 285–293
15. Zheng L, Baumann U, Reymond J-L (2004) An efficient one-step site-directed and site-saturation mutagenesis protocol. Nucleic Acids Res 32:e115–e115. https://doi.org/10.1093/nar/gnh110
16. Hollander I, Kunz A, Hamann PR (2008) Selection of reaction additives used in the preparation of monomeric antibody-calicheamicin conjugates. Bioconjug Chem 19:358–361. https://doi.org/10.1021/bc700321z

Chapter 2

Application of Next-Generation Maleimides (NGMs) to Site-Selective Antibody Conjugation

Maurício Morais, Nafsika Forte, Vijay Chudasama, and James R. Baker

Abstract

Site-selective antibody conjugation is widely recognized as a key strategy for the optimum construction of antibody–drug conjugates (ADCs). Achieving such bioconjugation directly onto native antibodies would represent the ideal solution, as it would afford greatly improved homogeneity whilst avoiding the need for genetic engineering, and even allow the repurposing of existing antibodies "off-the shelf." Here we describe a protocol for the use of next-generation maleimides (NGMs) for the selective modification of the four interchain disulfide bonds present in a typical IgG1 antibody format. These reagents retain the efficiency of classical maleimides whilst serving to rebridge each reduced disulfide bond, affording one attachment per disulfide. The approach is simple, uses readily available reagents, and generates robustly stable conjugates which are ideal for in vitro or in vivo applications. In addition to use in the construction of ADCs these reagents can also be used to develop antibody conjugates for imaging, bispecifics, and broadly for use across biology and medicine.

Key words Site-selective antibody conjugation, Antibody–drug conjugates (ADCs), Disulfide bridging, Next-generation maleimides (NGMs)

1 Introduction

Antibody conjugates represent one of the most diversely exploited classes of bioconjugates. They form the basis of immunoassays such as ELISAs, radioimmunoconjugates for imaging and therapy, and antibody–drug conjugates (ADCs) for the targeted delivery of cytotoxins. Approaches to antibody conjugation have been dominated by nonselective methods, most commonly targeting lysine residues which cover the surface of antibodies. On a typical antibody there are >80 lysine residues, and it has been estimated for ADCs that a desired average loading of 3–4 drugs per antibody leads to around 10^6 molecular species present in the resulting heterogeneous conjugates [1, 2]. Each component in such a mixture will have a different pharmacological profile; thus, this represents a far from optimal outcome [2, 3]. Alternatively cysteine

Sam Massa and Nick Devoogdt (eds.), *Bioconjugation: Methods and Protocols*, Methods in Molecular Biology, vol. 2033, https://doi.org/10.1007/978-1-4939-9654-4_2,

residues can be targeted as sites for attachment. There are four solvent accessible disulfide bonds in the major therapeutically relevant antibody isotype (IgG1), and these can be readily reduced in preference to the buried intrachain disulfides to generate eight cysteines. In practice, partial reduction is usually carried out, with subsequent conjugation affording products with drug-to-antibody ratios (DARs) of 0, 2, 4, 6, and 8 as the major components [4]. Such conjugates still represent heterogeneous mixtures. Notably the presence of unmodified antibody can inhibit the activity of the ADC, and the higher loaded species have been identified as having reduced stability [5, 6], and poorer outcomes in vivo due to their accelerated clearance [4, 7]. The use of classical maleimides as the favored approach for efficient cysteine conjugation has also been widely demonstrated to suffer a further limitation, as the conjugates are unstable over several days in vivo, due to their ability to undergo retro conjugate additions [8, 9].

Increased homogeneity in conjugates has been shown to offer the prospect of improved therapeutic windows [2, 7, 10, 11]; thus, site selectivity is considered a key component in the design of future ADCs [12, 13]. Strategies being explored include the use of genetic engineering to incorporate cysteine mutants, nonnatural amino acids, or enzymatic recognition sequences as handles for controlled drug loading [10, 12–14]. Site-selective methods which avoid this requirement for engineering can offer the further advantage of being directly applicable to native antibodies. To this end disulfide bridging is a leading strategy, in which the interchain disulfide bonds are targeted with reagents which reconnect the two cysteine residues [13, 15–19]. This approach targets antibody conjugates with a controlled loading of one drug per disulfide bond and thus a DAR of ~4 in IgG1s.

We have recently described the use of next-generation maleimides (NGMs) [20–22], which are able to achieve efficient disulfide bridging and thus represent an ideal platform for site-selective antibody conjugation (Fig. 1) [23–27]. These reagents retain the rapid kinetics of maleimide conjugation (the conjugation step occurs in <1 min) and are "locked" as robustly stable conjugates following quantitative hydrolysis after just 2 h in the conjugation buffer. Here we summarize a step-by-step guide to their use.

2 Materials

The optimum NGM reagents we suggest for antibody bioconjugation by disulfide bridging are dibromomaleimides (DBM) with a C-2 linker (DBM-C_2 reagent, Fig. 2) [28]. These reagents are readily synthesized in two steps, starting from commercially available dibromomaleic acid (e.g., Sigma-Aldrich). The synthesis involves initial condensation with glycine to generate DBM-C_2-

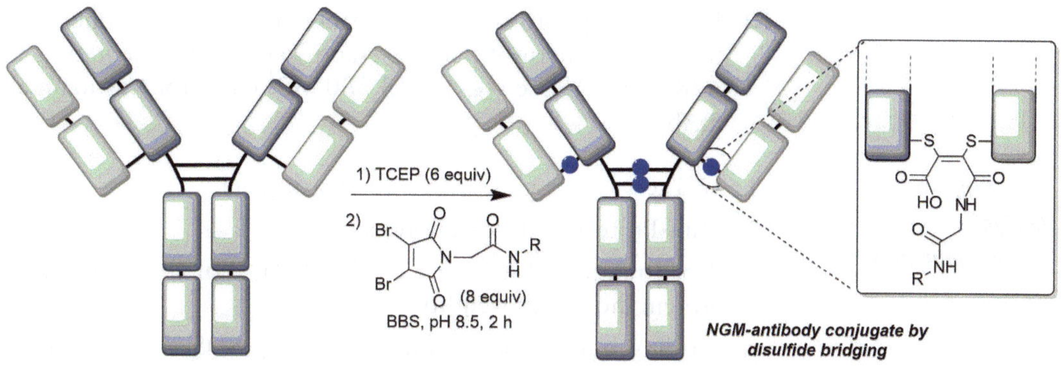

Fig. 1 Antibody conjugation by disulfide bridging using next-generation maleimides

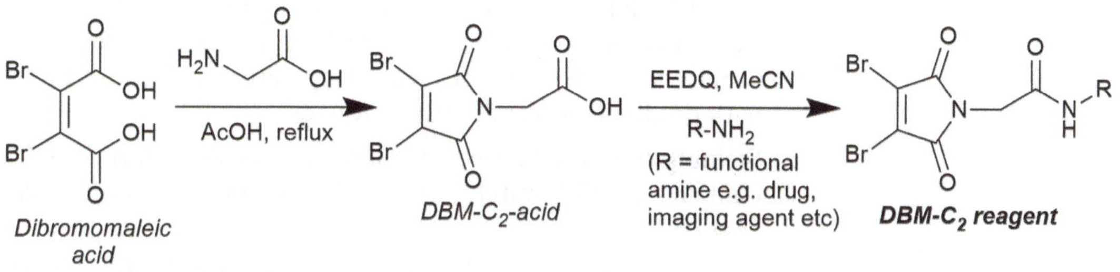

Fig. 2 Synthesis of the DBM-C_2 reagent

acid, which is then coupled with the functional amine of choice (drug, imaging agent, etc.). Details of this synthesis are described elsewhere [28], and are not the focus of this protocol chapter.

2.1 Antibody Reduction and Conjugation

1. Prepare buffer solutions with double-deionized water. Filter-sterilize (through 0.22 μm membrane) and degas them, prior to use (*see* **Note 1**).
2. Borate buffered saline (BBS) solution: 50 mM sodium borate, 50 mM NaCl, and 5 mM ethylenediaminetetraacetic acid (EDTA), at pH 8.5.
3. IgG1 antibody (e.g., lyophilized).
4. Tris(2-carboxyethyl)phosphine hydrochloride (TCEP–HCl).
5. DBM-C_2 reagent [28].
6. *N*,*N*-Dimethylformamide, anhydrous (DMF).
7. Vivaspin® 500 polyethersulfone (PES) membrane concentrators with a molecular weight cutoff (MWCO) of 10 kDa for ultrafiltration.
8. Centrifuge operating at 14,000 rcf at 20 °C (e.g., Eppendorf 5415R fixed angle rotor centrifuge).
9. Thermomixer for temperature and agitation controlled experiments.

10. UV spectrophotometer to determine the antibody concentration (e.g., ThermoScientific NanoDrop 2000C operating in A280 mode). Use $\varepsilon_{280} = 215{,}380\ M^{-1}\ cm^{-1}$ for trastuzumab mAb.

2.2 SDS-PAGE Analysis of the Conjugates

1. 12% acrylamide gels, with a 4% stacking gel.
2. Mini-Protean® Electrophoresis system.
3. Broad-range MW marker (10–250 kDa, BioLabs) to estimate protein molecular weights.
4. Loading buffer stock solution (5×): 1 g SDS, 3 mL glycerol, 6 mL 0.5 M Tris-HCl pH 6.8, 2 mg R-250 dye (*see* **Note 2**).
5. 10 mM dithiothreitol (DTT) solution in water (*see* **Note 3**).
6. SDS-PAGE running buffer (composition for 10× buffer): 30 g Tris base, 144 g glycine, 10 g SDS in 1 L double-deionized water. Store the buffer at room temperature and dilute to 1× with double-deionized water before use.
7. Staining solution: Coomassie blue (0.05% w/v) solution (1 L), contains 49.95% water, 40% methanol, and 10% acetic acid (AcOH).
8. Destaining solution (1 L, % v/v): 10% methanol, 10% AcOH, and 80% water.

2.3 LCMS Analysis of the Conjugates

1. 50 mM ammonium acetate buffer, pH 6.9.
2. HPLC-grade acetonitrile, water, and formic acid solvents.
3. 15,000 units/mL PNGase F in 20 mM Tris–HCl, 50 mM NaCl, 5 mM EDTA, pH 7.5 (New England BioLabs Inc.).
4. 0.22 μm centrifuge tube filters (Corning® Costar Spin-X®).
5. Agilent 6510 QTOF LC-MS system (Agilent, UK) with MassHunter software (version B.07.00) or alternative high accuracy LC-MS with deconvolution software.
6. PLRP-S column, 1000 Å, 8 μm, 150 mm × 2.1 mm.
7. Mobile phase of A (5% MeCN in aqueous 0.1% formic acid) and B (95% MeCN, 5% water, 0.1% formic acid).

3 Methods

The following procedure has been optimized for trastuzumab (an IgG1 antibody). For other antibodies the equivalents of reducing agent will have to be tested to determine the amount which is required for full reduction. For different antibody isotypes, which contain higher numbers of interchain disulfide bonds, the equivalents of the DBM-C2 reagent would also require increasing.

3.1 Antibody Reduction and Conjugation

1. Dissolve the lyophilized antibody in double-deionized water to a final concentration of 10 mg/mL. Buffer exchange it into BBS, via ultrafiltration (*see* **Note 4**). Determine the concentration and adjust it to 22.9 μM (3.32 mg/mL, *see* **Note 5**) with BBS (*see* **Note 6**).
2. Dissolve TCEP–HCl in BBS to make a 10 mM (2.87 mg/mL) solution (*see* **Note 7**).
3. Dissolve the DBM-C2 reagent in dry DMF to make a 10 mM solution (*see* **Note 8**).
4. To the antibody solution (22.9 μM, 100 μL) add TCEP (10 mM, 1.4 μL, 6 equiv.) and warm at 37 °C for 2 h, under mild agitation.
5. Add the DBM-C_2 reagent (10 mM, 1.8 μL, 8 equiv) to the reduced antibody solution and incubate at 20 °C for 5 min, under mild agitation.
6. Carry out ultrafiltration (10 kDa MWCO) with BBS to remove any remaining reagent (*see* **Note 9**).
7. Retain these conjugates in BBS buffer (22.9 μM) at 25 °C for 2 h to ensure complete hydrolysis of the maleimides generating robustly stable conjugates.

3.2 SDS-PAGE Analysis of the Conjugates

1. Take a 3 μL sample (~35 μM mAb) and dilute in 10 μL of water. Add 2 μL of the loading buffer and heat at 65 °C for 5 min to denature the sample.
2. For the reducing conditions take a 3 μL sample (~35 μM in total mAb) and dilute in 10 μL of water, then add 2 μL of the loading buffer (5×) along with 1 μL of the DTT solution (10 mM), vortex (10 s), and heat at 65 °C for 5 min.
3. Centrifuge the samples for 5 s and load (5 μL of each sample) onto the gel, along with the marker. Also include samples for the unmodified antibody and the reduced antibody.
4. Run the gel at constant current (30–35 mA) for 40 min in 1× SDS running buffer.
5. Following gel electrophoresis, separate the two plates with the use of a cassette opening lever. Carefully remove the gel and transfer it to a container with staining solution (enough to cover the gel). Incubate at room temperature for ca. 45 min (*see* **Note 10**). Discard the staining solution and rinse the gel twice with double-deionized water. Add destaining solution and incubate at RT, until a clear background is obtained (*see* **Note 11**).
6. The lanes generated in such a gel are shown in Fig. 3. The key observations are as follows: (1) the reduced antibody (lane 2) confirms that the TCEP procedure reduces all the interchain

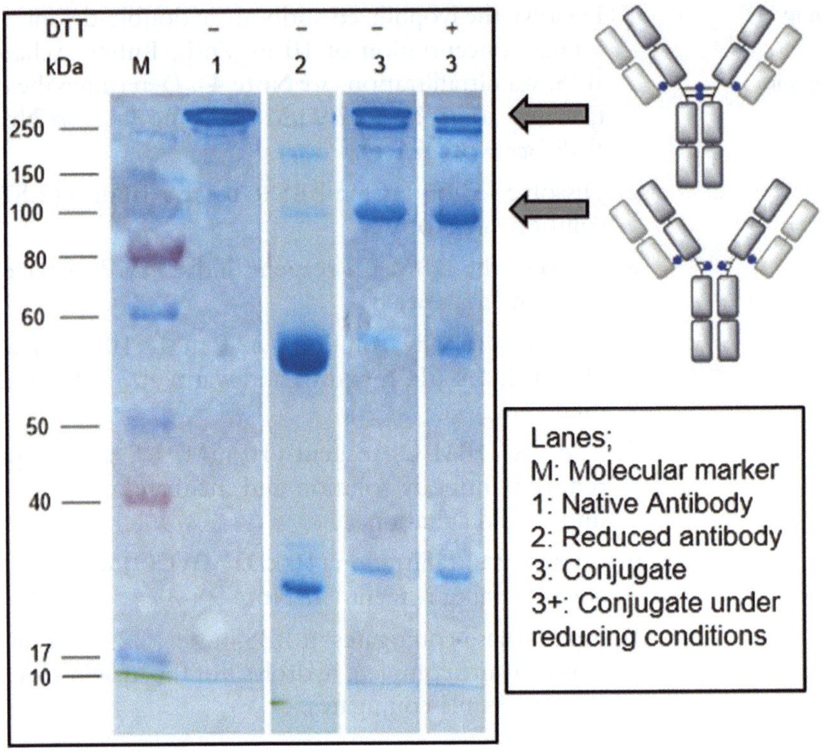

Fig. 3 SDS-PAGE analysis of DBM-C2 reagent antibody conjugation

disulfide bonds, to afford bands for the heavy chain (~50 kDa) and light chain (~25 kDa); (2) the DBM-C_2 reagent is efficient in bridging the cysteine residues generated (lane 3), reconnecting the heavy and light chains; (3) the conjugate is resistant to reducing conditions (lane 3+), confirming the installation of robustly stable maleamic acid bridges. NOTE: two dominant isomeric forms are observed in the conjugate, due to competing formation of inter- and intra-heavy chain bridges [28]. The denaturing conditions of SDS-PAGE separates these species, which are otherwise inseparable by native analytical or purification methods.

3.3 LC-MS Analysis of the Conjugates

1. Take a 20 μL aliquot of the antibody conjugate and exchange it into ammonium acetate buffer via ultrafiltration (10 kDa MWCO). Determine the concentration by UV/Vis absorbance and adjust it to 0.7 μM with ammonium acetate buffer.
2. Add PNGase F (1 μL) to the antibody conjugate and incubate at 37 °C for 2 h, to effect deglycosylation.
3. Filter the resultant solution though a 0.22 μm centrifuge tube filter to remove any residual particles, prior to LC-MS analysis.

4. 10 μL of the resultant solution is injected onto the PLRP column, which is maintained at 60 °C. The flow rate used is 300 μL/min and the gradient is as follows: 15% B for the first 2 min followed by increase to 32% B over 1 min, remained at 32% B for 1 min. Mobile phase B then increased to 50% over 10 min, with further increase of B to 95% over 4 min and maintained at 95% B for 2 min. The mobile phase B is changed to 15% B for the final 3 min.
5. Acquire mass spectra in positive electrospray ionization (ESI) mode using the *m/z* range 800–5000.
6. Convert the raw data to zero charge mass spectra using a maximum entropy deconvolution algorithm.

The LC-MS data for trastuzumab, reduced trastuzumab, and a trastuzumab DBM-C_2 conjugate are shown in Fig. 4, to highlight what is to be expected. In each case the total ion count (TIC), raw data, and deconvoluted spectrum are shown. The following key observations are made: (1) the PNGase F step is important to remove heterogeneity due to glycosylation, leading to well resolved spectra; (2) the reduced antibody is observed as separated heavy and light chains due to the denaturing conditions of the analytical method; (3) the antibody DBM–C_2 conjugate is observed as two dominant species, that is, the full antibody conjugate with four maleamic acid bridges and the half-antibody with two maleamic acid bridges. The latter species is derived from the presence of the intra heavy chain bridge, consistent with the SDS-PAGE. Under the highly acidic LC-MS conditions (pH 2–3), maleamic acid conjugates are sometimes observed to undergo a small amount of fragmentation that leads to loss of the payload, retaining a maleic acid bridge [28].

4 Notes

1. The term "degassing" refers to the process of removing O_2 from a solution by bubbling argon through it for 10 min.
2. SDS-PAGE loading buffers can be stored in aliquots at −20 °C for up to 12 months.
3. DTT solution should be prepared fresh each time.
4. If the antibody is already obtained in solution, buffer-exchange this directly into BBS buffer.
5. This exact concentration is not critical. The equivalents of TCEP and DBM-C_2 relative to the antibody must be

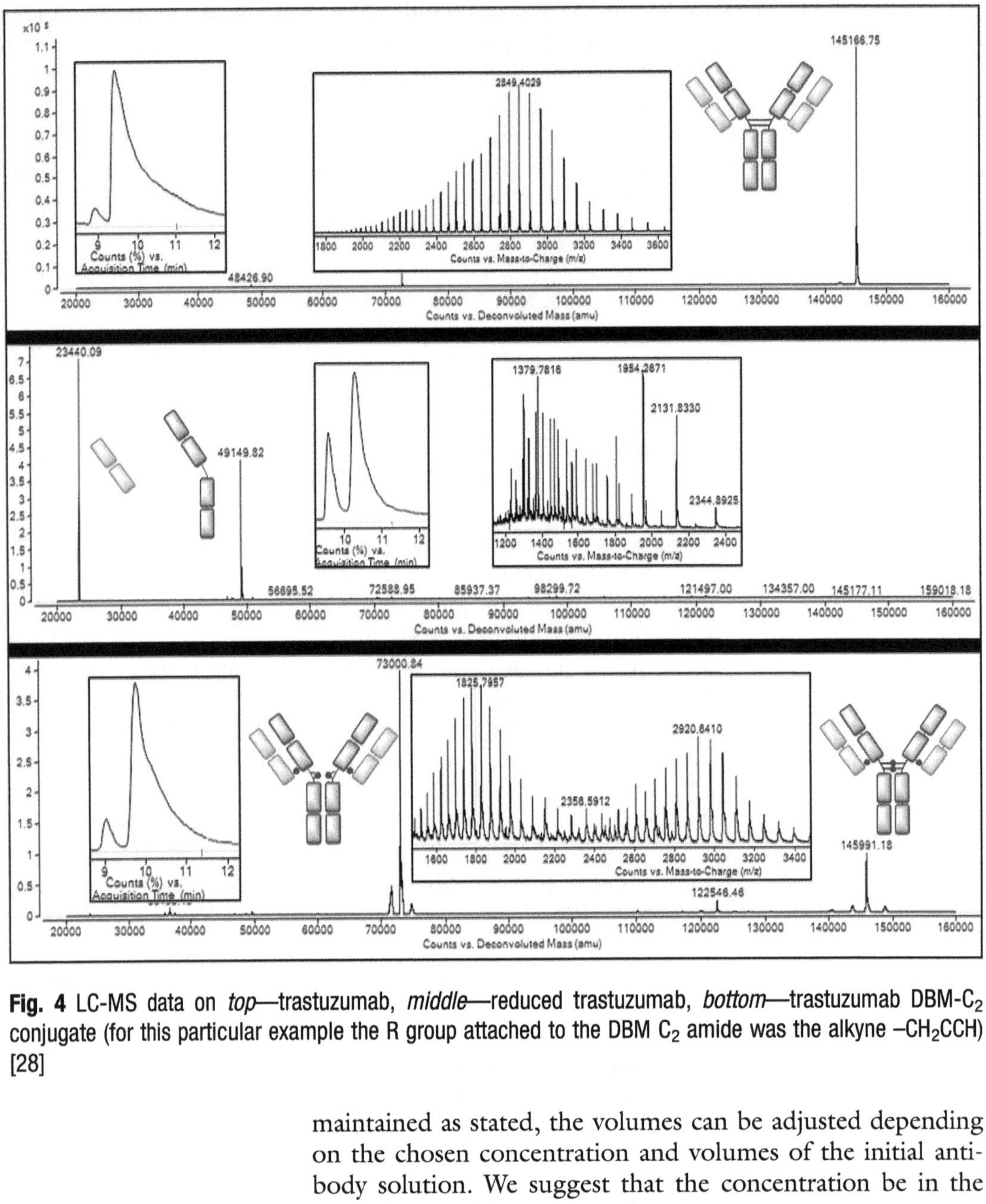

Fig. 4 LC-MS data on *top*—trastuzumab, *middle*—reduced trastuzumab, *bottom*—trastuzumab DBM-C_2 conjugate (for this particular example the R group attached to the DBM C_2 amide was the alkyne $–CH_2CCH$) [28]

maintained as stated, the volumes can be adjusted depending on the chosen concentration and volumes of the initial antibody solution. We suggest that the concentration be in the range 20–40 μM.

6. Antibody aliquots can be flash-frozen and stored at −80 °C.
7. The TCEP solution in BBS should be prepared fresh prior to use.
8. DBM-C_2 solution should be prepared fresh each time.
9. Ultrafiltration at this stage is not essential and can instead be carried out after the subsequent hydrolysis step for convenience.

10. Staining can be accelerated by microwaving the gel at full power for 1 min, letting it cool to room temperature and repeating for 1 min.
11. Destaining can be accelerated by microwaving the gel at full power for 1 min.

References

1. Wang LT, Amphlett G, Blattler WA, Lambert JM, Zhang W (2005) Structural characterization of the maytansinoid—monoclonal antibody immunoconjugate, huN901-DM1, by mass spectrometry. Protein Sci 14 (9):2436–2446
2. Junutula JR, Raab H, Clark S, Bhakta S, Leipold DD, Weir S et al (2008) Site-specific conjugation of a cytotoxic drug to an antibody improves the therapeutic index. Nat Biotechnol 26(8):925–932
3. Junutula JR, Flagella KM, Graham RA, Parsons KL, Ha E, Raab H et al (2010) Engineered thio-trastuzumab-DM1 conjugate with an improved therapeutic index to target human epidermal growth factor receptor 2-positive breast cancer. Clin Cancer Res 16 (19):4769–4778
4. Hamblett KJ, Senter PD, Chace DF, Sun MMC, Lenox J, Cerveny CG et al (2004) Effects of drug loading on the antitumor activity of a monoclonal antibody drug conjugate. Clin Cancer Res 10(20):7063–7070
5. Beckley NS, Lazzareschi KP, Chih HW, Sharma VK, Flores HL (2013) Investigation into temperature-induced aggregation of an antibody drug conjugate. Bioconjug Chem 24 (10):1674–1683
6. Adem YT, Schwarz KA, Duenas E, Patapoff TW, Galush WJ, Esue O (2014) Auristatin antibody drug conjugate physical instability and the role of drug payload. Bioconjug Chem 25(4):656–664
7. Boswell CA, Mundo EE, Zhang C, Bumbaca D, Valle NR, Kozak KR et al (2011) Impact of drug conjugation on pharmacokinetics and tissue distribution of anti-STEAP1 antibody-drug conjugates in rats. Bioconjug Chem 22(10):1994–2004
8. Alley SC, Benjamin DR, Jeffrey SC, Okeley NM, Meyer DL, Sanderson RJ et al (2008) Contribution of linker stability to the activities of anticancer immunoconjugates. Bioconjug Chem 19(3):759–765
9. Shen B-Q, Xu K, Liu L, Raab H, Bhakta S, Kenrick M et al (2012) Conjugation site modulates the in vivo stability and therapeutic activity of antibody-drug conjugates. Nat Biotechnol 30(2):184–189
10. Dennler P, Fischer E, Schibli R (2015) Antibody conjugates: from heterogeneous populations to defined reagents. Antibodies 4 (3):197–224
11. Jackson D, Atkinson J, Guevara CI, Zhang CY, Kery V, Moon SJ et al (2014) In vitro and in vivo evaluation of cysteine and site specific conjugated herceptin antibody-drug conjugates. PLoS One 9(1):14
12. Akkapeddi P, Azizi SA, Freedy AM, Cal P, Gois PMP, Bernardes GJL (2016) Construction of homogeneous antibody-drug conjugates using site-selective protein chemistry. Chem Sci 7 (5):2954–2963
13. Jackson DY (2016) Processes for constructing homogeneous antibody drug conjugates. Org Process Res Dev 20(5):852–866
14. Chudasama V, Maruani A, Caddick S (2016) Recent advances in the construction of antibody-drug conjugates. Nat Chem 8 (2):113–118
15. Kuan SL, Wang T, Weil T (2016) Site-selective disulfide modification of proteins: expanding diversity beyond the proteome. Chem Eur J 22(48):17112–17129
16. Bryant P, Pabst M, Badescu G, Bird M, McDowell W, Jamieson E et al (2015) In vitro and in vivo evaluation of cysteine rebridged trastuzumab-MMAE antibody drug conjugates with defined drug-to-antibody ratios. Mol Pharm 12(6):1872–1879
17. Badescu G, Bryant P, Bird M, Henseleit K, Swierkosz J, Parekh V et al (2014) Bridging disulfides for stable and defined antibody drug conjugates. Bioconjug Chem 25 (6):1124–1136
18. Maruani A, Smith MEB, Miranda E, Chester KA, Chudasama V, Caddick S (2015) A plug-and-play approach to antibody-based therapeutics via a chemoselective dual click strategy. Nat Commun 6:6645
19. Bahou C, Richards DA, Maruani A, Love EA, Javaid F, Caddick S et al (2018) Highly homogeneous antibody modification through optimisation of the synthesis and conjugation of

functionalised dibromopyridazinediones. Org Biomol Chem 16(8):1359–1366

20. Tedaldi LM, Smith MEB, Nathani R, Baker JR (2009) Bromomaleimides; new reagents for the selective and reversible modification of cysteine. Chem Commun (43):6583–6585
21. Smith MEB, Schumacher FF, Ryan CP, Tedaldi LM, Papaioannou D, Waksman G et al (2010) Protein modification, bioconjugation, and disulfide bridging using bromomaleimides. J Am Chem Soc 132(6):1960–1965
22. Ryan CP, Smith MEB, Schumacher FF, Grohmann D, Papaioannou D, Waksman G et al (2011) Tunable reagents for multi-functional bioconjugation: reversible or permanent chemical modification of proteins and peptides by control of maleimide hydrolysis. Chem Commun 47(19):5452–5454
23. Castaneda L, Maruani A, Schumacher FF, Miranda E, Chudasama V, Chester KA et al (2013) Acid-cleavable thiomaleamic acid linker for homogeneous antibody-drug conjugation. Chem Commun 49(74):8187–8189
24. Schumacher FF, Sanchania VA, Tolner B, Wright ZVF, Ryan CP, Smith MEB et al (2013) Homogeneous antibody fragment conjugation by disulfide bridging introduces 'spinostics'. Sci Rep 3:1525
25. Schumacher FF, Nunes JPM, Maruani A, Chudasama V, Smith MEB, Chester KA et al (2014) Next generation maleimides enable the controlled assembly of antibody-drug conjugates via native disulfide bond bridging. Org Biomol Chem 12(37):7261–7269
26. Nunes JP, Morais M, Vassileva V, Robinson E, Rajkumar VS, Smith ME et al (2015) Functional native disulfide bridging enables delivery of a potent, stable and targeted antibody-drug conjugate (ADC). Chem Commun 51 (53):10624–10627
27. Robinson E, Nunes JP, Vassileva V, Maruani A, Nogueira J, Smith MEB et al (2017) Pyridazinediones deliver potent, stable, targeted and efficacious antibody-drug conjugates (ADCs) with a controlled loading of 4 drugs per antibody. RSC Adv
28. Morais M, Nunes JPM, Karu K, Forte N, Benni I, Smith MEB et al (2017) Optimisation of the dibromomaleimide (DBM) platform for native antibody conjugation by accelerated post-conjugation hydrolysis. Org Biomol Chem 15(14):2947–2952

Chapter 3

Lysine Bioconjugation on Native Albumin with a Sulfonyl Acrylate Reagent

Maria J. Matos, Gonzalo Jiménez-Osés, and Gonçalo J. L. Bernardes

Abstract

This protocol details a novel bioconjugation strategy that uses a methanesulfonyl acrylate reagent that is directed to the most reactive lysine on human serum albumin, which enables the construction of chemically defined and stable bioconjugates. The reaction proceeds rapidly and a regioselective modification is achieved using a single molar equivalent of the reagent under biocompatible conditions (37 °C, pH 8.0). Importantly, the bioconjugate retains both the secondary structural content and function of the unmodified protein. During the reaction of the amino group of lysine and the sulfonyl acrylate reagent, methanesulfinic acid is released after the conjugate addition, which then generates an electrophilic acrylate moiety on the protein. This acrylate can be further used for site-specific protein labeling using a synthetic molecule bearing a reactive amine under biocompatible conditions (21 °C, pH 8.0).

Key words Lysine, Sulfonyl acrylate, Bioconjugation, Aza-Michael addition, Human serum albumin

1 Introduction

Protein-drug conjugates have been reported and evaluated in clinical trials [1, 2]. Compared to the parental small molecule drugs, protein–drug conjugates offer several advantages, including half-life extension, localization to a target tissue, avoidance of drug--drug interactions, and toxicity reduction [3–5]. Site- or residue-specific methods to install drugs on proteins simplify the interpretation of results and yield predictable conjugates [6]. However, chemical reactions that can proceed under conditions mild enough to maintain the structure and function of the modified proteins are limited. Even rarer are chemical strategies that can target a single site, leading to products with uniform properties and optimal function without the need for sequence engineering [7].

Conjugation to serum albumin (HSA) has emerged as a powerful approach for adjuvanting the immune response triggered by synthetic vaccines and extending the in vivo half-life of many small-molecule and peptide/protein drugs [8, 9]. However, HSA

Sam Massa and Nick Devoogdt (eds.), *Bioconjugation: Methods and Protocols*, Methods in Molecular Biology, vol. 2033, https://doi.org/10.1007/978-1-4939-9654-4_3,

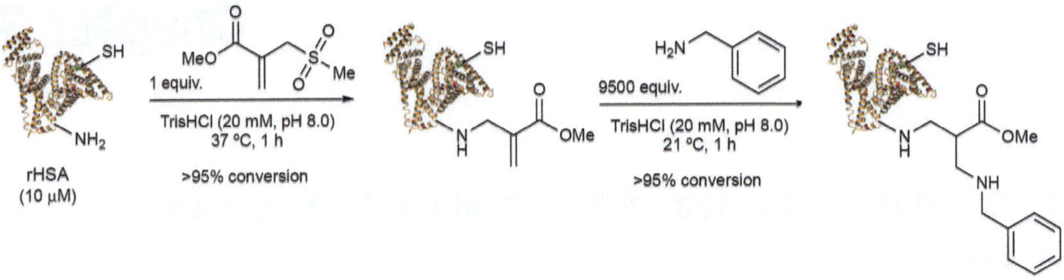

Fig. 1 Chemoselective and regioselective modification of lysine on rHSA. Briefly, the addition reaction of the most reactive lysine on the sulfonyl acrylate reagent is followed by spontaneous elimination of methanesulfinic acid. This restores a type 2 alkene that undergoes site-specific modification through aza-Michael addition of suitable synthetic molecules bearing an amine nucleophile, such as benzylamine

conjugation strategies, can often yield heterogeneous mixtures with inadequate pharmacokinetics, low efficacies, and variable safety profiles. To date, none of these compounds have displayed site specificity for a single lysine (site-selective bioconjugation). Barbas et al. first approached this goal in 2014 by showing that certain lysine residues in albumin reacted more rapidly with α,-β-unsaturated sulfonamides than other lysine residues [10].

The method reported therein explores a two-step strategy to build protein–drug conjugates: the first step consists of lysine chemoselective and regioselective chemical modification utilizing a methanesulfonyl acrylate aza-Michael acceptor; the second step then generates a type-2 alkene that can be further modified through a second aza-Michael ligation with amine- and hydroxylamine-containing molecules (Fig. 1) [11]. Many relevant conjugation reagents such as fluorescent probes, polyethylene glycol (PEG), carbohydrate, deoxyribonucleic acid (DNA), or drug derivatives are commercially available and feature a free reactive amine handle. This two-step, irreversible, and chemoselective approach can be directly used to modify several commercial proteins and antibodies bearing free lysine residues (can be confirmed by protein digestion and analysis via LC–MS/MS analysis—*see* Subheading 3.3), and in some cases even in the presence of potentially reactive cysteines. In this protocol, we describe the method for the modification of recombinant human serum albumin (rHSA) at a single lysine from potential 59 reactive lysines.

2 Materials

Prepare all reagents and solutions using ultrapure water (prepared by purifying deionized water at 25 °C to >15 MΩ/cm resistance and filtered through a 0.2-μm disc filter) and analytical grade reagents. Prepare all reagents at room temperature and store them at 4 °C (unless indicated otherwise). Use all commercially available reagents as received unless otherwise noted. Reagents used in the procedure are potentially dangerous, and appropriate care should be taken during their manipulation. The solid and liquid waste products generated should be disposed of appropriately, as defined locally.

2.1 Sulfonyl Acrylate Reagent

1. Methyl 2-(bromomethyl)acrylate 97%.
2. Methanol anhydrous, 99.8%.
3. Sodium methanesulfinate technical grade, 85%.
4. Ethyl acetate (distilled under CaH_2).
5. Hexane (distilled under CaH_2).
6. Sodium chloride saturated solution.
7. Magnesium sulfate anhydrous, ≥98.0%.
8. Sulfonyl acrylate reagent characterization:
   (a) Potassium permanganate ($KMnO_4$) staining solution for thin layer chromatography (TLC).
   (b) Silica gel high purity grade, Merck grade 9385 pore size 60 Å, 230–400 mesh particle size for flash column chromatography (5 cm diameter column).
   (c) Deuterated chloroform ($CDCl_3$) for nuclear magnetic resonance (^{1}H NMR and ^{13}C RMN).
   (d) Mobile phases for liquid chromatography–mass spectrometry (LC-MS): 95% aqueous acetonitrile with 0.05% formic acid and 10 mM ammonium acetate with 0.1% formic acid; 50% aqueous acetonitrile with 0.25% formic acid.

2.2 Reagents for Protein Modification

1. rHSA was kindly provided by Albumedix Limited. rHSA (59 lysines, 1 free cysteine, 17 disulfide bonds).
2. Methyl 2-((methylsulfonyl)methyl)acrylate (**1**, synthesized as described in Subheading 3.1).
3. Benzylamine ReagentPlus®, 99%.
4. *N,N*-Dimethylformamide anhydrous (DMF).

2.3 Solutions for Protein Modification

1. Formic acid, LC-MS grade.
2. Acetonitrile for high-performance liquid chromatography (HPLC ultragradient grade).
3. LC-MS solvent A for protein/antibody analysis: deionized H_2O + 0.1% formic acid.
4. LC-MS solvent B for protein/antibody analysis: 71% MeCN, 29% H_2O, 0.075% formic acid.
5. 20 mM Tris–HCl, pH 8.0.

(a) Prepare 800 mL of distilled water in a suitable container.
(b) Add 2.420 g of Tris base to the solution.
(c) Adjust solution to pH 8.0 using HCl (*see* **Notes 1** and **2**).
(d) Add distilled water until volume is 1 L.
(e) Filtrate under vacuum using Millipore Express® PLUS Membrane Filters (*see* **Note 3**).

6. 50 mM NaPi, pH 7.4.
   (a) Prepare 800 mL of distilled water in a suitable container.
   (b) Add 10.105 g of sodium phosphate dibasic (Na_2HPO_4) to the solution.
   (c) Add 1.697 g of sodium phosphate monobasic (NaH_2PO_4) to the solution.
   (d) Adjust solution to final desired pH using HCl or NaOH (*see* **Note 2**).
   (e) Add distilled water until volume is 1 L.
   (f) Filtrate under vacuum using Millipore Express® PLUS Membrane Filters (*see* **Note 3**).

2.4 Equipment

1. Bruker 400 MHz DPX-400 Dual Spectrometer.
2. LC-MS system for sulfonyl acrylate reagent characterization: Waters micromass ZQ instrument equipped, electrospray (ESI) ionization, with Waters 2795 HPLC and a Waters 2996 photodiode array detector. The separation technology is based on a 50 × 4.6 mm C18 column (currently a Phenomenex Kinetix solid core column).
3. LC–MS system for protein analysis: Xevo G2-S TOF mass spectrometer coupled to an Acquity UPLC system using an Acquity Q6 UPLC BEH300 C4 column (1.7 mm, 2.1 × 50 mm).
4. Chirascan spectrophotometer equipped with a Quantum TC125.
5. Vortex shaker suitable for short-time operation (touch function), activated by pressing shaker attachment or continuous operation.
6. Eppendorf thermomixer shaker incubator—heating/mixing/cooling.

2.5 Consumables

1. Flash-chromatography column available with ball joints (capacity 400 mL, useful for rapid, preparative separations).
2. TLC plates: Merck TLC Silica gel 60F254 glass plates (cut into rectangles of 5 × 15 cm).
3. NMR tubes: thin glass walled tubes (5 mm diameter).

4. Mass vials: polypropylene 12 × 32 mm screw neck vial, with cap and preslit PTFE/silicone septa 300 μL volume.
5. Quartz cuvette for circular dichroism: parameter 200–2500 nm spectral range, pathlength 10 mm, chamber volume 300 μL.

3 Methods

Carry out all procedures at room temperature, unless otherwise specified.

3.1 Synthesis of Sulfonyl Acrylate Reagent

1. Prepare a solution of 0.18 mL of methyl 2-(bromomethyl) acrylate (1.5 mmol) in 5 mL of methanol.
2. Treat the solution with 0.18 g of sodium methanesulfinate (1.5 mmol) portion wise over 10 min at room temperature (*see* **Note 4**).
3. Stir the solution at room temperature for a period of 1.25 h and concentrate the solution in vacuum by evaporating all the solvent.
4. The residue is taken up in water (10 mL) and extracted four times with ethyl acetate (4× 10 mL) using a liquid/liquid extraction flask. Liquid–liquid extraction is a very well-known method to separate compounds (usually the desired compound from impurities) based on their relative solubilities in two different immiscible liquids, usually water (polar) and an organic solvent (nonpolar).
5. Wash the combined ethyl acetate solution (40 mL) with saturated sodium chloride solution in an Erlenmeyer flask, dry over anhydrous magnesium sulfate, filter (fluted paper filter) and concentrate in vacuum to get a white solid.
6. Purify the solid residue by flash column chromatography (pre-packed until 16 cm high with silica gel—*see* Subheading 2.1) eluting with hexane–ethyl acetate 8:2, to get the desired product (Scheme 1) as a white solid (0.175 g, 65% yield) (*see* **Notes 5** and **6**).
7. Analyze reaction mixtures by analytical TLC on TLC silica gel plates (*see* Subheading 2.5). Visualization is accomplished with UV light (254 nm) or $KMnO_4$ staining solution.

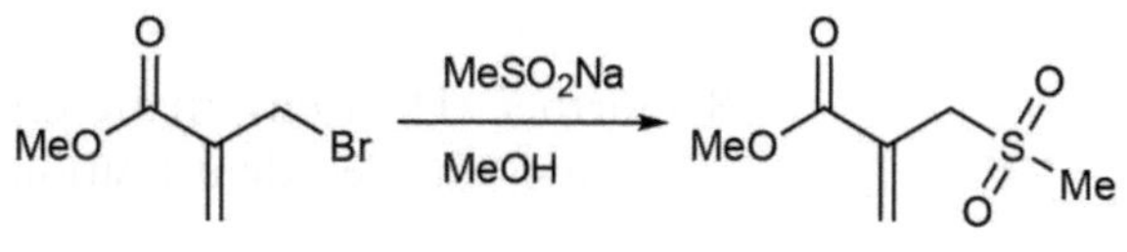

Scheme 1 Synthesis of **1**

8. ^{1}H NMR and ^{13}C NMR spectra are recorded on a Bruker 400 MHz DPX-400 Dual Spectrometer in deuterated $CDCl_3$ as a solvent using tetramethylsilane as an internal standard. To prepare the sample, 5 mg of the product are dissolved in deuterated $CDCl_3$ and transferred to an RMN tube. Chemical shifts are reported in parts per million (ppm) on the δ scale from tetramethylsilane (NMR descriptions: s, singlet).
9. Mass spectroscopy is performed using a Waters micromass ZQ (LC–MS) instrument (*see* Subheading 2.4). This system is an automated service utilizing electrospray (ESI) ionization. To prepare the sample, 1 mg of the product is dissolved in methanol and transferred to a mass vial (*see* Subheading 2.5).
10. The mobile phases are described in Subheading 2.1. The separation technology is based on a 50 × 4.6 mm C18 column (currently a Phenomenex Kinetix solid core column) (*see* **Note 7**). The system runs using 50% aqueous acetonitrile with 0.25% formic acid as mobile phase and can measure accurate masses from 150 Da to 1500 Da.
11. Characterization of compound **1**: Melting point 64.9–66.7 °C. ^{1}H NMR (400 MHz, $CDCl_3$): 2.89 (s, 3H, CH_3), 3.81 (s, 3H, OCH_3), 4.05 (s, 2H, CH_2), 6.15 (s, 1H, CH_2), 6.62 (s, 1H, CH_2). ^{13}C NMR (100 MHz, $CDCl_3$): 40.5, 52.7, 56.4, 128.8, 134.2, 165.8. DEPT: 40.5, 52.7, 56.4, 134.2. IR (ATR): 1712 (CO) cm^{-1}. LC-MS *m/z* (ESI): 179 (MH^+). High-resolution mass spectrometry (HRMS) calcd. For $C_6H_{10}O_4S$ (MH^+): 179.0300; found: 179.0347. Data is consistent with that previously reported [12].

3.2 rHSA Conjugation with 1

1. Prepare a solution of 9.4 μL of 20 mM Tris–HCl, pH 8.0 and 3.3 μL of DMF (*see* **Note 8**) in an eppendorf.
2. Add a 26.6 μL aliquot of a stock solution of 15.05 μM rHSA (*see* **Notes 9** and **10**).
3. Vortex the resulting mixture for 10 s (*see* **Note 11**).
4. Add 0.7 μL of a 0.56 mM solution of **1** (1 equiv.) in DMF (Fig. 2, *see* **Notes 12** and **13**).
5. Shake the reaction for 1, 2, and 24 h, at 37 °C (600 rpm).
6. Analyze a 10 μL aliquot at each reaction time by LC–MS (*see* **Note 14**).
7. Complete conversion to the expected product is observed after 1 h (calculated mass, 66,536 Da; observed mass, 66,536 Da, Fig. 3).
8. Further rHSA enzymatic digestion followed by LC–MS/MS analysis enables identification of the modified lysine residue. In

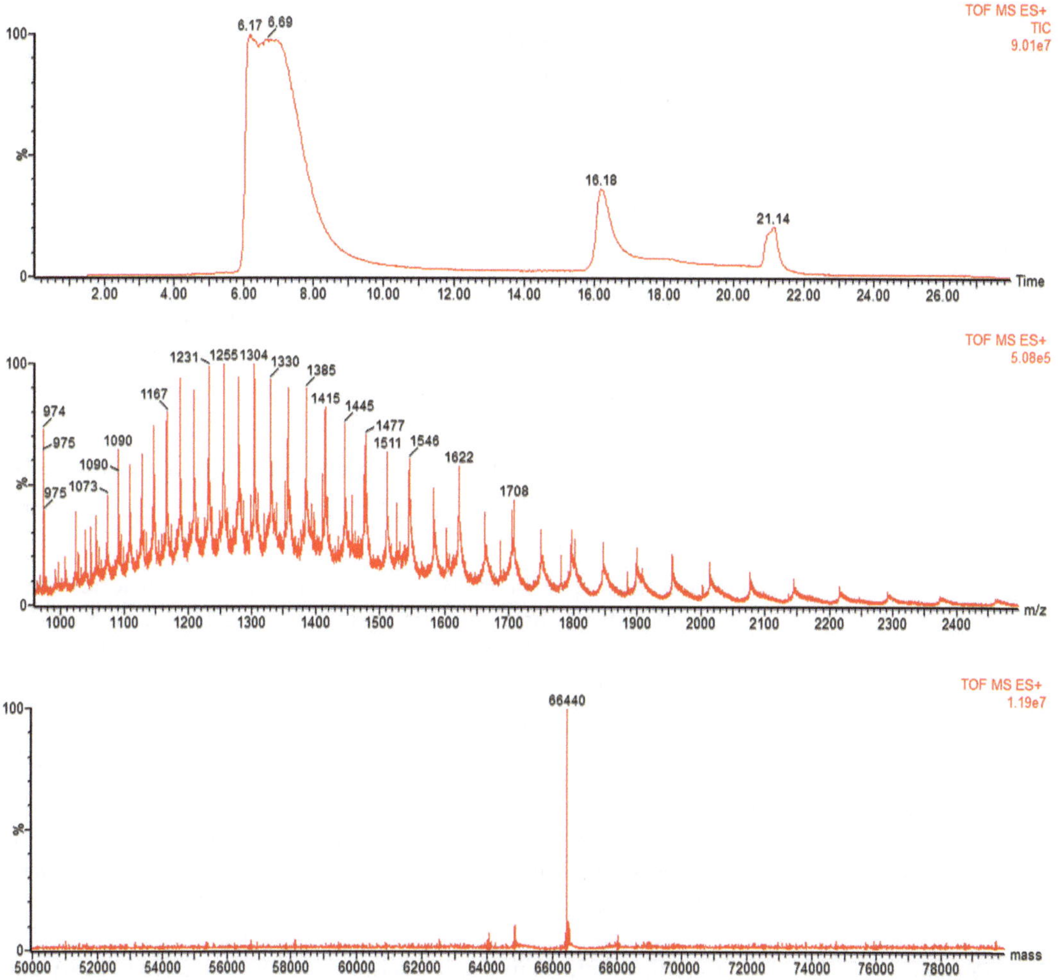

Fig. 2 Typical analysis of rHSA conjugation. The total ion chromatogram, combined ion series and deconvoluted spectra are shown

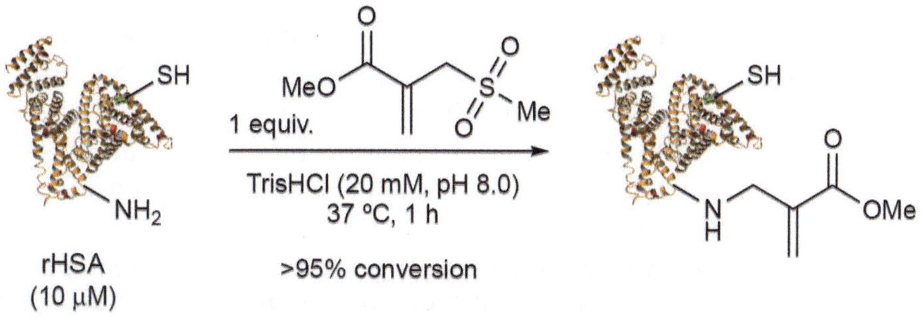

Fig. 3 Scheme of reaction for the bioconjugation rHSA with **1**

the case of rHSA, we found that the modification occurred in the peptide G**K**KLVAASQAALGL (modified residue underlined and in bold, total of 91% sequence coverage) as observed in the MS/MS spectrum, which corresponds to lysine 573 [11].

3.3 Protein Analysis by LC-MS

1. LC–MS is performed on a Xevo G2-S (*see* Subheading 2.4).
2. LC-MS solvents A and B (*see* Subheading 2.3) are used as the mobile phase at a flow rate of 0.2 mL/min. The gradient is programmed as follows: 72% A to 100% B after 25 min then 100% B for 2 min and after that 72% A for 18 min (*see* **Note 15**).
3. The electrospray source is operated with a capillary voltage of 2.0 kV and a cone voltage of 40 V.
4. Nitrogen is used as the desolvation gas at a total flow of 850 L/h.
5. Mass vials described in Subheading 2.5 are used for the measurements.
6. Total mass spectra are reconstructed from the ion series using the MaxEnt algorithm preinstalled on MassLynx software (v. 4.1 from Waters) according to the manufacturer's instructions.
7. To obtain the ion series described, the major peak(s) of the chromatogram are selected for integration and further analysis.
8. The total ion chromatogram, combined ion series and deconvoluted spectra are shown for the rHSA (Fig. 4). Identical analyses are carried out for all the conjugation reactions performed.

3.4 Circular Dichroism

1. Circular dichroism spectroscopy is used to analyze rHSA and rHSA–**1** secondary structure in solution (Fig. 5). The different types of secondary structure—α-helix, β-sheet, and random coil—all have characteristic circular dichroism spectra in the far-UV region of the spectrum (190–250 nm).
2. Concentrate the samples to 10 nM in 50 mM NaPi, pH 7.4 using a Vivaspin® 500, 30,000 MWCO PES (Sartorius) column (*see* **Note 16**).
3. Record circular dichroism measurements using a Chirascan spectrophotometer equipped with a Quantum TC125 temperature control unit (25 °C).
4. Acquire the data in a 0.1 cm path length with a response time of 1 s, a per-point acquisition delay of 5 ms and a prescan and postscan delay of 50 ms.
5. Spectra are an averaged over three scans, in a wavelength range from 200 nm to 260 nm, and the spectrum from a blank sample containing only buffer was subtracted from the averaged data.

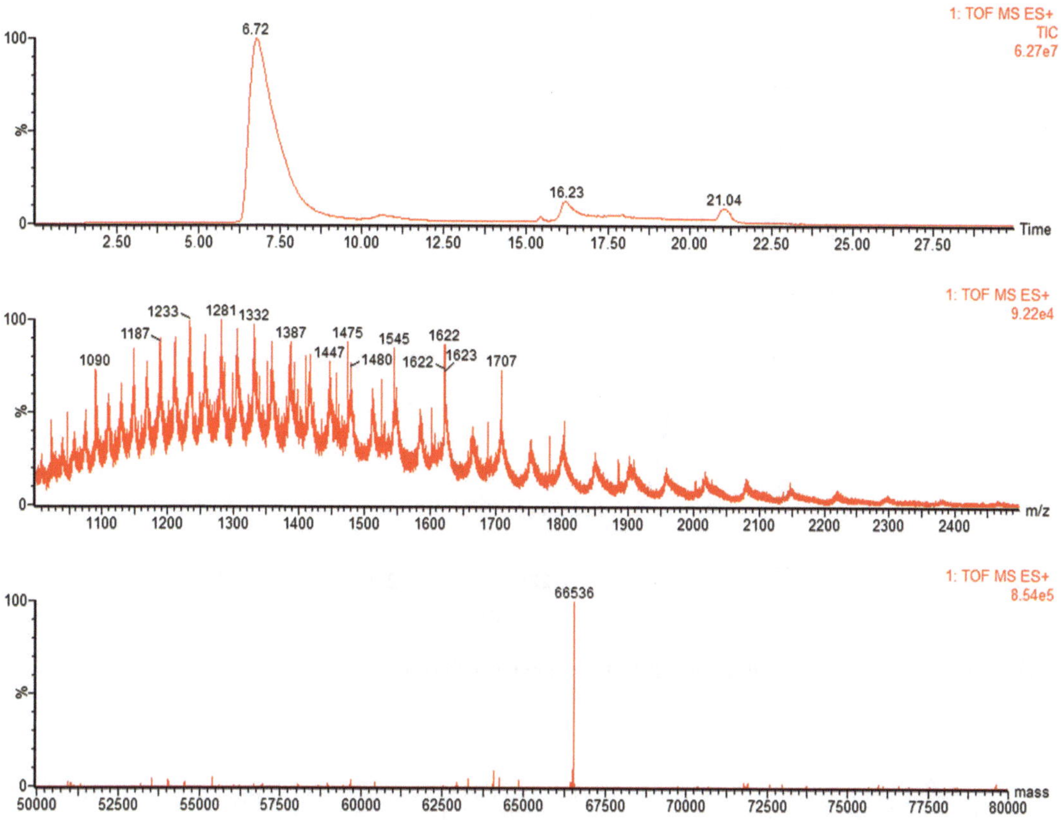

Fig. 4 Combined ion series and deconvoluted mass spectrum of the reaction of rHSA (10 μM) with 1 equiv. of **1** after 1 h at 37 °C. Identical data is obtained at 2 and 24 h

6. Circular dichroism analysis of rHSA and rHSA–**1** showed no alterations in secondary structural content, which reflects the mildness and efficiency of the conjugation process.

3.5 rHSA–1 Conjugation with Benzylamine

1. Thaw a 20 μL aliquot of rHSA–**1** (10 μM) in 20 mM Tris–HCl, pH 8.0 buffer (*see* **Notes 17** and **18**).
2. Add 0.2 μL of 9.5 M solution of benzylamine (9500 equiv., in DMF) (Fig. 6, *see* **Note 19**).
3. Vortex the resulting mixture for 10 s.
4. Shake the mixture for an additional hour at 21 °C (600 rpm).
5. Remove small molecules from the reaction mixture by loading the sample onto a Zeba Spin Desalting Column previously equilibrated with 20 mM Tris–HCl, pH 8.0 buffer (*see* **Note 20**). The sample is eluted via centrifugation for 2 min at 1500 × *g*.
6. Analyze a 10 μL aliquot by UPLC-MS (described in Subheading 3.2).

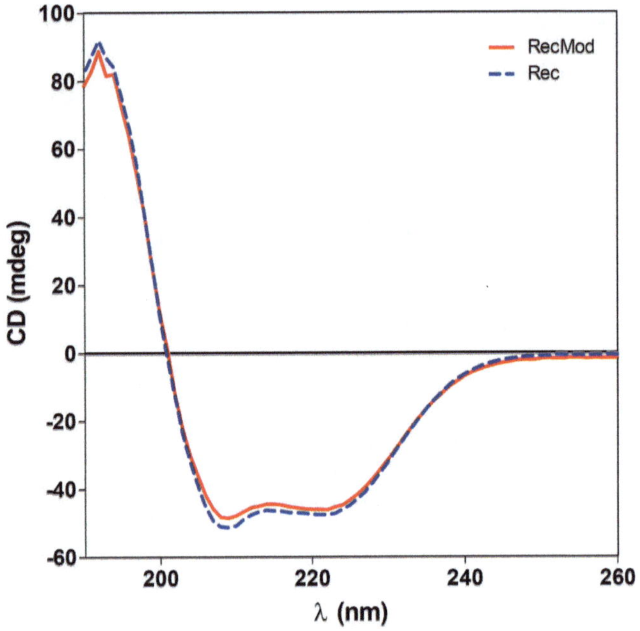

Fig. 5 Structural analysis of rHSA and rHSA–**1** by circular dichroism

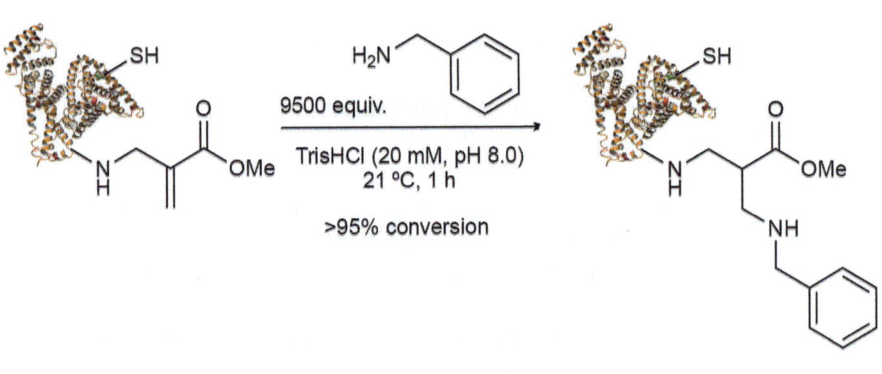

Fig. 6 Scheme of reaction for the bioconjugation rHSA–**1** with benzylamine

7. Complete conversion to the expected product is observed after 1 h (calculated mass, 66,645; observed mass, 66,643, Fig. 7).

4 Notes

1. The Tris–HCl buffer should be strictly 20 mM and pH 8.0. The reaction is very sensitive to the presence of salts, ionic strength and pH. The use of 20 mM Tris–HCl, pH 8.0 buffer is crucial to achieve complete conversion to the elimination product.

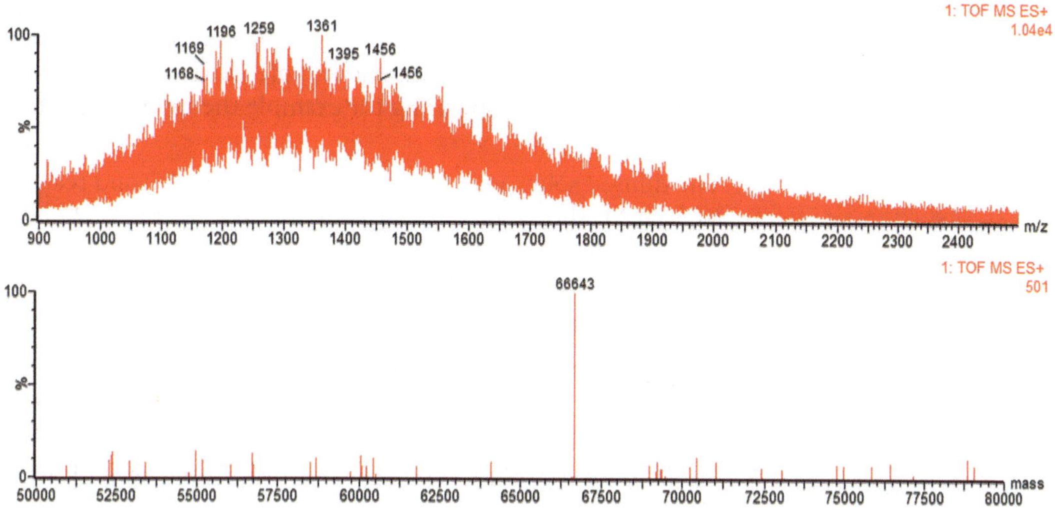

Fig. 7 Combined ion series and deconvoluted mass spectrum of the reaction of rHSA–**1** (10 μM) with 9500 equiv. of benzylamine in 20 mM Tris–HCl, pH 8.0 Tris–HCl with 10% DMF after 1 h at 21 °C

2. Concentrated HCl (12 M) can be used at first to narrow the gap from the starting pH to the required pH. From then on it would be better to use a series of HCl (i.e., 6 and 1 M) with lower ionic strengths to avoid a sudden drop in pH below the required pH.
3. When the buffers are not being used every day, consider to storage them at 4 °C to avoid degradation. Also, take aliquots and avoid using buffers prepared by other users. Once prepared and filtrated, the buffers can be stored and used for up to 1 month at room temperature (15–25 °C).
4. Sodium methanesulfinate must be added portionwise for the reaction to be complete.
5. Methyl 2-((methylsulfonyl)methyl)acrylate (**1**) can be stored as a solid at −20 °C for at least 4 months.
6. Toxicity information for this compound is not available: assume that it is toxic. Always handle the compound with gloves and work in an appropriately ventilated environment.
7. There are several methods available enabling the user to produce mass spectra for compounds up to 2 kDa in positive and negative modes of ionization. In some cases, a Waters LCT Premier combined with an Agilent 1100 autosampler can also be used.
8. The use of 10% DMF ensures the complete solubility of the reagent.

9. rHSA (100 mg/mL stock solution) can be stored at 4 °C for up to 6 months. Consider running a sample control before starting the experiments. Check the sample concentration before starting the labeling by using for example the Bradford protein assay.
10. The diluted solutions of both protein (15.05 μM) and reagent (0.56 mM) should be fresh (storage for no longer than 24 h, at 4 °C, before use). Stock solutions (1 M) of this reagent should be freshly prepared on the day of the experiment.
11. Vortex all the solutions for them to be homogeneous.
12. More than one single modification can be observed when using more than 1 equivalent of reagent. Concentrations of stock solutions should be accurate. To avoid errors, prepare a concentrated stock solution (1 M) and take aliquots from it to prepare diluted ones.
13. If the reaction is scaled up, the linker should be gradually dropped into the reaction solution while shaking. Both reaction and bioconjugate analysis are similar up to 150 μM/500 μL (highest concentration/volume tested).
14. Centrifuging the samples prior to the run helps remove insoluble particles that can damage the HPLC column (2000 × *g*, 10 s, room temperature).
15. For best results, freshly prepare all HPLC solvents and purge the LC-MS system before analysis of protein samples. Results can be slightly different based on the equipment.
16. Vivaspin® 500 columns are ideal for small molecules elimination, buffer exchange and concentration of the protein/bioconjugate samples. Follow the manufacturer's instructions.
17. For further use of rHSA–**1** conjugate, reagent excess can be removed by loading the sample onto a Zeba spin desalting column, preequilibrated with 20 mM Tris–HCl, pH 8.0. Perform the purification eluting the sample following the manufacturer's instructions. Check the sample concentration after this procedure using for example the Bradford protein assay.
18. The rHSA–**1** conjugate can be stored at −20 °C for up to a month.
19. Conditions must be optimized for different amines. Terminal primary and secondary amines can be used.
20. The huge excess of benzylamine may interfere with the resolution of the LC-MS spectra and also damage the HPLC column. Small molecules should be removed from the reaction mixture following the described method.

Acknowledgments

M.J.M. thanks Xunta da Galicia and Galician Plan of Research, Innovation and Growth 2011–2015 (Plan I2C, ED481B 2014/086-0 and ED481B 2018/007). G.J.O. thanks D.G.I. MINECO/FEDER (CTQ2015-70524-R and RYC-2013-14706 grants). G.J.L.B. is a Royal Society University Research Fellow (URF\R\180019).

References

1. Vhora I, Patil S, Bhatt P, Misra A (2015) Protein- and peptide-drug conjugates: an emerging drug delivery technology. Adv Protein Chem Struct Biol 98:1–55
2. Flygare JA, Pillow TH, Aristoff P (2013) Antibody-drug conjugates for the treatment of cancer. Chem Biol Drug Des 81:113–121
3. Boutureira O, Bernardes GJL (2015) Advances in chemical protein modification. Chem Rev 115:2174–2195
4. Bernardim B, Cal PMSD, Matos MJ, Oliveira BL, Martínez-Sáez N, Albuquerque IS, Perkins E, Corzana F, Burtoloso ACB, Jiménez-Osés G, Bernardes GJL (2016) Stoichiometric and irreversible cysteine-selective protein modification using carbonylacrylic reagents. Nat Commun 7:13128
5. Agarwal P, Bertozzi CR (2015) Site-specific antibody–drug conjugates: the nexus of bioorthogonal chemistry, protein engineering, and drug development. Bioconjug Chem 26:176–192
6. Krall N, da Cruz FP, Boutureira O, Bernardes GJL (2016) Site-selective protein-modification chemistry for basic biology and drug development. Nat Chem 8(2):103–113
7. Freedy AM, Matos MJ, Boutureira O, Corzana F, Guerreiro A, Akkapeddi P, Somovilla VJ, Rodrigues T, Nicholls K, Xie B, Jiménez-Osés G, Brindle KM, Neves AA, Bernardes GJL (2017) Chemoselective installation of amine bonds on proteins through aza-Michael ligation. J Am Chem Soc 139 (50):18365–18375
8. Larsen MT, Kuhlmann M, Hvam ML, Howard KA (2016) Albumin-based drug delivery: harnessing nature to cure disease. Mol Cell Ther 4:3
9. Matos MJ, Labão-Almeida C, Sayers C, Dada O, Tacke M, Bernardes GJL (2018) Synthesis and biological evaluation of homogeneous thiol-linked NHC*-Au-albumin and -Trastuzumab bioconjugates. Chem Eur J 24: 12250–12253
10. Asano S, Patterson JT, Gaj T, Barbas CF (2014) Site-selective labeling of a lysine residue in human serum albumin. Angew Chem Int Ed Engl 53:11783–11786
11. Matos MJ, Oliveira BL, Martínez-Sáez N, Guerreiro A, Cal PMSD, Bertoldo J, Maneiro M, Perkins E, Howard J, Deery MJ, Chalker JM, Corzana F, Jiménez-Osés G, Bernardes GJL (2018) Chemo- and regioselective lysine modification on native proteins. J Am Chem Soc 140:4004–4017
12. Vazquez ML, Mueller RA, Talley JJ, Getman DP, DeCrescenzo GA, Sun ET (1998) Sulfonylalkanoylamino hydroxyethylamino sulfanamides useful as retroviral protease inhibitors. Google Patents

Acknowledgments

Chapter 4

Engineering Dual Variable Domains for the Generation of Site-Specific Antibody–Drug Conjugates

Alex R. Nanna and Christoph Rader

Abstract

Site-specific antibody–drug conjugate (ADC) technologies are highly desirable for the production of therapeutics with well-defined biochemical and pharmacological characteristics. We have developed a strategy to produce site-specific ADCs using a highly reactive lysine residue embedded in a dual-variable-domain (DVD) format. Here we provide protocols for the engineering, expression, and purification of the DVDs used for this strategy. We also provide a protocol for DVD–drug conjugation and describe methods for their biochemical characterization, including a catalytic assay to monitor conjugation efficiency.

Key words Antibody–drug conjugate, Homogeneous, Site-specific conjugation, Dual-variable-domain, Drug-to-antibody ratio, Catalytic antibody

1 Introduction

Antibody–drug conjugates (ADCs) have established themselves as one of the most promising classes of cancer therapeutics. The strategy of conjugating highly cytotoxic drugs to antibodies has already been clinically validated with four FDA approved ADCs and more than 60 in clinical trials [1, 2]. Due to the high specificity of monoclonal antibodies, an ADC's cytotoxic payload is selectively delivered to target tumor cells thereby reducing off-target toxicity and broadening their therapeutic indices. All of the FDA approved and several ADCs in clinical trials use nonspecific conjugation chemistries to attach the drug to the antibody via lysine (Lys, i.e., ado-trastuzumab emtansine (Kadcyla®), inotuzumab ozogamicin (Besponsa®), and gemtuzumab ozogamicin (Mylotarg®)) or cysteine (Cys, i.e., brentuximab vedotin (Adcetris®)) residues already present. These methods result in heterogeneous mixtures containing ADCs with varying drug-to-antibody ratios (DARs), each species of which has distinct activity and toxicity profiles [3]. To overcome these issues, several site-specific conjugation technologies have been developed to prepare homogeneous ADCs

Sam Massa and Nick Devoogdt (eds.), *Bioconjugation: Methods and Protocols*, Methods in Molecular Biology, vol. 2033,
https://doi.org/10.1007/978-1-4939-9654-4_4,

[4]. These ADCs have a discrete number of drugs attached at designated sites on the antibody and have more predicable pharmacological properties [5].

We have developed a site-specific conjugation technology utilizing a buried Lys residue embedded in a dual-variable-domain (DVD) [6]. Although Lys conjugation is highly random with ~40 surface exposed Lys residues being modified using traditional conjugation chemistry, h38C2 is a humanized anti-hapten monoclonal antibody that contains a highly reactive Lys residue. Due to the Lys being buried in an 11 Å deep hydrophobic pocket [7], the Lys is more nucleophilic (pKa ~6) at physiological pH and is capable of reacting specifically and irreversibly with β-lactam functionalized drugs. In addition to this unique reactivity, the Lys is catalytic and capable of catalyzing aldol and retro-aldol reactions. This antibody has been used to prepare chemically programmed antibodies after being conjugated to β-lactam functionalized peptide or peptidomimetic targeting molecules [8] and chemically programmed bispecific antibodies that are capable of recruiting T-cells toward target tumor cells [9]. Because h38C2 only binds small molecules, an additional variable domain was incorporated at the *N*-terminus to prepare DVDs capable of binding a target of interest while still containing the reactive Lys residue needed for conjugation. We showed that conjugation to a β-lactam containing monomethylauristatin F (MMAF) derivative proceeds in a single step using physiological conditions (PBS, pH 7.4) in 4 h. The resulting DVD-ADCs were homogeneous with only a single Lys being modified and shown to be highly potent in vitro and in vivo [6].

Here we provide a protocol for the cloning, expression, and purification of the anti-HER2 DVD. We incorporate a short spacer sequence (ASTKGP) between the inner (h38C2) and outer (trastuzumab) variable domains (Fig. 1). This spacer sequence is derived

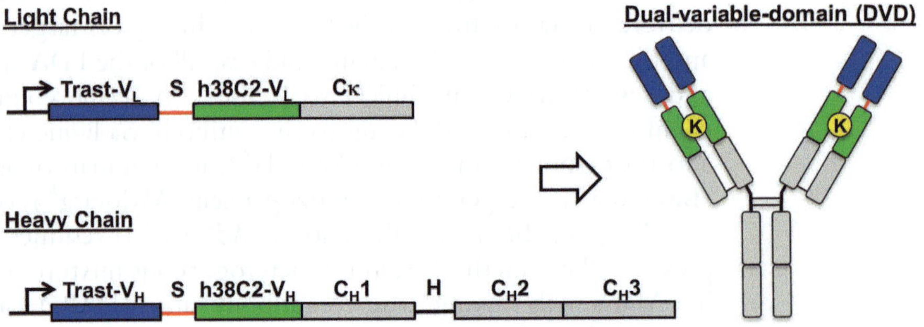

Fig. 1 Expression cassette for anti-HER2 DVD. The expression cassettes of the anti-HER2 DVD light and heavy chains are shown. The DVD incorporates the variable domains of trastuzumab (blue) and h38C2 (green), which are joined via a short spacer sequence (S = ASTKGP, red). The light chain includes a kappa (Cκ) constant domain and the heavy chain three constant domains (C_H1, C_H2, and C_H3, grey) from human IgG1. The hinge region (H) joins C_H1 to C_H2 and C_H3. Coexpression of the two cassettes (arrow) leads to the formation of the DVD, which contains a reactive Lys residue (yellow circle) in the h38C2 variable domain

a

```
GCTAGC
ATGGACTGGACTTGGAGAATCCTGTTTCTCGTAGCTGCTGCAACTGGAGCACACTCCGATATCCAGATGACCCAGTCCCCGAGCTCCCTGTCCGCCTCTGTGGGC 111
 M  D  W  T  W  R  I  L  F  L  V  A  A  A  T  G  A  H  S  D  I  Q  M  T  Q  S  P  S  S  L  S  A  S  V  G
GATAGGGTCACCATCACCTGCCGTGCCAGTCAGGATGTGAATACTGCTGTAGCCTGGTATCAACAGAAACCAGGAAAAGCTCCGAAACTACTGATTTACTCGGCA 216
 D  R  V  T  I  T  C  R  A  S  Q  D  V  N  T  A  V  A  W  Y  Q  Q  K  P  G  K  A  P  K  L  L  I  Y  S  A
TCCTTCCTCTACTCTGGAGTCCCTTCTCGCTTCTCTGGATCCAGATCTGGGACGGATTTCACTCTGACCATCAGCAGTCTGCAGCCGGAAGACTTCGCAACTTAT 321
 S  F  L  Y  S  G  V  P  S  R  F  S  G  S  R  S  G  T  D  F  T  L  T  I  S  S  L  Q  P  E  D  F  A  T  Y
TACTGTCAGCAACATTATACTACTCCTCCCACGTTCGGACAGGGTACCAAGGTGGAGATCAAAGCCAGCACCAAGGGACCTGAGCTGCAGATGACCCAGAGCCCC 426
 Y  C  Q  Q  H  Y  T  T  P  P  T  F  G  Q  G  T  K  V  E  I  K  A  S  T  K  G  P  E  L  Q  M  T  Q  S  P
AGCAGCCTGAGCGCCAGCGTGGGCGACAGAGTGACCATCACCTGTCGGAGCAGCCAGAGCCTGCTGCACACCTACGGCAGCCCCTACCTGAACTGGTATCTGCAG 531
 S  S  L  S  A  S  V  G  D  R  V  T  I  T  C  R  S  S  Q  S  L  L  H  T  Y  G  S  P  Y  L  N  W  Y  L  Q
AAGCCCGGCCAGTCCCCCAAGCTGCTGATCTACAAGGTGTCCAACCGGTTCAGCGGCGTGCCCAGCAGATTCAGCGGCAGCGGCTCCGGCACCGACTTCACCCTG 636
 K  P  G  Q  S  P  K  L  L  I  Y  K  V  S  N  R  F  S  G  V  P  S  R  F  S  G  S  G  S  G  T  D  F  T  L
ACCATCAGCTCCCTGCAGCCCGAGGACTTCGCCGTGTACTTCTGTAGTCAAGGCACCCATCTGCCCTACACCTTCGGCGGAGGCACCAAGGTGGAAATCAAACGG 741
 T  I  S  S  L  Q  P  E  D  F  A  V  Y  F  C  S  Q  G  T  H  L  P  Y  T  F  G  G  G  T  K  V  E  I  K  R
ACCGTGGCCGCCCCCAGCGTCTTCATCTTCCCGCCCAGCGACGAGCAGCTGAAGTCGGGCACGGCCAGCGTGGTGTGCCTCCTGAACAACTTCTACCCCCGCGAG 846
 T  V  A  A  P  S  V  F  I  F  P  P  S  D  E  Q  L  K  S  G  T  A  S  V  V  C  L  L  N  N  F  Y  P  R  E
GCGAAGGTCCAGTGGAAGGTGGACAACGCCCTGCAGAGCGGGAACAGCCAGGAGAGCGTGACCGAGCAGGACTCGAAGGACAGCACCTACAGCCTCAGCAGCACC 951
 A  K  V  Q  W  K  V  D  N  A  L  Q  S  G  N  S  Q  E  S  V  T  E  Q  D  S  K  D  S  T  Y  S  L  S  S  T
CTGACGCTGAGCAAGGCCGACTACGAGAAGCACAAGGTCTACGCCTGCGAGGTGACCCACCAGGGGCTCTCCAGCCCCGTGACCAAGAGCTTCAACCGGGGCGAG 1056
 L  T  L  S  K  A  D  Y  E  K  H  K  V  Y  A  C  E  V  T  H  Q  G  L  S  S  P  V  T  K  S  F  N  R  G  E
TGCTGA                                                                                                   1062
 C  *
CTCGAG
```

b

```
GCTAGC
ATGGACTGGACTTGGAGAATCCTGTTTCTCGTAGCTGCTGCAACTGGAGCACACTCCGAGGTTCAGCTGGTGGAGTCTGGCGGTGGCCTGGTGCAGCCAGGGGGC 111
 M  D  W  T  W  R  I  L  F  L  V  A  A  A  T  G  A  H  S  E  V  Q  L  V  E  S  G  G  G  L  V  Q  P  G  G
TCACTCCGTTTGTCCTGTGCAGCTTCTGGCTTCAACATTAAAGACACCTATATACACTGGGTGCGTCAGGCCCCGGGTAAGGGCCTGGAATGGGTTGCAAGGATT 216
 S  L  R  L  S  C  A  A  S  G  F  N  I  K  D  T  Y  I  H  W  V  R  Q  A  P  G  K  G  L  E  W  V  A  R  I
TATCCTACGAATGGTTATACTAGATATGCCGATAGCGTCAAGGGCCGTTTCACTATAAGCGCAGACACATCCAAAAACACAGCCTACCTGCAGATGAACAGCCTG 321
 Y  P  T  N  G  Y  T  R  Y  A  D  S  V  K  G  R  F  T  I  S  A  D  T  S  K  N  T  A  Y  L  Q  M  N  S  L
CGTGCTGAGGACACTGCCGTCTATTATTGTTCTAGATGGGGAGGGGACGGCTTCTATGCTATGGACTACTGGGGTCAAGGAACCCTGGTCACCGTCTCCTCGGCC 426
 R  A  E  D  T  A  V  Y  Y  C  S  R  W  G  G  D  G  F  Y  A  M  D  Y  W  G  Q  G  T  L  V  T  V  S  S  A
AGCACCAAGGGACCTGAGGTGCAGCTGGTGGAATCTGGCGGCGGACTGGTGCAGCCTGGCGGCAGCCTGAGACTGTCTTGCGCCGCCAGCGGCTTCACCTTCAGC 531
 S  T  K  G  P  E  V  Q  L  V  E  S  G  G  G  L  V  Q  P  G  G  S  L  R  L  S  C  A  A  S  G  F  T  F  S
GTGAAGGGCCGGTTCACCATCAGCCGGGACAACAGCAAGAACACCCTGTACCTGCAGATGAACAGCCTGCGGGCCGAGGACACCGGCATCTACTACTGCAAGACC 636
 V  K  G  R  F  T  I  S  R  D  N  S  K  N  T  L  Y  L  Q  M  N  S  L  R  A  E  D  T  G  I  Y  Y  C  K  T
TACTTCTACAGCTTCAGCTACTGGGGCCAGGGCACCCTGGTGACAGTGTCTAGCGCGTCGACCAAGGGCCCATCGGTCTTCCCCCTGGCACCCTCCTCCAAGAGC 741
 Y  F  Y  S  F  S  Y  W  G  Q  G  T  L  V  T  V  S  S  A  S  T  K  G  P  S  V  F  P  L  A  P  S  S  K  S
ACCTCTGGGGGCACAGCGGCCCTGGGCTGCCTGGTCAAGGACTACTTCCCCGAACCTGTGACGGTCTCGTGGAACTCAGGCGCCCTGACCAGCGGCGTGCACACC 846
 T  S  G  G  T  A  A  L  G  C  L  V  K  D  Y  F  P  E  P  V  T  V  S  W  N  S  G  A  L  T  S  G  V  H  T
TTCCCGGCTGTCCTACAGTCCTCAGGACTCTACTCCCTCAGCAGCGTGGTGACCGTGCCCTCCAGCAGCTTGGGCACCCAGACCTACATCTGCAACGTGAATCAC 951
 F  P  A  V  L  Q  S  S  G  L  Y  S  L  S  S  V  V  T  V  P  S  S  S  L  G  T  Q  T  Y  I  C  N  V  N  H
AAGCCCAGCAACACCAAGGTGGACAAGAGAGTTGAGCCCAAATCTTGTGACAAAACTCACACATGCCCACCGTGCCCAGCACCTGAACTCCTGGGGGGACCGTCA 1056
 K  P  S  N  T  K  V  D  K  R  V  E  P  K  S  C  D  K  T  H  T  C  P  P  C  P  A  P  E  L  L  G  G  P  S
GTCTTCCTCTTCCCCCCAAAACCCAAGGACACCCTCATGATCTCCCGGACCCCTGAGGTCACATGCGTGGTGGTGGACGTGAGCCACGAAGACCCTGAGGTCAAG 1161
 V  F  L  F  P  P  K  P  K  D  T  L  M  I  S  R  T  P  E  V  T  C  V  V  V  D  V  S  H  E  D  P  E  V  K
TTCAACTGGTACGTGGACGGCGTGGAGGTGCATAATGCCAAGACAAAGCCGCGGGAGGAGCAGTACAACAGCACGTACCGTGTGGTCAGCGTCCTCACCGTCCTG 1266
 F  N  W  Y  V  D  G  V  E  V  H  N  A  K  T  K  P  R  E  E  Q  Y  N  S  T  Y  R  V  V  S  V  L  T  V  L
CACCAGGACTGGCTGAATGGCAAGGAGTACAAGTGCAAGGTCTCCAACAAAGCCCTCCCAGCCCCCATCGAGAAAACCATCTCCAAAGCCAAAGGGCAGCCCCGA 1371
 H  Q  D  W  L  N  G  K  E  Y  K  C  K  V  S  N  K  A  L  P  A  P  I  E  K  T  I  S  K  A  K  G  Q  P  R
GAACCACAGGTGTACACCCTGCCCCCATCCCGGGAGGAGATGACCAAGAACCAGGTCAGCCTGACCTGCCTGGTCAAAGGCTTCTATCCCAGCGACATCGCCGTG 1476
 E  P  Q  V  Y  T  L  P  P  S  R  E  E  M  T  K  N  Q  V  S  L  T  C  L  V  K  G  F  Y  P  S  D  I  A  V
GAGTGGGAGAGCAATGGGCAGCCGGAGAACAACTACAAGACCACGCCTCCCGTGCTGGACTCCGACGGCTCCTTCTTCCTCTATAGCAAGCTCACCGTGGACAAG 1581
 E  W  E  S  N  G  Q  P  E  N  N  Y  K  T  T  P  P  V  L  D  S  D  G  S  F  F  L  Y  S  K  L  T  V  D  K
AGCAGGTGGCAGCAGGGGAACGTCTTCTCATGCTCCGTGATGCATGAGGCTCTGCACAACCACTACACGCAGAAGAGCCTCTCCCTGTCCCCGGGTGCCTGA    1683
 S  R  W  Q  Q  G  N  V  F  S  C  S  V  M  H  E  A  L  H  N  H  Y  T  Q  K  S  L  S  L  S  P  G  A  *
CTCGAG
```

Fig. 2 DNA and amino acid sequence of the anti-HER2 DVD. (**a**) Using custom DNA synthesis, the variable light chain domain of anti-human HER2 mAb trastuzumab was linked to the variable light chain domain of h38C2 via an ASTKGP linker. The light chain incorporates a human kappa constant domain (Cκ) at the *C*-terminus of the h38C2 variable domain. (**b**) Using custom DNA synthesis, the variable heavy chain domain of anti-human HER2 mAb trastuzumab was linked to the variable heavy chain domain of h38C2 via an ASTKGP linker. The heavy chain incorporates constant domains of human IgG1 (C_H1-hinge-C_H2-C_H3) at the *C*-terminus of the h38C2 variable domain. Flanking sequences with *NheI* and *XhoI* restriction sites at the 5′ and 3′ ends, respectively, are shown in bold italics. Underlined amino acid sequences indicate signal peptide, ASTKGP linker, and the reactive Lys residue located in the heavy chain. The Lys residue found at the *C*-terminus of the heavy chain was replaced with an alanine

from the *N*-terminal 6 amino acids of human C_H1 and was found to generate DVDs that fully retain h38C2 Lys reactivity. The DVD was cloned with a human IgG1 heavy chain and κ light chain (Fig. 2). After transient expression in HEK 293 Phoenix-ECO

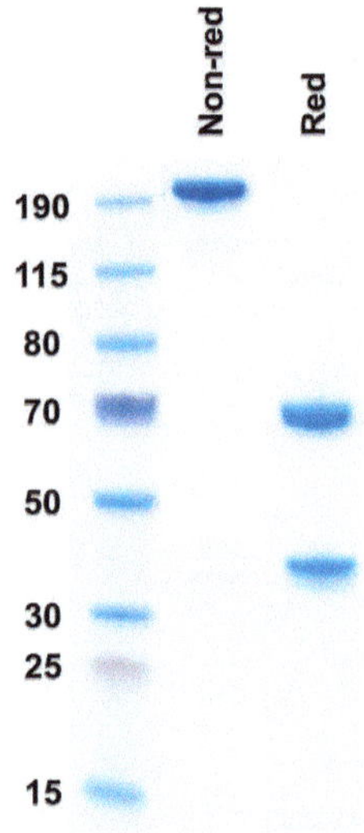

Fig. 3 SDS-PAGE analysis of anti-HER2 DVD purified by Protein A affinity chromatography. For each lane, 3 μg of protein in NuPAGE LDS Sample Buffer with (Red) or without (Non-red) NuPAGE Sample Reducing Reagent was separated by electrophoresis on a 1.5-mm, 10-well NuPAGE Novex 4–12% Bis-Tris Gel using NuPAGE MES SDS Running Buffer followed by staining with PageBlue Protein Staining Solution with Coomassie G-250. A molecular weight standard (PageRuler Plus Prestained Protein Ladder; Thermo Fisher Scientific) is given on the left in kDa. Both the non-reduced (expected band ~200 kDa) and reduced (heavy chain ~63 kDa, light chain ~36 kDa) samples indicate high purity

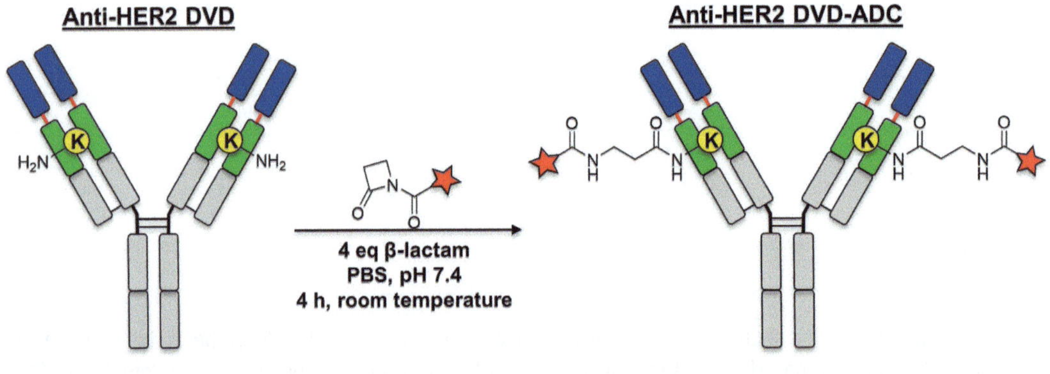

Fig. 4 Scheme for the conjugation of β-lactam MMAF to anti-HER2 DVD. The conjugation proceeds using 4 equivalents (eq) of β-lactam MMAF with respect to DVD in 4 h at room temperature at physiological pH (pH 7.4). The reactive Lys residue (yellow circle) reacts with the β-lactam by opening the ring and forming an amide bond. Since there are two Lys residues per DVD, the drug-to-antibody ratio (DAR) is 2

cells and purification using Protein A, anti-HER2 DVD was obtained with a yield of ~10 mg/L in high purity (Fig. 3). This protocol also describes the method for selective drug conjugation using β-lactam MMAF to prepare anti-HER2 DVD-ADC (Fig. 4) and a catalytic assay used to monitor and assess drug conjugation efficiency (Fig. 5). The detailed description for the production,

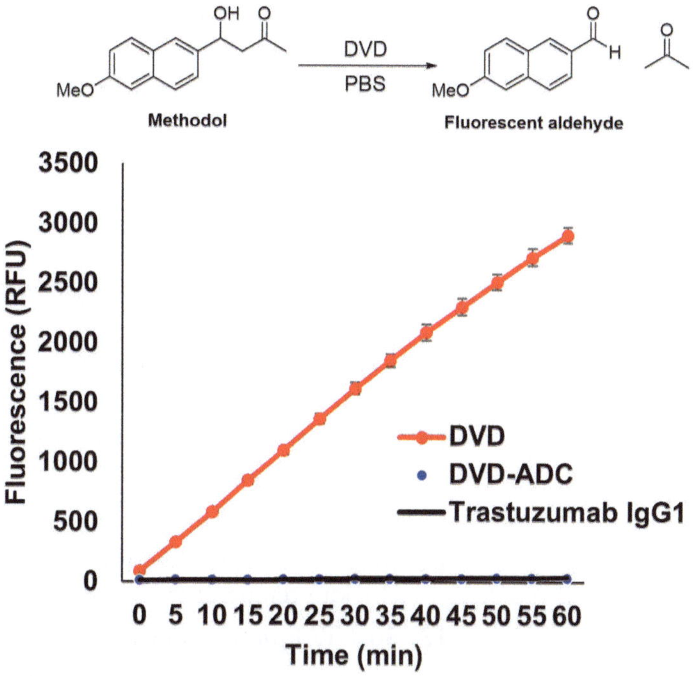

Fig. 5 Methodol assay to assess drug conjugation efficiency. The Lys residue present in the anti-HER2 DVD is capable of performing retro-aldol reactions. The signal corresponds to the formation of the fluorescent aldehyde when methodol is used as a substrate. The signal is reported in relative fluorescent units (RFU; mean ± SD of triplicates). The anti-HER2 DVD-ADC (blue) has completely lost catalytic activity indicating complete and specific conjugation at the Lys residue. Unconjugated anti-HER2 DVD (red) is used as a positive control and trastuzumab IgG1 (black) as a negative control

conjugation, and analysis of the anti-HER2 DVD-ADC we provide here has been successfully applied to >10 DVD-ADCs that target other cell surface antigens.

2 Materials

2.1 Cloning of pCEP4/Anti-HER2 DVD Plasmids

1. Cell spreader, sterile.
2. Phusion High-Fidelity (HF) DNA polymerase (20 U/μL) and 5× Phusion HF buffer (New England Biolabs), store at −20 °C.
3. 10 mM total dNTP mixture.
4. Anti-HER2 DVD gBlocks (Integrated DNA Technologies) (*see* Fig. 2).
5. *Nhe*I-HF/*Xho*I double cut and purified pCEP4 vector.
6. anti-HER2 DVD PCR primers:

Forward (5′-TTACCGGCTAGCATGGACTGGACTTGGA GAATCCTGTTTCTCGTAGCTGCTGCAACTGGA-3′).

Heavy Reverse (5′-GTCATCTCGAGTCAGGCACCCGGGG ACAGGGA-3′).

Light Reverse (5′-TAACTCTCGAGTCAGCACTCGCCCC GGTTG-3′).

7. Razor blades.
8. 14-mL round-bottom tubes with snap caps (Thermo Fisher Scientific).
9. *Nhe*I-HF (20 U/μL) and 10× CutSmart buffer (New England Biolabs), store at −20 °C.
10. *Xho*I (20 U/μL) and 10× CutSmart buffer (New England Biolabs), store at −20 °C.
11. QIAprep Spin Miniprep Kit (Qiagen).
12. Ambion nuclease-free water (Thermo Fisher Scientific), store at room temperature.
13. 50× TAE buffer (Quality Biological).
14. UltraPure agarose (Thermo Fisher Scientific).
15. 6× DNA gel loading dye solution (Thermo Fisher Scientific).
16. E.Z.N.A. Gel Extraction Kit (Omega Bio-Tek).
17. E.Z.N.A. Cycle Pure Kit (Omega Bio-Tek).
18. NanoDrop instrument (Thermo Fisher Scientific).
19. LB medium: Dissolve 20 g LB broth powder (Lennox) in 1 L water, autoclave, add antibiotics immediately upon reaching room temperature, and store at 4 °C.
20. LB-agar carbenicillin plates: Combine 1.5 g agar with 100 mL of LB media (no antibiotics), autoclave, add antibiotics immediately upon cooling (do not let agar solidify), dispense ~10 mL per plate, let cool until agar solidifies, and store sealed at 4 °C.
21. One Shot TOP10 Chemically Competent *E. coli cells* (Thermo Fisher Scientific) store at −80 °C.
22. SOC medium, store at room temperature.
23. 100 μg/μL carbenicillin: dissolve 1 g carbenicillin disodium in 10 mL water, sterilize by filtration through a 0.22-μm filter, and store in 1 mL aliquots at −20 °C.
24. 100% (v/v) molecular grade glycerol.
25. T4 DNA ligase (5 U/μL) and 5× T4 DNA ligase reaction buffer (Thermo Fisher Scientific), store at −20 °C.
26. DNA sequencing primers:

pCEP forward (5′-AGCAGAGCTCGTTTAGTGAACCG-3′),
ASTKGP forward (5′-GCCAGCACCAAGGGACCT-3′).
EBV reverse (5′-GTGGTTTGTCCAAACTCATC-3′).

2.2 Production of Anti-HER2 DVD in Mammalian Cells

1. pCEP4/anti-HER2 DVD light chain plasmid.
2. pCEP4/anti-HER2 DVD heavy chain plasmid.
3. Cell growth medium: High glucose DMEM supplemented with 10% (v/v) fetal bovine serum (FBS), 100 μg/mL streptomycin, and 100 U/mL penicillin (all from Thermo Fisher Scientific), store at 4 °C.
4. Antibody production medium: High glucose DMEM supplemented with 100 μg/mL streptomycin, 100 U/mL penicillin (all from Thermo Fisher Scientific), store at 4 °C.
5. HEK 293 Phoenix-ECO (American Tissue Culture Collection #CRL-3214).
6. PEI transfection reagent: dissolve 100 mg polyethylenimine (MW~25,000 Da) (Polysciences) in 100 mL water, adjust pH to 7.0. Aliquot and store at −20 °C.
7. 10-mL disposable polystyrene pipettes, sterile, individually packaged.
8. 50-mL conical centrifuge tubes.
9. Opti-MEM Reduced Serum Medium (Thermo Fisher Scientific), store at 4 °C.
10. 500-mL 0.45-μm Stericup Filter Unit (Millipore).
11. HiTrap Protein A HP columns (GE Healthcare), store at 4 °C.
12. Peristaltic P-1 pump (GE Healthcare).
13. DPBS (1×), Dulbecco's phosphate buffered saline, no calcium and no magnesium.
14. Protein A elution buffer: 0.5 M acetic acid.
15. Neutralization buffer: 1 M Tris–HCl, pH 9.0.
16. Column storage buffer: 20% (v/v) ethanol.
17. PD-10 Desalting Columns (GE Healthcare).
18. 15-mL Amicon Ultra Centrifugal Filter Devices with a 30-kDa MWCO membrane (EMD Millipore).
19. UV photometer instrument with disposable cuvettes.
20. NuPAGE Novex 4–12% Bis-Tris Gel 1.5 mm, 10 well (Thermo Fisher Scientific), store at 4 °C.
21. NuPAGE MES SDS Buffer Kit with MES SDS Running Buffer, Sample Reducing Agent, Antioxidant, and LDS Sample Buffer (Thermo Fisher Scientific).
22. PageBlue Protein Staining Solution with Coomassie G-250 (Thermo Fisher Scientific).

2.3 Conjugation to β-Lactam-Reactive Compounds

1. DPBS (1×), Dulbecco's phosphate buffered saline, no calcium and no magnesium.
2. 10 mM β-lactam MMAF [6] in DMSO, store at −20 °C.
3. NAP-10 columns (GE Healthcare).
4. 4-mL Amicon Ultra Centrifugal Filter Devices with a 30-kDa MWCO membrane (EMD Millipore).

2.4 Methodol Catalytic Assay

1. 10 mM methodol [10] solution in 100% biological grade ethanol (Thermo Fisher Scientific).
2. DPBS (1×), Dulbecco's phosphate buffered saline, no calcium and no magnesium.
3. Falcon 96-Well Clear Flat Bottom TC-treated Culture Microplate (Thermo Fisher Scientific).
4. SpectraMax M5 instrument (Molecular Devices) with SoftMax Pro software.

3 Methods

3.1 Cloning of pCEP4/Anti-HER2 DVD Plasmids

1. Set up two 50-μL PCR reactions containing the following:
   (a) 10 ng of anti-HER2 heavy or light chain gBlock.
   (b) 1.0 μL of 10 mM total dNTP mixture.
   (c) 2.5 μL of 10 μM forward primer.
   (d) 2.5 μL of 10 μM reverse primer.
   (e) 10 μL of 5× Phusion Polymerase reaction buffer.
   (f) Nuclease-free water to make the final volume 49.5 μL.
   (g) 0.5 μL Phusion HF DNA polymerase (20 U/μL).
2. Vortex and centrifuge the reaction tube.
3. Run the PCR using the following program: 30 s at 98 °C for denaturation, followed by 27 cycles of 10 s denaturation at 98 °C, 30 s annealing at 57 °C, 40 s of extension at 72 °C, followed by 7 min final extension at 72 °C (*see* **Note 1**).
4. Directly add 10 μL of 6× DNA gel loading dye solution, and separate by electrophoresis on a 1% (w/v) agarose gel in 1× TAE buffer.
5. Identify the PCR products (~1.0 kb for light chain and ~1.8 kb for heavy chain).
6. Excise the DNA fragments using a razor blade and purify by E.Z.N.A Gel Extraction Kit.
7. Set up a 35-μL *Nhe*I-HF/*Xho*I double digestion reaction containing 3.5 μL 10× CutSmart buffer, 1 μL *Nhe*I-HF (20 U/μL), 1 μL *Xho*I (20 U/μL), and 30 μL of purified DNA fragment

solution (eluted using 30 μL nuclease-free water for final step of E.Z.N.A. Gel Extraction Kit). Incubate at 37 °C for 3 h. Purify the digested DNA fragments using the E.Z.N.A. Cycle Pure Kit.

8. Set up two 10-μL ligation reactions containing the following:
   (a) 50 ng *Nhe*I-HF/*Xho*I double-cut and purified pCEP4.
   (b) 30 ng *Nhe*I-HF/*Xho*I double-cut and purified anti-HER2 DVD light or heavy chain.
   (c) 2.0 μL 5× T4 DNA ligase buffer.
   (d) Nuclease-free water to make the final volume 9.5 μL.
   (e) 0.5 μL ExpressLink T4 DNA Ligase (5 U/μL) and incubate reaction at room temperature for 10 min.
9. Add 1 μL of above reaction sample to 25 μL One Shot TOP10 Chemically Competent *E. coli* and incubate the cells on ice for 30 min, followed by 30 s at 42 °C, and then place immediately back on ice for 2 min. Add 200 μL of SOC medium to the reaction tube and shake at 37 °C and 225 rpm for 1 h.
10. Spread the 200 μL on an LB-agar carbenicillin plate and incubate for 16 h at 37 °C.
11. Randomly pick four colonies from the plate and inoculate in 5 mL LB medium with 100 μg/mL carbenicillin in four 14-mL round-bottom tubes with snap caps, and grow culture at 37 °C and 250 rpm for 14–16 h.
12. Prepare glycerol stocks by diluting 500 μL of each culture with 500 μL 100% (v/v) glycerol and store at −80 °C.
13. Using the remaining cultures, isolate the plasmid DNA with the QIAprep Spin Miniprep Kit with 30 μL sterile Millipore ultrapure water for elution. Measure the concentration of each plasmid using a NanoDrop instrument.
14. Confirm sequences using sequencing primers (*see* **Note 2**).
15. Store confirmed pCEP4/anti-HER2 DVD plasmids at −20 °C.

3.2 Production of Anti-HER2 DVD in Mammalian Cells

1. To initiate plasmid preparation, scrape the frozen bacterial glycerol stock containing pCEP4/anti-HER2 DVD light chain or pCEP4/anti-HER2 DVD heavy chain plasmid using a sterile pipette tip, inoculate each in 500 mL LB medium with 100 μg/mL carbenicillin in a 2-L Erlenmeyer flask, and shake at 37 °C and 250 rpm for 14–16 h. Isolate the plasmid DNA with a QIAGEN HiSpeed Plasmid Maxi Kit following the manufacturer's instructions (*see* **Note 3**).
2. The day before transfection, prepare ten 140-cm^2 cell culture dishes and seed 1.0×10^7 HEK 293 Phoenix-ECO cells into each dish with 20 mL cell growth medium. Grow cells for

18–24 h in a cell culture incubator with humidified atmosphere containing 5% CO_2 at 37 °C (*see* **Note 4**).

3. Dilute 10 × 15 μg (Σ 150 μg) pCEP4/anti-HER2 DVD light chain plasmid and 10 × 15 μg (Σ 150 μg) pCEP4/anti-HER2 DVD heavy chain plasmid with 20 mL of Opti-MEM medium in a 50-mL conical centrifuge tube. Subsequently add 10 × 60 μL (Σ 600 μL) PEI transfection reagent directly to the diluted plasmid, mix gently using a 10-mL pipette, and incubate at room temperature for 15 min (*see* **Note 5**).
4. Carefully add 2.0 mL transfection complex dropwise to each dish of cells and gently mix the medium (*see* **Note 6**).
5. Incubate the cells in the CO_2 incubator at 37 °C for 12–14 h.
6. Aspirate growth medium and add 20 mL antibody production medium to each dish and incubate the cells in the cell culture incubator at 37 °C for 72 h.
7. Collect the medium, store at 4 °C, add 20 mL fresh medium for antibody production, and incubate cells for another 72 h.
8. Repeat **step 7**.
9. Collect the third supernatant. The total volume of the three harvested supernatants is ~ 600 mL.
10. To purify total antibody by Protein A affinity chromatography, spin the pooled harvests at 10,000 × *g* for 20 min at 4 °C, pour the supernatant into a 500-mL 0.45-μm Stericup Filter Unit, and filter under vacuum.
11. Load the ~600-mL filtrate directly on a PBS preequilibrated prepacked 1-mL HiTrap Protein A HP column using a peristaltic P-1 pump at 1 mL/min at 4 °C (*see* **Notes 7** and **8**).
12. Wash the column with 20 mL of PBS at 1 mL/min.
13. Prepare twelve 1.5-mL microcentrifuge tubes containing 0.35 mL 1 M Tris–HCl (pH 9.0) and place them under the column to collect the elution at 0.65 mL/tube. Elute the column with 0.5 M acetic acid at a flow rate of 1 mL/min (*see* **Note 9**).
14. Wash the column with 15 mL of 0.5 M acetic acid at a flow rate of 1 mL/min.
15. Wash the column with 15 mL water followed by 15 mL 20% (v/v) ethanol at a flow rate of 1 mL/min. Store the column in 20% (v/v) ethanol at 4 °C.
16. Identify the fractions that contain the eluted protein by measuring the absorbance at 280 nm with a disposable cuvette in a UV photometer (*see* **Note 10**).

17. Set the two most concentrated fractions aside and concentrate the remaining fractions to 0.5 mL using a 4-mL Amicon Ultra Centrifugal Filter Device with a 30-kDa MWCO membrane.
18. Prepare a PD-10 Desalting Column by equilibrating with 25 mL DPBS (1×) (*see* **Note 11**).
19. Load all 2.5 mL of the eluted protein solution onto the PD-10 column. Let the samples enter the packed bed completely. Discard the flow-through.
20. Elute the column with 6.0 mL DPBS (1×) and collect twelve fraction of 0.5 mL each.
21. Identify the fractions that contain the eluted protein by measuring the absorbance at 280 nm with a disposable cuvette in a UV photometer (*see* **Note 10**).
22. Determine the concentration of the anti-HER2 DVD by measuring the absorbance at 280 nm. Use this absorbance to calculate the protein concentration based on the estimation that 1 mg/mL of an immunoglobulin protein gives an absorbance of 1.4. Dilute or concentrate the protein to a suitable storage concentration (typically 1–5 mg/mL), aliquot, and store at −80 °C.
23. Analyze the purity and integrity of the anti-HER2 DVD protein on an SDS-PAGE gel under reducing and non-reducing conditions following Coomassie staining.

3.3 Conjugation to β-Lactam Reactive Compounds

1. Concentrate the anti-HER2 DVD prepared in Subheading 3.2 to 10 mg/mL (50 μM) in DPBS (1×) using a 4-mL Amicon Ultra Centrifugal Filter Device with a 30-kDa MWCO membrane.
2. Add 4 equivalents of β-lactam MMAF per antibody equivalent using the 10 mM DMSO drug solution and vortex immediately (*see* **Note 12**).
3. Centrifuge briefly and incubate at room temperature for 4 h.
4. Wash and equilibrate a NAP-10 column with DPBS (1x), add the reaction mixture to the column, let the sample enter the packed bed completely, and discard flow-through (*see* **Notes 13** and **14**).
5. Elute the column with 2.8 mL DPBS and collect 200 μL fractions.
6. Identify the fractions that contain the eluted protein by measuring the absorbance at 280 nm with a disposable cuvette in a UV photometer (*see* **Note 15**).
7. Determine the concentration of the anti-HER2 DVD-ADC by measuring the absorbance at 280 nm. Use this absorbance to calculate the protein concentration based on the estimation

that 1 mg/mL of an immunoglobulin protein gives an absorbance of 1.4. Dilute or concentrate the protein to a suitable storage concentration (typically 1–3 mg/mL), aliquot, and store at 4 °C for short term or −80 °C for long term storage.

3.4 Methodol Catalytic Assay

1. Prepare 310 μL of each antibody sample (0.2 mg/mL, 1 μM) by diluting in DPBS.
2. Dispense 98 μL of each sample in triplicate in a 96-well plate (*see* **Note 16**).
3. Prepare a SpectraMax M5 instrument (Molecular Devices) by setting the wavelength of excitation (λ_{ext}) to 330 nm and wavelength of emission (λ_{em}) to 452 nm. Set the instrument to record every 5 min for 1 h. Add 2 μL of methodol (10 mM in 100% biological grade ethanol) to each well using a multichannel and read plate immediately (*see* **Note 17**).
4. Determine the signal by normalizing to 98 μL DPBS with 2 μL of the methodol solution added.

4 Notes

1. When ordering gBlock fragments, care should be taken to make the DNA sequences of the variable domains substantially different. This step is necessary to prevent amplification of only the inner domain during PCR. A codon optimization tool should be used to convert the desired protein sequence to the gBlock (e.g., www.idtdna.com/CodonOpt). Alternative PCR conditions can also be used to prevent this issue assuming the annealing temperature of the primers are sufficiently high: 30 s at 98 °C for denaturation, followed by 27 cycles of 10 s denaturation at 98 °C and 30 s annealing at 72 °C, and lastly a 7 min final extension at 72 °C.
2. When performing sequencing with the standard sequencing primers (pCEP forward and EBV reverse), the heavy chain is too long for the inner portion to be accurately and completely sequenced. Thus, an additional sequencing should be performed with ASTKGP forward and EBV reverse primers for complete coverage of the heavy chain sequence.
3. Use 15 mL buffer P1, 15 mL buffer P2, and 15 mL buffer P3 for 500-mL culture. After adding buffer P3, mix thoroughly, then centrifuge the mixture at 3000 × *g* for 15 min before pouring into the barrel of the QIAfilter Cartridge. pCEP4 is a low copy vector and elution of the DNA with 1 mL warm (65 °C) water typically yields a plasmid concentration from 600 to 900 μg/mL.

4. Other HEK cells can be used for this transfection, such as 293 c18 cells (American Tissue Culture Collection). However, we observed higher yields with HEK 293 Phoenix-ECO when compared to 293 c18 cells.
5. The amount of PEI used for each plate must not exceed 60 μL due to its cytotoxicity. The PEI solution can be stored at −20 °C for up to 1 year without significant loss of transfection efficiency. All working batches of PEI should be tested using a GFP containing plasmid to ensure high transfection efficiency before large-scale transfections are performed. A PEI stock solution (5 mg/mL) can be made by dissolving all PEI powder (2 g) into 0.2 M HCl (400 mL) and storing at −20 °C. To prepare PEI transfection reagent, dilute this stock solution to 1 mg/mL using water, adjust pH to 7.0, and store at −20 °C. Use Polysciences PEI (MW~25,000 Da) with catalog number 23966. Other sizes of PEI do not work as efficiently as this one.
6. HEK 293 Phoenix-ECO cells adhere very weakly on the plate so care should be taken to add the solution gently. A mark should be written on each plate so the media is harvested and replaced from the same location. This technique will minimize disturbing the cells resulting in higher protein yields.
7. The binding capacity of a 1-mL HiTrap Protein A HP column is 20 mg IgG1. Thus, the loading volume depends on the expected yield of antibody. To ensure all the antibody is captured, the flow-through should be saved, stored at 4 °C, and analyzed for the presence of target antibody using ELISA or flow cytometry. If substantial protein is detected, the flow-through should be filtered again through a 0.45-μm Stericup Filter Unit before loading.
8. For large scale protein expression (>1 L supernatant), HiTrap Protein A columns often clog. Multiple columns should be used to prevent this or gravity columns packed with Protein A can be used (e.g., rProtein A Gravitrap (GE Healthcare)).
9. After Protein A elution and immediate neutralization, the antibody solution should be kept on ice or can be stored at 4 °C overnight with caution (precipitation has been observed when yield is high) before buffer-exchanging into DPBS.
10. Alternatively, the protein containing fractions can be determined using a NanoDrop instrument (absorbance set to 280 nm).
11. If a large amount of protein is being purified with multiple Protein A columns, multiple PD-10 columns should be used. Ensure that the loading volume does not exceed 2.5 mL for each PD-10 column. The pH of DPBS is reported to range from 7.0 to 7.3. However, the DPBS is used directly as provided and is not adjusted.

12. For smaller reactions, the 10 mM β-lactam MMAF can be diluted to 1 mM using DPBS and used immediately. The final amount of DMSO in the reaction should not exceed 5% (v/v). Increasing the amount of DMSO decreases the conjugation efficiency.
13. The methodol assay described in Subheading 3.4, can be used to confirm 100% conjugation before purification using the NAP-10 column. To perform this assay, dilute the crude reaction to 1 μM and perform the assay as described in Subheading 3.4.
14. Depending on the volume, a NAP-5 (GE Healthcare) or PD-10 can be used for this purification.
15. Later fractions should be collected with caution since they may contain excess free compound. A volume collection guide is provided with the NAP-5, NAP-10, and PD-10 columns and is dependent on the input volume.
16. Black-walled 96-well plates with a clear bottom (e.g., Corning Black 96-Well Clear bottom Flat Bottom TC-treated Culture Microplate (Thermo Fisher Scientific)) can be used to increase sensitivity.
17. The assay is highly sensitive, so a multichannel pipette should be used to ensure all the methodol is added at the same time in each row of the plate.

References

1. Beck A, Goetsch L, Dumontet C, Corvaia N (2017) Strategies and challenges for the next generation of antibody-drug conjugates. Nat Rev Drug Discov 16:315–337
2. Carter PJ, Lazar GA (2017) Next generation antibody drugs: pursuit of the 'high-hanging fruit'. Nat Rev Drug Discov 17:197–223
3. Hamblett KJ, Senter PD, Chace DF, Sun MM, Lenox J, Cerveny CG, Kissler KM, Bernhardt SX, Kopcha AK, Zabinski RF, Meyer DL, Francisco JA (2004) Effects of drug loading on the antitumor activity of a monoclonal antibody drug conjugate. Clin Cancer Res 10:7063–7070
4. Agarwal P, Bertozzi CR (2015) Site-specific antibody–drug conjugates: the nexus of bioorthogonal chemistry, protein engineering, and drug development. Bioconjug Chem 26:176–192
5. Panowski S, Bhakta S, Raab H, Polakis P, Junutula JR (2014) Site-specific antibody drug conjugates for cancer therapy. MAbs 6:34–45
6. Nanna AR, Li X, Walseng E, Pedzisa L, Goydel RS, Hymel D, Burke TR Jr, Roush WR, Rader C (2017) Harnessing a catalytic lysine residue for the one-step preparation of homogeneous antibody-drug conjugates. Nat Commun 8:1112
7. Rader C, Turner JM, Heine A, Shabat D, Sinha SC, Wilson IA, Lerner RA, Barbas CF (2003) A humanized aldolase antibody for selective chemotherapy and adaptor immunotherapy. J Mol Biol 332:889–899
8. Rader C (2014) Chemically programmed antibodies. Trends Biotechnol 32:186–197
9. Walseng E, Nelson CG, Qi J, Nanna AR, Roush WR, Goswami RK, Sinha SC, Burke TR Jr, Rader C (2016) Chemically programmed bispecific antibodies in diabody format. J Biol Chem 291:19661–19673
10. List B, Barbas CF 3rd, Lerner RA (1998) Aldol sensors for the rapid generation of tunable fluorescence by antibody catalysis. Proc Natl Acad Sci U S A 95:15351–15355

Chapter 5

Efficient Production of Homogeneous Lysine-Based Antibody Conjugates Using Microbial Transglutaminase

Jared L. Spidel and Earl F. Albone

Abstract

Random conjugation of chemical linkers to endogenous lysines or cysteines within an antibody yields a heterogeneous mixture of conjugates with various drug-to-antibody ratios. One approach for generating homogeneous antibody conjugates utilizes enzymatic transfer of payloads onto a specific glycan or amino acid residue. Microbial transglutaminase (MTG) is an enzyme that catalyzes the formation of a stable isopeptide bond between a glutamine and a lysine. We have previously identified and reported several sites throughout the antibody structure where an engineered lysine is sufficient for transfer of a glutamine-based substrate onto the antibody. Whereas other enzymatic transfer strategies typically require significant antibody engineering to either modify the *N*-glycans or introduce a multi-amino acid enzyme recognition site, the lower contextual specificity of MTG for lysines allows just a single lysine point mutation in an antibody to be efficiently transamidated. Here we describe the molecular positioning of these single engineered lysine residues and the conjugation conditions for producing homogeneous antibody conjugates exemplified using azido- and auristatin F-based acyl donor substrates.

Key words Antibody–drug conjugate, Site-specific conjugation, Transglutaminase, Lysine, Acyl donor, Acyl acceptor

1 Introduction

Site-specific conjugation methods can alleviate the heterogeneous complexities of random chemical conjugation to native cysteine or lysine residues when producing antibody–drug conjugates (ADCs). Chemical or biochemical methods unique and specific to the desired target residue include thiol-based conjugation to an engineered unpaired cysteine [1–5], incorporation of nonnatural amino acids with unique chemically reactive groups [6], and site-specific enzymatic modification. Several site-specific enzymatic modifications requiring multiple-amino acid substitutions or peptide tags have been described [7–16]. In particular, utilization of microbial transglutaminase (MTG) isolated from the bacterium *Streptomyces mobaraensis* is an attractive conjugation enzyme due to its ability to

Sam Massa and Nick Devoogdt (eds.), *Bioconjugation: Methods and Protocols*, Methods in Molecular Biology, vol. 2033, https://doi.org/10.1007/978-1-4939-9654-4_5,

function under a wide range of pH, salt, and temperature conditions, as well as its low cost.

Transglutaminases are a family of proteins found in organisms ranging from bacteria to humans that catalyze the formation of a stable isopeptide bond between the γ-carboxyamide group (acyl donor) of a glutamine and the ε-amino group (acyl acceptor) of a lysine (reviewed in refs. 17–19). MTG has been used extensively to crosslink proteins together for various applications within the food and biotechnology industries including production of homogeneous ADCs where typically glutamine residues on an engineered antibody are transamidated by an acyl acceptor substrate [11–16]. Deglycosylation of IgG1 either enzymatically by PNGase F or via the heavy chain point mutation N297Q produces acyl donor sites at Gln295 or both Gln295 and Gln297, respectively [12, 13, 15]. Antibodies can also be engineered at the N or C terminus as well as various internal positions with the four amino acid acyl-donor peptide LLQG [14].

We recently demonstrated an alternative method to produce MTG-assisted, site-specific conjugates using glutamine-based acyl donor substrates to transamidate a lysine acyl acceptor site introduced by a single point mutation or single C-terminal amino acid extension on an otherwise wild-type antibody (e.g., fully glycosylated) [20]. Favorable acyl acceptor sites generated via lysine point mutations were found in each constant region of the antibody, particularly at heavy chain (HC) positions 135, 136, 193, 221, 252, 297, 445, and 447, and on the kappa light chain (LC) positions 201 and 202 (EU numbering; Fig. 1). While we

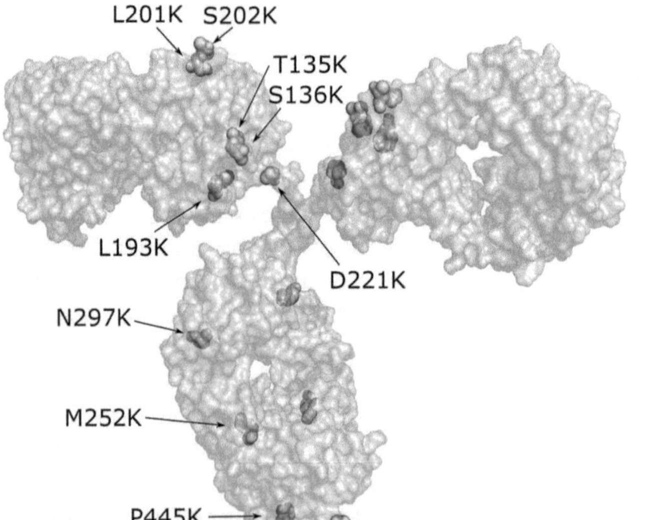

Fig. 1 Transamidation sites are highlighted in the 1HZH crystal structure of human IgG1

found some sequence requirements for optimal transamidation (no adjacent acidic or proline residues), no clear consensus sequence emerged to predict the most optimal sites to engineer an acyl acceptor site. Our results did suggest that the structure surrounding a putative MTG substrate residue likely influences its transamidation. The diversity of engineered acyl acceptor sites allows flexibility when designing ADCs by not only pairing these sites to generate drug-to-antibody ratios (DARs) greater than 2, but also integrating them into antibody fragments.

2 Materials

2.1 Mutagenesis

1. QuikChange Lightning Site-Directed Mutagenesis Kit (Agilent).
2. XL-10 Gold cells (Agilent).
3. LB agar plates with appropriate antibiotic selection (e.g., 100 μg/mL ampicillin or 50 μg/mL kanamycin).
4. LB medium with appropriate antibiotic selection (e.g., 100 μg/mL ampicillin or 50 μg/mL kanamycin).
5. Small- and large-scale plasmid DNA purification kits (e.g., QIAprep Spin Miniprep Kit and Plasmid Plus Maxi Kit (Qiagen)).
6. Thermal cycler.
7. SOC medium.
8. 1.5-mL Eppendorf tubes.
9. 17 × 100 mm Falcon Snap Cap tubes.
10. 1-L Erlenmeyer flasks.

2.2 Cell Culture and Transfection

1. Mammalian cell counter (e.g., Vi-CELL XR (Beckman Coulter)).
2. HC expression plasmid DNA and LC expression plasmid DNA.
3. Humidified incubator at 37 °C with 8% CO_2.
4. Expi293™ Expression System Kit (ThermoFisher) containing ExpiFectamine™ 293 Reagent, Enhancer 1, and Enhancer 2.
5. Expi293™ Expression Medium (ThermoFisher).
6. Polycarbonate Erlenmeyer cell culture flasks.
7. 50 mL conical tubes.
8. Opti-MEM® I Reduced Serum (ThermoFisher).
9. Centrifuge and rotor (e.g., Avanti® J-26 XP and J-LITE® JLA-8.1000 Rotor (Beckman Coulter)).

10. Centrifuge bottles (e.g., 1000 mL J-Lite PC-1000, 95 × 191 mm (Beckman Coulter)).
11. Nalgene Rapid-Flow Sterile Disposable Filter Units with PES Membrane, 0.2 μm (ThermoFisher).

2.3 Antibody Purification

1. ÄKTA FPLC equipped with Frac 950 fraction collector (GE Healthcare).
2. M5 multi-mode spectrophotometer (Molecular Devices).
3. HiTrap MabSelect protein A 5 mL column (GE Healthcare).
4. Equilibration buffer: 20 mM sodium phosphate, 10 mM EDTA, pH 7.2.
5. 0.1 M glycine, pH 2.9.
6. 1 M Tris–HCl, pH 8.0.
7. Dulbecco's [D]PBS: 0.2 g/L KCl, 0.2 g/L KH_2PO_4, 8 g/L NaCl, 1.15 g/L Na_2HPO_4, pH 7.2.
8. Slide-A-Lyzer™ G2 Dialysis Cassette, 20K MWCO (ThermoFisher).
9. Amicon® Ultra 15 mL Centrifugal Filter, 30,000 NMWL (Millipore Sigma).

2.4 ZQG Substrates and Synthesis

1. ZQG substrates: Z-Gln-Gly-OH (Bachem), Z-Gln-Gly-CAD-biotin (Zedira), Z-Gln-Gly-PEG_3-N_3 (Zedira).
2. Waters Alliance e2695 HPLC with Acquity SQ single quadrupole detector and Acquity 2998 pulsed diode array (PDA) detector.
3. Agilent Zorbax SB-C18 9.4 mm × 150 mm 5 μm pore size HPLC column.
4. Vacuum.
5. Pentafluorophenol.
6. *N*,*N*′-dimethylformamide (DMF).
7. EDAC–HCl.
8. Compressed N_2 gas.
9. Diethyl ether.
10. Methanol.
11. Azidopropylamine (Click Chemistry Tools) or other primary amine-containing compound (*see* **Note 1**).
12. 1,2-Ethylenediamine or other alkyl diamine compound (*see* **Note 2**).
13. *N*-hydroxysuccinimide (NHS)-PEG_2-auristatin F (AuF) (Concortis Biosystems) or other NHS-based payload (*see* **Note 3**).
14. Triethylamine.

15. Acetonitrile (Fluka) (*see* **Note 4**).
16. Water (18 MΩ or greater with volatile organic compounds (VOC) removed using in-line C-18 cartridge).
17. Formic acid (Sigma) (*see* **Note 4**).

2.5 MTG Reaction

1. Microbial transglutaminase (Zedira).
2. DPBS.
3. Superdex 200 Increase 10/300 gel filtration column (GE Healthcare).
4. AKTA FPLC equipped with Frac 900 fraction collector (GE Healthcare).
5. 1.0 mL injection loop.
6. Spectrophotometer (e.g., the M5 multi-mode spectrophotometer (Molecular Devices)).

2.6 ESI-MS

1. Waters Alliance e2695 HPLC with Acquity SQD single quadrupole detector and Acquity 2998 PDA detector.
2. Proteomix RP-1000 column (5 μm, 4.6 mm diameter or 50–150 mm length) (Sepax).
3. Acetonitrile.
4. Formic acid.

3 Methods

3.1 Mutagenic Introduction of Acyl Acceptor Sites

1. Select the IgG codons to mutate to an acyl acceptor site(s). Ten acyl acceptor sites were identified as efficient sites of transamidation in human IgG (*see* Subheading 1) [20] (*see* **Note 5**).
2. Design complementary forward and reverse primers to change the wild-type codon to either lysine codon flanked with 15 nucleotides on either side of the lysine codon homologous to the plasmid. If introducing a L448 codon between the K447 and stop codons, the primer will consist of codons 442–447, L448 codon, and the stop codon plus 12 nucleotides homologous to the 3′ untranslated region of the plasmid.
3. Set up a QuikChange Lightning site-directed mutagenesis reaction by combining 2.5 μL of 10× reaction buffer, 0.5 μL of dNTP mix, 0.75 μL of QuikSolution, 62.5 ng of each forward and reverse primer, 5 ng of template DNA, and water for a final volume of 24.5 μL.
4. After mixing, add 0.5 μL of QuikChange Lightning Enzyme.
5. Add the reaction to a thermal cycler and all reaction to proceed using the following parameters:

Segment	Cycles	Temperature	Time
1	1	95 °C	2 min
2	18	95 °C 60 °C 68 °C	20 s 10 s 30 s/kb of plasmid
3	1	68 °C	5 min

6. Upon completion of the reaction, add 1 μL of *Dpn*I to the reaction, mix, and incubate for 5 min at 37 °C.
7. Thaw XL10-Gold ultracompetent cells on ice for several minutes.
8. Combine 50 μL of the cells to a prechilled Eppendorf tube and 2 μL of the *Dpn*I-treated QuikChange reaction to the cells.
9. Incubate on ice for 30 min.
10. Incubate the transformation at 42 °C for 30 s, then chill on ice for 2 min.
11. Add 250 μL of room temperature SOC medium to the transformation and incubate at 37 °C for 60 min with shaking at 225–250 rpm.
12. Plate 50–300 μL of the transformation on LB agar plates with the appropriate antibiotic.
13. Incubate overnight in a 37 °C incubator.
14. Grow several colonies in 3–5 mL LB medium plus antibiotic in a 17 × 100 mm Falcon Snap Cap tube overnight at 37 °C with shaking at 225–250 rpm.
15. Isolate plasmid DNA using a plasmid miniprep kit such as QIAprep Spin Miniprep Kit from Qiagen, reserving 0.25 mL of the overnight culture and storing at 4 °C.
16. Sequence the plasmid to confirm the presence of the desired mutation and no undesired mutations.
17. Inoculate 150 mL LB medium plus antibiotic in a 1-L flask with 50 μL of the reserved overnight culture and incubate overnight at 37 °C with shaking at 225–250 rpm.
18. Extract plasmid DNA using a maxiprep kit such as Plasmid Plus Maxi Kit from Qiagen.

3.2 Transfection and Stable Cell Line Generation

1. Culture and expand a sufficient amount of Expi293™ cells to transfect a 1-L culture (*see* **Note 6**).
2. The day prior to transfection, count cells and seed a 2.8-L polycarbonate Erlenmeyer cell culture flask with 1.7×10^9 cells in 850 mL Expi293™ Expression Medium and incubate

in a humidified incubator at 37 °C, 8% CO_2 with shaking at 125 rpm.

3. On the day of transfection, count cells from **step 2** and seed 2.5×10^9 cells in 850 mL Expi293™ Expression Medium in a 2.8-L polycarbonate Erlenmeyer cell culture flask.
4. Combine and mix 333–667 μg of the HC plasmid, 333–667 μg of the LC plasmid, and 50 mL Opti-MEM® I Reduced Serum in a 50-mL conical tube.
5. Combine and mix 2.7 mL of ExpiFectamine™ 293 Reagent and 50 mL Opti-MEM® I Reduced Serum in a 50-mL conical tube. Incubate 5 min at room temperature.
6. Mix the ExpiFectamine™ 293 Reagent solution with the DNA solution in a 125-mL polycarbonate Erlenmeyer cell culture flask and incubate an additional 20–30 min at room temperature.
7. Add the DNA-ExpiFectamine™ 293 Reagent complex to the cells.
8. Incubate in a humidified incubator at 37 °C, 8% CO_2 with shaking at 125 rpm.
9. The following day, add 5 mL of ExpiFectamine™ 293 Transfection Enhancer 1 and 50 mL of ExpiFectamine™ 293 Transfection Enhancer 2 to the flask.
10. Monitor the viability of the culture and harvest the supernatant after the culture reaches <50% viable cells.
11. Centrifuge the cells >15,000 × *g* for 60–90 min.
12. Filter the supernatant using a Nalgene Rapid-Flow sterile disposable filter unit with a 0.2 μm PES membrane.
13. Store supernatant at 4 °C or −20 °C until purification.

3.3 Antibody Purification

1. Using an ÄKTA FPLC, equilibrate a HiTrap MabSelect column with 10 column volumes (CV) of equilibration buffer.
2. Load the sample, followed by washing unbound material with 10 CV of equilibration buffer.
3. Elute the sample using 5 CV of 0.1 M glycine, pH 2.9.
4. Pool the fractions containing the antibody.
5. Dialyzed in DPBS using an MWCO 20K Slide-A-Lyzer.
6. Concentrate the antibody to >10 mg/mL using an Amicon® Ultra 15 mL Centrifugal Filter, 30,000 NMWL.
7. Determine protein concentration by absorbance at 280 nm using a molar extinction coefficient (ε) of 210,000 M^{-1} cm^{-1}.

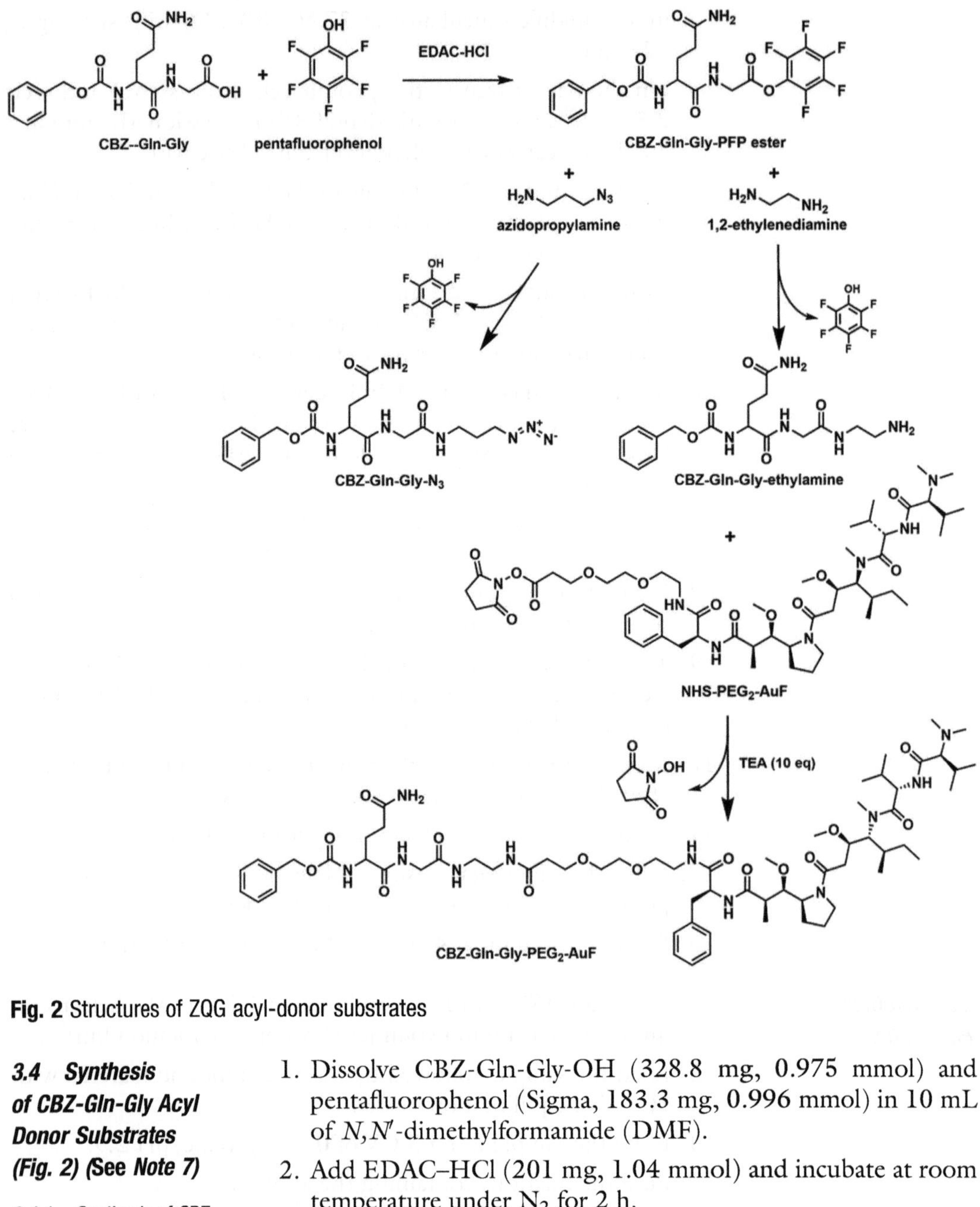

Fig. 2 Structures of ZQG acyl-donor substrates

3.4 Synthesis of CBZ-Gln-Gly Acyl Donor Substrates (Fig. 2) (See Note 7)

3.4.1 Synthesis of CBZ-Gln-Gly-Pentafluorophenyl Ester (ZQG-PFP)

1. Dissolve CBZ-Gln-Gly-OH (328.8 mg, 0.975 mmol) and pentafluorophenol (Sigma, 183.3 mg, 0.996 mmol) in 10 mL of *N,N′*-dimethylformamide (DMF).
2. Add EDAC–HCl (201 mg, 1.04 mmol) and incubate at room temperature under N_2 for 2 h.
3. Add 100 mL of diethyl ether cooled on dry ice to the reaction and precipitate overnight at −80 °C.
4. Collect the crude product by centrifugation and recrystallize with 20 mL of 60 °C methanol.
5. Rinse the final product with cold diethyl ether and dry over a stream of N_2 and to final dryness under vacuum.

6. Analyze final product by electrospray ionization-mass spectrometry (ESI-MS) (direct infusion of 2 ppm solution in 50% acetonitrile in 0.1% formic acid) *m/z* 504.0 ([M+H], 86%), 526.0 ([M+Na], 100%), 542.0 ([M+K], 22%).

3.4.2 Synthesis of CBZ-Gln-Gly-Azide (ZQG-N_3)

1. Dissolve ZQG-PFP (21.24 mg, 4.22×10^{-5} mol) and azidopropylamine (42.2 μL of a 0.91 M stock solution in DMF, 3.84×10^{-5} mol) in 0.42 mL final volume of DMF.
2. Stir the reaction under N_2 overnight at room temperature.
3. Purify the product by HPLC using a 0.1% formic acid in H_2O (A)/0.1% formic acid in acetonitrile (B) mobile phase with a gradient of 1–70% B from 1.5–14 min at 5 mL/min flow rate with peak collection using mass triggering at 380.4 Da ([M +H]) (window of ±2 Da).
4. Dry the product in vacuo.
5. Analyze the product by ESI-MS with direct infusion of 2 ppm solution of compound in 25% acetonitrile in water with 0.1% formic acid—*m/z* 420.2 ([M+H], 100%), 442.1 ([M+Na], 32%).

3.4.3 Synthesis of CBZ-Gln-Gly-PEG$_2$-Auristatin F (ZQG-AuF)

1. To synthesize ZQG-NH_2, dissolve ZQG-PFP (22.2 mg, 4.37×10^{-5} mol) in 0.85 mL DMF and add 1,2-ethylenediamine (2.3×10^{-5} L, 3.5×10^{-4} mol). Mix.
2. Stir the reaction under N_2 overnight at room temperature.
3. Purify the product by HPLC using a 0.1% formic acid in H_2O (A)/0.1% formic acid in acetonitrile (B) mobile phase with a gradient of 1–70% B from 1.5–14 min at 5 mL/min flow rate with peak collection using mass triggering at 380.4 Da ([M+H]) (window of ±2 Da).
4. Dry the product in vacuo.
5. Analyze purity of final product by ESI-MS with direct infusion of 2 ppm solution of compound in 25% acetonitrile in water with 0.1% formic acid- *m/z* 380.1 ([M+H], 100%).
6. Dissolve ZQG-NH_2 (3.8 mg, 1.01×10^{-5} mol) and NHS-PEG$_2$-AuF (10.3 mg, 1.03×10^{-5} mol, Concortis Biosystems) in 0.2 mL DMF.
7. Add triethylamine (14 μL, 1×10^{-4} mol) and incubate under N_2 overnight at room temperature.
8. Purify the product by HPLC using a 0.1% formic acid in H_2O (A)/0.1% formic acid in acetonitrile (B) mobile phase with a gradient of 1–70% B from 1.5–14 min at 5 mL/min flow rate with peak collection using mass triggering at 634.3 Da ([M+H]2+) (window of ±2 Da).
9. Dry the product in vacuo.

10. Analyze the product by ESI-MS with direct infusion of 2 ppm solution of compound in 25% acetonitrile in water with 0.1% formic acid—*m/z* 634.0 ([M+H]2+,100%), 645.1 ([M+Na]2+, 45%), 1267.0 ([M+H], 16%).

3.5 MTG Reaction

1. Dilute the acyl acceptor antibody >2.5 mg/mL in DPBS (*see* **Note 8**).
2. Dissolve the acyl donor in appropriate diluent (*see* **Note 9**).
3. Add MTG to the diluted antibody at a final concentration of 0.4 U/mg antibody (*see* **Note 10**).
4. Add the acyl donor substrate (e.g., ZQG-N_3 from Subheading 3.4.2 or ZQG-AuF from Subheading 3.4.3) at a 50:1 acyl donor–antibody ratio (*see* **Note 11**).
5. Mix the reaction immediately (*see* **Note 12**).
6. Incubate overnight at room temperature (*see* **Note 13**).
7. Connect a Superdex 200 Increase 10/300 gel filtration column to an ÄKTA FPLC and equilibrate with 2 column volumes (CV) of 1× DPBS at a flow rate of 1 mL/min.
8. Using a 1 mL injection loop, inject 1 mL of transglutaminase reaction mixture onto the equilibrated gel filtration column. Use smaller injection loops if lower volumes are to be purified. Use repeated injections or a larger column if >1 mL is to be purified.
9. Collect peaks at 280 nm. Aggregated material will elute in the void volume (8–10 min), while monomeric antibody species will elute from 13–15 min. Transglutaminase will elute at approx. 18 min.
10. Pool monomeric antibody-containing fractions. Determine protein concentration by absorbance at 280 nm using a molar extinction coefficient (ε) of 210,000 M^{-1} cm^{-1}.
11. Analyze conjugation efficiency by determining DARs using either hydrophobic-interaction chromatography–HPLC (HIC-HPLC) and/or reversed-phase HPLC (RP-HPLC) with in-line mass detection for confirmation of identity of conjugated species.

4 Notes

1. Other primary and secondary amine-based molecules can also be utilized to add other function groups to the CBZ-Gln-Gly building block using CBZ-Gln-Gly-PFP.
2. We utilized ethylenediamine to minimize the size of the alkyl spacer between the glycine residue and AuF. Other alkyl- or PEG-containing diamines could be used as well.

3. NHS-PEG_2-AuF is used as a model compound. Other NHS-based payloads can be utilized as well.
4. For chemicals used in LC-MS use LC-MS-grade chemicals.
5. Not all sites are equally transamidated with a variety of acyl donors; therefore it is important to test multiple acyl acceptor sites to optimize the efficiency of transamidation for any particular acyl donor.
6. The transfection protocol herein details a 1-L transfection; however, the transfection protocol can be proportionally scaled to the desired volume.
7. Synthesis of two acyl donor substrates is exemplified following similar protocols outlined in Fig. 2. In short, an amine-reactive acyl donor building block CBZ-Gln-Gly-PFP-ester is synthesized and then reacted with amine-based substrates to produce the final acyl donor substrate (as with ZQG-N_3) or an amine-based intermediate for reacting with NHS-based substrates (as with ZQG-PEG_2-AuF).
8. The optimal antibody concentration can vary with the solubility of different antibodies. A 2.5 mg/mL ratio is an ideal starting concentration, but high and lower concentrations can be used.
9. Water is preferred. In cases where the substrate is not soluble in water, use DMSO. If DMSO is used, dissolve to a minimum concentration of 50 mM to minimize the amount of DMSO being added to the reaction (*see* **Note 11**).
10. The optimal MTG–antibody ratio can vary. A 0.4 U:mg ratio is an ideal starting concentration, but high and lower concentrations of MTG can be used.
11. The optimal molar ratio of acyl donor to antibody can vary with different acyl donors. A 50:1 ratio is an ideal starting ratio. Do not exceed 5% DMSO in the reaction if the acyl donor substrate is dissolved in DMSO.
12. It is critical to mix immediately if the acyl donor substrate has limited solubility in aqueous solutions as slow dispersion through the aqueous environment can result in precipitation of the substrate.
13. MTG is active over a wide range of temperatures and we have successfully performed the reaction from 4 °C to 45 °C. The optimal reaction time can be tested empirically. We recommend an overnight reaction time to ensure the reaction goes to completion.

Acknowledgments

We thank Ben Vaessen for his work optimizing the MTG reaction conditions, Dr. Xin Cheng for ESI-MS analysis, and Sara Jacob for antibody purification.

References

1. Albone EF, Spidel JL, Cheng X, Park YC, Jacob S, Milinichik AZ, Vaessen B, Butler J, Kline JB, Grasso L (2017) Generation of therapeutic immunoconjugates via residue-specific conjugation technology (RESPECT) utilizing a native cysteine in the light chain framework of Oryctolagus cuniculus. Cancer Biol Ther 18 (5):347–357
2. Spidel JL, Albone EF, Cheng X, Vaessen B, Jacob S, Milinichik AZ, Verdi A, Kline JB, Grasso L (2017) Engineering humanized antibody framework sequences for optimal site-specific conjugation of cytotoxins. MAbs 9:1–9
3. Junutula JR, Raab H, Clark S, Bhakta S, Leipold DD, Weir S, Chen Y, Simpson M, Tsai SP, Dennis MS et al (2008) Site-specific conjugation of a cytotoxic drug to an antibody improves the therapeutic index. Nat Biotechnol 26(8):925–932
4. Chen XN, Nguyen M, Jacobson F, Ouyang J (2009) Charge-based analysis of antibodies with engineered cysteines: from multiple peaks to a single main peak. MAbs 1 (6):563–571
5. Gomez N, Ouyang J, Nguyen MD, Vinson AR, Lin AA, Yuk IH (2010) Effect of temperature, pH, dissolved oxygen, and hydrolysate on the formation of triple light chain antibodies in cell culture. Biotechnol Prog 26 (5):1438–1445
6. Hallam TJ, Smider VV (2014) Unnatural amino acids in novel antibody conjugates. Future Med Chem 6(11):1309–1324
7. Beerli RR, Hell T, Merkel AS, Grawunder U (2015) Sortase enzyme-mediated generation of site-specifically conjugated antibody drug conjugates with high in vitro and in vivo potency. PLoS One 10(7):e0131177
8. Madej MP, Coia G, Williams CC, Caine JM, Pearce LA, Attwood R, Bartone NA, Dolezal O, Nisbet RM, Nuttall SD et al (2012) Engineering of an anti-epidermal growth factor receptor antibody to single chain format and labeling by Sortase A-mediated protein ligation. Biotechnol Bioeng 109(6):1461–1470
9. Swee LK, Guimaraes CP, Sehrawat S, Spooner E, Barrasa MI, Ploegh HL (2013) Sortase-mediated modification of alpha-DEC205 affords optimization of antigen presentation and immunization against a set of viral epitopes. Proc Natl Acad Sci U S A 110 (4):1428–1433
10. Drake PM, Albers AE, Baker J, Banas S, Barfield RM, Bhat AS, de Hart GW, Garofalo AW, Holder P, Jones LC et al (2014) Aldehyde tag coupled with HIPS chemistry enables the production of ADCs conjugated site-specifically to different antibody regions with distinct in vivo efficacy and PK outcomes. Bioconjug Chem 25 (7):1331–1341
11. Josten A, Haalck L, Spener F, Meusel M (2000) Use of microbial transglutaminase for the enzymatic biotinylation of antibodies. J Immunol Methods 240(1–2):47–54
12. Mindt TL, Jungi V, Wyss S, Friedli A, Pla G, Novak-Hofer I, Grunberg J, Schibli R (2008) Modification of different IgG1 antibodies via glutamine and lysine using bacterial and human tissue transglutaminase. Bioconjug Chem 19 (1):271–278
13. Jeger S, Zimmermann K, Blanc A, Grunberg J, Honer M, Hunziker P, Struthers H, Schibli R (2010) Site-specific and stoichiometric modification of antibodies by bacterial transglutaminase. Angew Chem Int Ed Engl 49 (51):9995–9997
14. Strop P, Liu SH, Dorywalska M, Delaria K, Dushin RG, Tran TT, Ho WH, Farias S, Casas MG, Abdiche Y et al (2013) Location matters: site of conjugation modulates stability and pharmacokinetics of antibody drug conjugates. Chem Biol 20(2):161–167
15. Dennler P, Chiotellis A, Fischer E, Bregeon D, Belmant C, Gauthier L, Lhospice F, Romagne F, Schibli R (2014) Transglutaminase-based chemo-enzymatic conjugation approach yields homogeneous antibody-drug conjugates. Bioconjug Chem 25(3):569–578
16. Siegmund V, Schmelz S, Dickgiesser S, Beck J, Ebenig A, Fittler H, Frauendorf H, Piater B, Betz UA, Avrutina O et al (2015) Locked by design: a conformationally constrained transglutaminase tag enables efficient site-specific

conjugation. Angew Chem Int Ed Engl 54 (45):13420–13424

17. Yokoyama K, Nio N, Kikuchi Y (2004) Properties and applications of microbial transglutaminase. Appl Microbiol Biotechnol 64 (4):447–454
18. Strop P (2014) Versatility of microbial transglutaminase. Bioconjug Chem 25(5):855–862
19. Kieliszek M, Misiewicz A (2014) Microbial transglutaminase and its application in the food industry. A review. Folia Microbiol (Praha) 59(3):241–250
20. Spidel JL, Vaessen B, Albone EF, Cheng X, Verdi A, Kline JB (2017) Site-specific conjugation to native and engineered lysines in human immunoglobulins by microbial transglutaminase. Bioconjug Chem 28(9):2471–2484

Chapter 6

Engineering Antibodies with C-Terminal Sortase-Mediated Modification for Targeted Nanomedicine

Rania A. Hashad, Jaclyn L. Lange, Natasha C. W. Tan, Karen Alt, and Christoph E. Hagemeyer

Abstract

The current advances in nanoengineered materials coupled with the precise targeting capability of recombinant antibodies can create nanoscale diagnostics and therapeutics which show enhanced accumulation and extended retention at a target tissue. Smaller antibodies such as single-chain variable fragments (scFv) preserve the selective and strong binding of their parent antibody to their antigen with the benefits of low immunogenicity, more efficient tissue penetration and easy introduction of functional residues suitable for site-specific conjugation. This is of high importance as nonspecific antibody modification often involves attachment to free cysteine or lysine amino acids which may reside in the active site, leading to reduced antigen binding.

In this chapter, we outline a facile and versatile chemoenzymatic approach for production of targeted nanocarrier scFv conjugates using the bacterial trans-peptidase Sortase A (Srt A). Srt A efficiently mediates sequence-specific peptide ligation under mild conditions and has few undesirable side reactions. We first describe the production, purification and characterization of Srt A enzyme and a scFv construct which targets activated platelets, called $scFv_{anti\text{-}GPIIb/IIIa}$. Following this, our protocol illustrates the chemoenzymatic modification of the antibody at the C-terminus with an orthogonal click chemistry linker. This avoids any random attachment to the biologically active antigen binding site of the antibody. Finally, we describe the modification of a nanoparticle surface with scFv attachment via two methods: (1) direct Sortase-mediated conjugation; or (2) a two-step system which consists of scFv Sortase-mediated conjugation followed by strain promoted azide-alkyne cycloaddition. Finally, methodology is described to assess the successful assembly of targeted particles.

Key words Nanomedicine, Recombinant antibodies, Targeting, Sortase A, Chemoenzymatic, Site-specific bioconjugation

1 Introduction

Posttranslational modification of antibody fragments for biological and medical applications is a critical step toward more potent therapeutics and high-sensitivity diagnostics. Typically, lysine or cysteine residues are targeted for those conjugation strategies, which if in close proximity to the active site, can lead to antibody

Sam Massa and Nick Devoogdt (eds.), *Bioconjugation: Methods and Protocols*, Methods in Molecular Biology, vol. 2033, https://doi.org/10.1007/978-1-4939-9654-4_6,

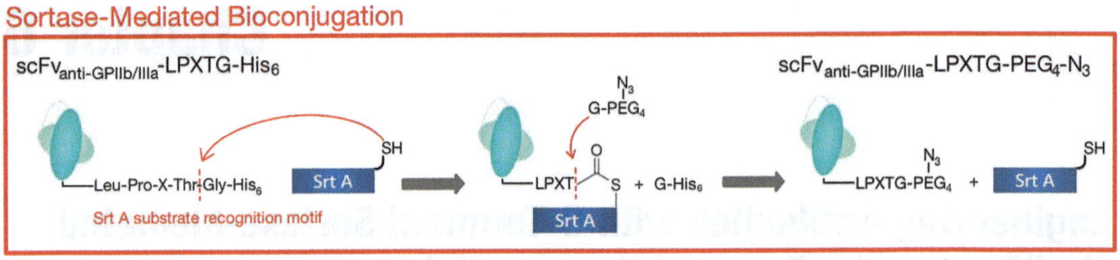

Fig. 1 Graphical scheme of the Sortase-mediated scFv modification. Site-specific modification of the scFv-LPXTG by the transpeptidase Sortase A. Sortase A recognizes the substrate with an LPXTG motif, cleaving the peptide bond between the threonine and glycine and causing a thioacyl intermediate. The modified glycine links to the targeting scFv via peptide linkage

inactivation [1]. This is even more pertinent when smaller scFvs are the targeting entity. These scFvs are composed of a variable heavy and a variable light chain of an immunoglobulin, connected by an adaptable linker and typically obtained by phage display or cloning from existing hybridomas cell lines [2, 3].

Bioconjugation of scFv with the enzyme Sortase A (Srt A), a bacterial transpeptidase from *Staphylococcus aureus*, ensures binding at the C-terminal thereby avoiding compromise of the active site of smaller antibody fragments [4]. This is achieved through recognition of a specific LPXTG peptide sequence which is cloned into the scFv framework at the C-terminal. Once identified, the Srt A will cleave the motif between the threonine and glycine amino acids via a catalytic cysteine. Nucleophilic attack of the carboxyl substituent of the threonine by an incoming glycine-based reaction partner generates an amide bond (Fig. 1) [5]. This reaction proceeds at physiological temperature and pH [5].

Through addition of specific tags, purification and confirmation of the scFvs can be achieved after posttranslational modification. Efficient separation of unreacted scFvs can be facilitated through addition of a hexahistidine affinity purification tag downstream of the cleavage site, while a second polypeptide protein tag makes use of an internal validation site in flow cytometry and immunohistochemistry techniques such as western blot analysis.

Attaching these modified scFvs to nanomaterial can be achieved with high yield and site specificity either using direct Sortase-mediated conjugation to the nanoparticle or via orthogonal click chemistry. Click chemistry refers to high yielding, modular, easily purified, stereospecific and simple chemical reactions [6]. The most famous of these, and the one that has defined the entire field since delineation in the seminal paper by Sharpless et al., is the copper-catalyzed azide–alkyne cycloaddition [6]. A particularly useful variant of click chemistry is strain-promoted azide–alkyne click chemistry. This system uses biorthogonal reactants that are inert to their surrounding biological milieu, that proceed without the copper catalyst and do not disrupt delicate quaternary protein structures of antibodies and scFvs [7]. Due to its small size and inert nature, the azide functionality is preferably conjugated to scFvs and the strained alkyne group to the nanoparticle surface. Based on its higher reaction rate compared to other strain-promoted substituents [8] the click reactive bicyclo-[6.1.0]-nonyne (BCN) substituent was selected in this protocol for tethering to a nanoparticle.

Nanoparticles are submicron (10–100 nm), highly monodisperse and multifunctional structures that allow for a high level of functional capability [9]. Any nanoparticle formulation is amenable to the following procedure, the only requirement being simple surface modification to yield an amine or BCN functional group available for further conjugation. High payload and therapeutic capacity can be engineered within the structure, wherein the nanoparticle contains a chemotherapy agent such as docetaxel or a combined treatment with radioactive isotopes for radionuclide therapy [10, 11]. There is also scope for multi-drug delivery systems and a corresponding decrease in the adverse effects of therapy using nanoparticle technology [12].

Platelets play an important role in thrombosis and hemostasis [13] and in recent years have also been recognized as major players in inflammation [14]. In this protocol, we will describe the production and purification of a highly specific scFv against activated platelets targeting the GPIIb/IIIa integrin complex ($scFv_{anti\text{-}GPIIb/IIIa}$). With approximately 50,000–80,000 copies on the surface of each platelet, this complex is an exquisite target for highly specific enrichment. The $scFv_{anti\text{-}GPIIb/IIIa}$ specifically binds to the activated form of this integrin complex thereby further reducing the background blood pool binding to nonactivated platelets. We can use this specificity to diagnosis thrombosis and inflammation in a variety of disease models [15–19].

In summary, we here present two protocols for conjugation of the anti-platelet $scFv_{anti\text{-}GPIIb/IIIa}$ to a nanoparticle. The first involves direct Srt A-mediated conjugation of the scFv to the nanoparticle, whereas the second involves a two-step click reaction with Srt A-mediated conjugation of the anti-platelet $scFv_{anti\text{-}GPIIb/IIIa}$ to a BCN-modified nanoparticle.

2 Materials

2.1 Protein Expression and Purification (Srt A/Targeting and Mutated Control scFv-LPXTG)

1. pET20b plasmid (Novagen) to express His_6-tagged SrtA in *E. coli*.
2. pET20b plasmid (Novagen) to express His_6- and LPXTG-tagged scFvs in *E. coli* (*see* **Note 1**).
3. BL21 (DE3) pLys competent bacteria.
4. LB medium: 10 g/L Bacto tryptone, 5 g/L yeast extract, 10 g/L NaCl.

5. 1000× Ampicillin stock solution: Dissolve at 100 mg/L in water.
6. Ampicillin-supplemented LB agar: Dissolve ampicillin (100 μg/mL) and 1.5% (w/v) agarose in autoclaved LB medium then pour into petri plates and leave to dry 1 h in sterile conditions.
7. Glycerol 50% (v/v): Dissolve in nuclease-free water.
8. 1 M Isopropyl β-D-thiogalactopyranoside (IPTG): Dissolve in water.
9. Bacterial lysis buffer (e.g., BugBuster® Master Mix (Novagen)).
10. Protease Inhibitor (e.g., cOmplete™ EDTA-free Protease Inhibitor (Roche)).
11. Binding/wash buffer 1: 50 mM Tris–HCl pH 7.5, 150 mM NaCl.
12. Binding/wash buffer 2: 50 mM Tris–HCl pH 7.5, 150 mM NaCl, 20 mM imidazole.
13. Elution buffer: 50 mM Tris–HCl pH 7.5, 150 mM NaCl, 250 mM imidazole.
14. Ni-NTA Superflow columns for FPLC system (Qiagen).
15. Centrifugal Filters for protein purification and concentration (e.g., Amicon® Ultra 0.5 mL Centrifugal Filters).
16. Spectrometer (e.g., Direct Detect spectrometer (Merck Millipore)).

2.2 Two-Pot Synthesis of Targeted Nanoparticles

2.2.1 Srt A-Mediated Azide-Functionalization scFv-LPXTG

1. Purified Srt A enzyme from Subheading 3.1.
2. Purified scFv-LPXTG$_{\text{anti-GPIIb/IIIa}}$ from Subheading 3.1.
3. 500 μM azide-functionalized glycine linker (G-PEG$_4$-Az, purchased from a standard peptide supplier, e.g., Wuxi Nordisk Biotech Co). Dissolve in DMSO.
4. Srt A reaction buffer: 50 mM Tris–HCl pH 7.5, 150 mM NaCl. Adjust to pH 8.
5. 20 mM $CaCl_2$ stock solution
6. Nuvia™ IMAC Resin: Use 25 μL for every 1 mg His-tagged protein in the reaction mixture.
7. Srt A reaction wash buffer: 30 mM sodium dihydrogen phosphate, 300 mM NaCl. Adjust to pH 7.
8. Phosphate-buffered saline (PBS), pH 7.4: 137 mM NaCl, 2.7 mM KCl, 10 mM Na_2HPO_4, 1.8 mM KH_2PO_4.
9. Thermomixer (e.g., Eppendorf Thermomixer® C).

2.2.2 Functional Assessment of the Modified scFvs

1. 12% SDS-PAGE Gels (e.g., NuPAGE Bis-Tris Gels) is purchased from Thermo Fisher
2. Coomassie blue stain: Dissolve 1.25 g of brilliant blue R in a mixture of methanol (200 mL), water (250 mL), and acetic acid (50 mL).
3. Destaining solution: Mix water, ethanol, and acetic acid at a ratio of 6:3:1.
4. BCN-functionalized near infrared (NIR) fluorescent dye (e.g., BCN—Dylight 800 dye (BCN-DL800) purchased from Synaffix®).
5. Odyssey CLX® imaging system LI-COR Biosciences.

2.2.3 Click-Reactive Nanoparticle Preparation

1. Amine surface-functionalized nanoparticles (e.g., gold nanoparticles (Cat No 765341 from Sigma-Aldrich) or protein nanoparticles (custom made in the lab)) are used in PBS buffer.
2. 500 μM amine and click bireactive cross-linker (NHS-PEG$_x$-BCN from BroadPharm), dissolved in DMSO.
3. Centrifugal filters for protein purification and concentration (e.g., Amicon® Ultra 0.5 mL centrifugal filters).
4. PBS.
5. Azide-functionalized FITC dye (e.g., Alexa Fluor™ 488 Azide (Az-AF488) ThermoFisher Cat no. A10266).

2.3 Targeted Nanoparticle (scFv-NP) Assembly Using Click Chemistry

1. BCN-surface-modified nanoparticles (BCN-NPs) from Subheading 3.2.3.
2. Azide-functionalized scFv (scFv-N_3) from Subheading 3.2.1.
3. Centrifugal filters for protein purification and concentration (e.g., Amicon® Ultra 0.5 mL centrifugal filters).
4. PBS.

2.4 Evaluation of Target Specific Binding for scFv$_{anti\text{-}GPIIb/IIIa}$-Functionalized NPs and Controls via Flow Cytometry

1. Platelets collected from whole blood (10^5 platelet/mL).
2. PBS.
3. PBS with Ca and Mg: PBS + 1 mM $CaCl_2$ + 0.5 mM $MgCl_2$.
4. 200 μM 10× Adenosine diphosphate (ADP) solution purchased from Merck® CAS Number 20398-34-9: Dissolve in water.
5. PE anti-human CD62-P Antibody (e.g., CD62P (P-Selectin) Monoclonal Antibody (Psel.KO2.3), PE, eBioscience™: Prepare a 1:20 dilution in PBS).
6. Alexa Fluor™ 488 Azide (Az-AF488).
7. Flow cytometry platform (e.g., BD FACS Calibur™).

2.5 One Pot scFv-Nanoparticle Coupling Using Srt A

1. ScFv-LPXTG from Subheading 3.1.
2. *tert*-Butyloxycarbonyl-protected GGGWW peptides (e.g., BOC-GGGWW-COOH purchased from a standard peptide supplier, e.g., Wuxi Nordisk Biotech Co.).
3. PyBOP–DIPEA–DMSO mixture: dissolve 2 mg of PyBOP and 1 mg of DIPEA in 100 μL of DMSO.
4. 50% (v/v) trifluoroacetic acid (TFA) in dichloromethane (DCM).
5. 10% (v/v) trifluoroacetic acid (TFA) in dichloromethane (DCM).
6. Sortase reaction buffer: 50 mM Tris–HCl pH 7.5, 150 mM NaCl. Adjust to pH 8.
7. PBS.
8. Centrifugal filters for protein purification and concentration (e.g., Amicon® Ultra 0.5 mL centrifugal filters).

3 Methods

3.1 Protein Expression and Purification (Srt A/Targeting scFv-LPXTG)

Using recombinant DNA technology, the DNA constructs encoding both Srt A and the targeting/mutated scFv-LPXTG are cloned into the expression plasmid pET20b, and transfected into competent BL21 *E. coli* bacteria (*see* **Note 2**). Bacterial cultivation, protein expression, and purification steps are performed according to the outlined protocol. Note that all the DNA constructs used encode for hexahistidine-tagged proteins, which enables affinity chromatography-based purification with Ni-NTA Superflow columns using Tris-based buffers.

1. Transform DNA plasmid encoding for scFv-LPXTG and SrtA separately into BL21 *E. coli* cells and plate on ampicillin-supplemented LB agar. Pick single colonies and allow to grow overnight in ampicillin-supplemented LB medium till saturation.
2. For long-term storage of transformed *E. coli* cultures, create a glycerol stock of cells. Add overnight culture to 50% glycerol in a 1:1 ratio and freeze at −80 °C in cryo vials (*see* **Note 3**).
3. Start a preculture of a single clone of transformed *E. coli* cells containing scFv-LPXTG or Srt A into 10 mL of ampicillin-supplemented LB medium and leave to grow overnight at 37 °C while shaking at 250 rpm.
4. Cultivate 10 mL of overnight preculture into 1 L of ampicillin-supplemented LB medium and allow to grow until an optical density of 0.8 at 600 nm is reached, using the same temperature and shaking conditions outlined in **step 3** in Subheading 3.1.

5. Induce Srt A and scFv-LPXTG expression by adding 1 mM of IPTG to the culture and incubate for a further 6 h at 37 °C while shaking at 250 rpm.
6. Harvest bacteria by centrifugation in 250 mL spin bottles for 15 min at 6000 × *g* and 4 °C. Discard the supernatant and weigh the cell pellet. Per gram of pellet, resuspend in 3 mL of BugBuster master mix and cOmplete™ Protease Inhibitor. Then, incubate at room temperature for 15 min while shaking gently (*see* **Note 4**).
7. Clarify the cell lysate by centrifugation at 15,000 × *g* for 30 min at 4 °C and collect the supernatant for product purification using Ni-NTA Superflow columns on the FPLC system.
8. Equilibrate the Ni-NTA column with 10 mL of wash buffer. Set the detector to 280 nm and flow rate to 1 mL/min.
9. Apply the clear protein supernatant which will cause a UV absorbance peak. Then, wash the column with wash buffer 1 until UV absorbance returns to baseline.
10. Unspecific protein in the lysate is washed away by applying wash buffer 2 until all undesired protein is removed.
11. Elute the desirable proteins from the column by applying elution buffer. Collect the peak fractions.
12. Dialyze against PBS at 4 °C to remove excess imidazole.
13. Analyze the protein concentration of the purified proteins using spectrometry (*see* **Note 5**).
14. Analyze protein purity on a Coomassie-stained 12% SDS-PAGE gel. A minimal purity of 95% is recommended.

3.2 Two-Pot Synthesis of Targeted Nanoparticles

The principle of the two pot-reaction to generate functionalized nanoparticles (NPs) is illustrated in Fig. 2.

3.2.1 Srt A-Mediated Azide-Functionalization scFv-LPXTG

In Srt A-mediated bioconjugation, the molar ratios of the reactants should be adjusted to optimize the reaction yield which is dependent on the scFv utilized. Increasing the amount of the glycine nucleophilic probe or using a sterically hindered nucleophilic probe helps to drive the reaction to completion and avoids hydrolysis of the product. In this outlined protocol, the reaction proceeds at maximum efficiency when $scFv_{anti\text{-}GPIIb/IIIa}$-LPXTG, glycine-based nucleophile, and Srt A enzyme are used in a 1:3:3 molar ratio, respectively. Purification of Srt A modified-scFv from unreacted mixture is performed using Nuvia™ IMAC Resin based on the affinity for the hexahistidine purification tag in the construct.

1. In Srt A reaction buffer mix scFv-LPXTG (15 μM) with azide-functionalized glycine linker (45 μM), Srt A (45 μM) and

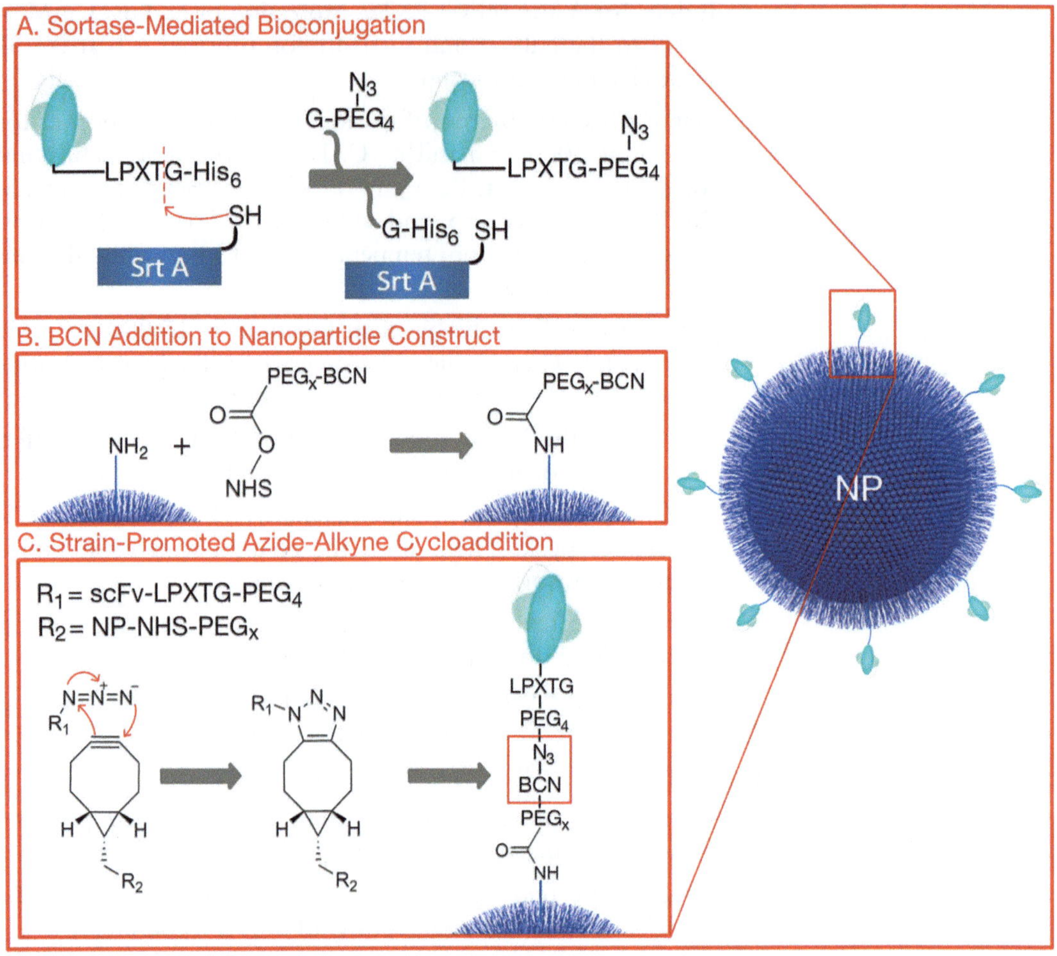

Fig. 2 Graphical scheme of the two-pot nanoparticle functionalization system. (**a**) Site-specific functionalization of $scFv_{anti\text{-}GPIIb/IIIa}$-LPXTG-$His_6$ with G-PEG_4-N_3 and (**b**) modification of the NP with NHS-PEG_x-BCN to a targeted nanosystem using bioenzymatic and click conjugation techniques (**c**)

$CaCl_2$ (0.5 mM). Incubate for 4–5 h at 37 °C, 750 rpm on ThermoMixer®.

2. For purification of hence-functionalized scFv-N_3 from starting unconjugated reactants, use a volume of Nuvia™ IMAC Resin equivalent to the total amount of his-tagged proteins in the reaction mixture. Wash the resin twice in 3× volume of Sortase wash buffer (*see* Subheading 2.2.1).
3. To the washed Nuvia™ IMAC Resin add the reaction mixture and incubate on gentle rotation at 4 °C for 30 min.
4. Centrifuge for 5 min at 13,500 × *g* and collect scFv-N_3 in the supernatant.
5. Dialyze against PBS overnight at 4 °C to isolate the purified scFv-N_3.

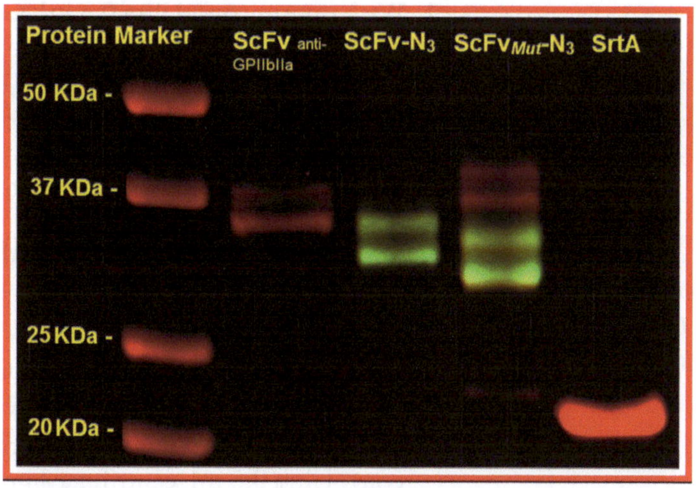

Fig. 3 Coomassie brilliant blue-stained SDS-PAGE gels. Demonstration of successful Srt A-modified $scFv_{anti\text{-}GPIIb/IIIa}$-$N_3$ post-click chemistry with BCN-DL800. Lane 1 Marker, Lane 2 $scFv_{anti\text{-}GPIIb/IIIa}$, Lane 3 N_3-linker modified $scFv_{anti\text{-}GPIIb/IIIa}$, Lane 4 N_3-linker-modified $scFv_{mut}$ control antibody, Lane 5 Sortase A enzyme. When viewed using the Odyssey LICOR system, the Coomassie brilliant blue stain appears red and the BCN-DL800 stain appears green

3.2.2 Functional Assessment of the Modified scFvs

Verification of Srt A-mediated modification efficiency can be performed via click-conjugation of scFv-N_3 to a cycloalkyne-functionalized NIR fluorophore (BCN-DL800) and analysis through SDS-PAGE. An example of such analysis is shown in Fig. 3.

1. Mix scFv-N_3 and BCN-DL800 in a molar ratio of 1:3, respectively.
2. Leave the reaction for 3 h at 4 °C.
3. Analyze the reaction by using a 12% SDS-PAGE.
4. Measure the protein band fluorescent intensity at 800 nm by using the Odyssey imaging system (*see* **Notes 5** and **6**, Subheading 2.2.2).

3.2.3 Click-Reactive Nanoparticle Preparation

Attachment of a BCN-click group to the surface of amine functionalized nanoparticles (NPs) precedes a click reaction to the Srt A modified scFv-N_3. We here describe the addition of an amine reactive bi-functional cross linker for attachment of the BCN click groups to the NP's surface:

1. Incubate NPs suspension in PBS (10^5/mL) with NHS-PEG_x-BCN in a ratio of 1:4 amine–NHS for 3 h at room temperature (*see* **Notes 7** and **8** for determination of amine groups per particle).

2. Wash on Amicon® Ultra 0.5 mL Centrifugal Filters three times with PBS.
3. Verification of click group attachment to NP surface can be performed via fluorescent labeling with a corresponding azide-functionalized fluorophore (same conditions as for the attachment of the click modified scFv to the particles).

3.3 Targeted Nanoparticle (scFv-NP) Conjugation Using Click Chemistry

A spontaneous click reaction drives the conjugation of targeting scFv-N_3 with BCN-functionalized nanoparticle according to the following steps:

1. Incubate surface-modified NPs-BCN suspended in PBS with a threefold molar excess of scFv-N_3 at 4 °C overnight.
2. Isolate scFv-N_3-conjugated nanoparticles by washing on Amicon® Ultra 0.5 mL centrifugal filters three times with PBS. The smaller scFv-N_3 will be in the flow-through.

Flow cytometry is a technique that allows to confirm the binding of $scFv_{anti\text{-}GPIIb/IIIa}$-functionalized nanoparticles to its target (GPIIb/IIIa integrin complex) on activated platelets. An example of such analysis is shown in Fig. 4.

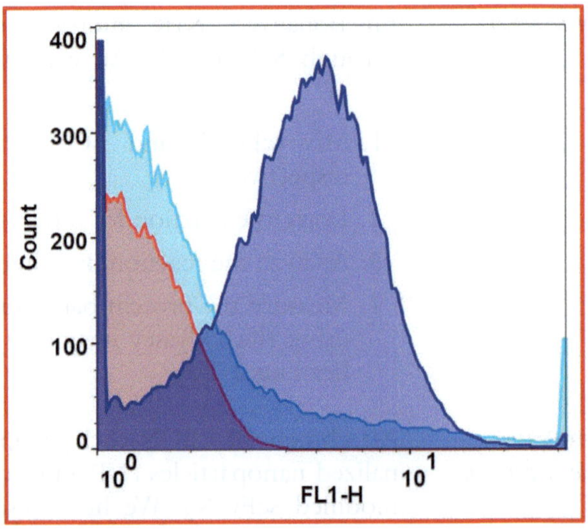

Fig. 4 Flow cytometry histogram. Overlay of the FACS histograms for platelets activated with 20 μM ADP and incubated with $scFv_{anti\text{-}GPIIb/IIIa}$-NP (dark blue) compared to $scFv_{mut}$-NP (red) (NP conjugated to control antibody). Target NP show a clear binding in comparison to the control. As additional control, activated platelets were incubated with an excess of nonconjugated $scFv_{anti\text{-}GPIIb/IIIa}$ for 30 min to block the antigen target before adding $scFv_{anti\text{-}GPIIb/IIIa}$-NP (light blue); no binding was observed, confirming specific binding of the functionalized NP to activated platelets

3.4 Evaluation of Target Specific Binding for $scFv_{anti\text{-}GPIIb/IIIa}$-Functionalized NPs via Flow Cytometry

1. Fluorescent labeling of targeted scFv-NPs with Az-AF488 is performed for 3 h in the dark at 4 °C followed by a wash with PBS using Amicon® Ultra 0.5 mL centrifugal filters. The density of free BCN groups on the nanoparticle surface is typically higher than what can be occupied with the larger scFv compared to a small dye. This allows for azide-labeled fluorophores to be clicked onto the extra BCN moieties on the surface that are not occupied by the scFv-N_3 as a result of the steric hindrance.
2. Platelet Rich Plasma (PRP) is separated from a whole blood sample and diluted 20-fold in PBS with Ca^{2+} (1 mM) and Mg^{2+} (0.5 mM) (*see* **Note 9**).
3. Activate platelets by 5 min incubation at room temperature with 20 μM ADP. Confirmation of successful platelet activation can be performed using CD62P-PE (GPIIb/IIIa-specific) fluorescent antibody on a flow cytometry device (*see* Subheading 2.4).
4. To a final concentration of 10^5 platelets/mL, AF-488 labeled $scFv_{anti\text{-}GPIIb/IIIa}$-NP is added at a concentration of 10^4 particles/mL and incubated at 4 °C for 15 min in the dark. Fix cell samples and leave in the dark till they are analyzed on a flow cytometer.

3.5 One Pot scFv-Nanoparticle Coupling Using Srt A

Nanoparticles with a glycine functionalized surface can be directly labeled with scFv-LPXTG by a Srt A-mediated transpeptidase reaction. The nanoparticle surface must have a glycine Srt A nucleophilic probe. This can be introduced using, for example, tert-Butyloxycarbonyl-protected GGGWW peptides (BOC-GGGWW-COOH) via carboxyl activation after which the t-Boc protection group can be removed. A one-pot scFv-nanoparticle bioconjugation can be performed as schematically illustrated in Fig. 5.

1. To a particle suspension (1.3×10^8 particles/mL) add BOC-GGGWW-COOH at a final concentration of 3.8 mM and 40 μL of PyBOP–DIPEA–DMSO mixture. Leave for 2 h at room temperature while shaking at 70 rpm.
2. Wash the particles on Amicon® Ultra 0.5 mL centrifugal filters three times with deionized water.
3. De-protect the t-Boc group by adding trifluoroacetic acid and leave stirring at room temperature for 30 min.
4. Wash with DCM three times, then incubate with 10% (v/v) TFA in DCM (*see* Subheading 2.5) for 20 min at room temperature.
5. Wash twice with water, then twice with Srt A reaction buffer.

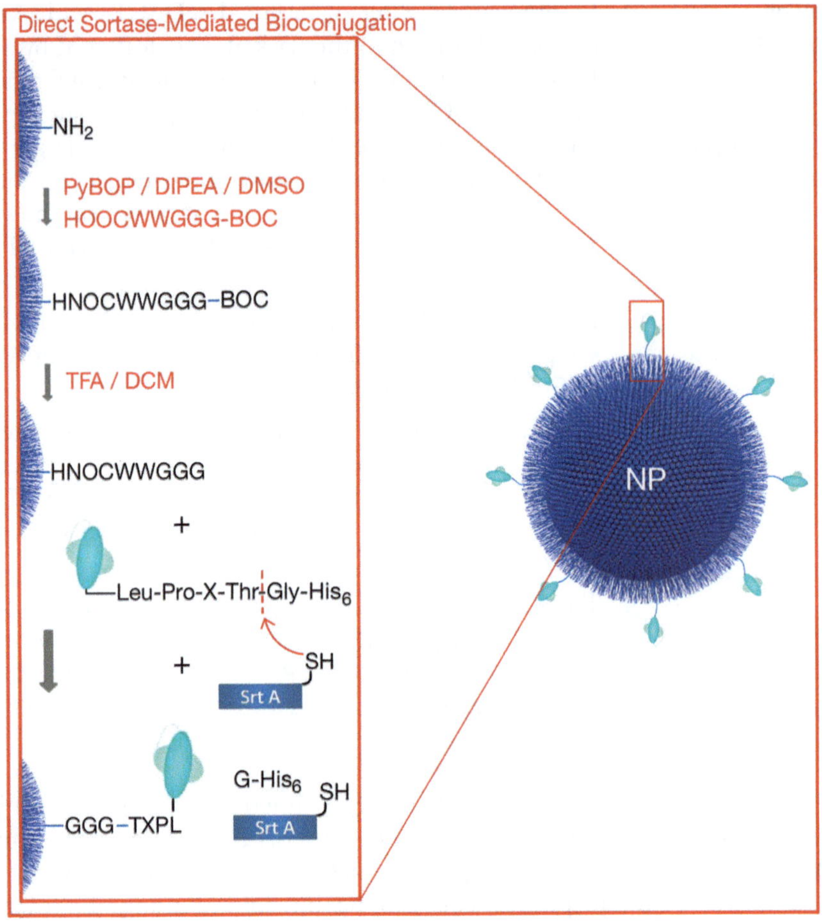

Fig. 5 Graphical scheme of the one-pot system. NP with amine surface groups are modified with tert-Butyloxycarbonyl-protected GGGWW peptides via carboxyl activation and deprotected. The particles are site-specifically labeled with scFv-LPXTG by Sortase A-mediated transpeptidase reaction

6. Resuspend the nanoparticles in Srt A coupling buffer containing scFv-LPXTG (10 μM), Srt A enzyme (30 μM) and $CaCl_2$ (0.5 mM). Incubate for 5 h at 37 °C while shaking.
7. Remove excess scFv and Srt A by washing the nanoparticles three times in PBS.

4 Notes

1. Any targeting scFv with a Sortase specific LPXTG recognition motif can be used for the Sortase-mediated bioconjugation to succeed. The results in this protocol are produced using scFv-LPXTG as the targeting ligand for activated platelets and a mutated $scFv_{mut}$-LPXTG as a negative control.

2. Depending on the protein of interest, other types of transfection and protein production systems can be used accordingly.
3. A glycerol stock of transformed *E. coli* cells is stable for years; however, repetitive freeze and thaw cycles may reduce the shelf life. To recover bacteria from the frozen glycerol stock, use a sterile pipette tip to scrape some of the frozen bacteria from the top and add directly to your culture medium. Do not allow the glycerol stock to fully thaw.
4. BugBuster® Master Mix is a bacterial lysis buffer combining BugBuster® Protein Extraction Reagent with Benzonase® Nuclease and rLysozyme™ Solution in one convenient reagent allowing for maximum recovery of active soluble protein.
5. *Staphylococcus aureus* Sortase enzyme (Srt A) is produced efficiently at a yield of 10–20 mg/L of culture by the protocol described here while scFv-LPXTG can be obtained at a yield of 1–2 mg/L of culture.
6. $ScFv_{anti\text{-}GPIIb/IIIa}$ is expressed with a C-terminal Srt A tag. Downstream of the tag an affinity tag (e.g., His_6-tag) is included, identical to the one present in the recombinant Srt A enzyme, to facilitate the clean-up of the final conjugate. Upon Srt A-mediated conjugation, the affinity tag is cleaved from the targeting agent. This is apparent in the SDS-PAGE gel electrophoresis performed on the targeting scFv before and after Srt A-mediated azide labeling, where a slight reduction in the molecular weight is seen. Please note that this change in size depends on the molecular weight of the tags after the LPXTG tag and the molecular weight of the linker added.
7. The chain length of the PEG cross-linker can be varied to optimize the click functional group availability on the nanoparticle surface for scFv coupling.
8. The molar concentration of amino groups available in protein-based NP is essential for achieving a successful stoichiometric reaction. This is variable for every type of NP and for commercial particles is typically given by the provider or can be determined via the calculation of the molar concentration of protein used in NP preparation and counting the number of particles in suspension.
9. Collected blood from healthy volunteers who have not taken any drugs known to affect platelet function for at least 14 days prior to the study.

References

1. Agarwal P, Bertozzi CR (2015) Site-specific antibody–drug conjugates: the nexus of bioorthogonal chemistry, protein engineering, and drug development. Bioconjug Chem 26 (2):176–192
2. Huston JS, McCartney J, Tai MS, Mottola-Hartshorn C, Jin D, Warren F, Keck P, Oppermann H (1993) Medical applications of single-chain antibodies. Int Rev Immunol 10 (2–3):195–217

3. Verhaar MJ, Keep PA, Hawkins RE, Robson L, Casey JL, Pedley B, Boden JA, Begent RHJ, Chester KA (1996) Technetium-99m radiolabeling using a phage-derived single-chain Fv with a C-terminal cysteine. J Nucl Med 37 (5):868–872
4. Marraffini LA, DeDent AC, Schneewind O (2006) Sortases and the art of anchoring proteins to the envelopes of gram-positive bacteria. Microbiol Mol Biol Rev 70(1):192–221
5. Hagemeyer CE, Alt K, Johnston APR, Such GK, Ta HT, Leung MKM, Prabhu S, Wang X, Caruso F, Peter K (2015) Particle generation, functionalization and sortase A-mediated modification with targeting of single-chain antibodies for diagnostic and therapeutic use. Nat Protocols 10(1):90–105
6. Kolb HC, Finn MG, Sharpless KB (2001) Click chemistry: diverse chemical function from a few good reactions. Angew Chem Int Ed 40 (11):2004–2021
7. Jewett JC, Bertozzi CR (2010) Cu-free click cycloaddition reactions in chemical biology. Chem Soc Rev 39(4):1272–1279
8. Schieber C, Bestetti A, Lim JP, Ryan AD, Nguyen T-L, Eldridge R, White AR, Gleeson PA, Donnelly PS, Williams SJ, Mulvaney P (2012) Conjugation of transferrin to azide-modified CdSe/ZnS core–shell quantum dots using cyclooctyne click chemistry. Angew Chem Int Ed 51(42):10523–10527
9. Walkey CD, Chan WCW (2012) Understanding and controlling the interaction of nanomaterials with proteins in a physiological environment. Chem Soc Rev 41 (7):2780–2799
10. Cui J, De Rose R, Alt K, Alcantara S, Paterson BM, Liang K, Hu M, Richardson JJ, Yan Y, Jeffery CM, Price RI, Peter K, Hagemeyer CE, Donnelly PS, Kent SJ, Caruso F (2015) Engineering poly(ethylene glycol) particles for improved biodistribution. ACS Nano 9 (2):1571–1580
11. Naguib YW, Rodriguez BL, Li X, Hursting SD, Williams RO III, Cui Z (2014) Solid lipid nanoparticle formulations of docetaxel prepared with high melting point triglycerides: in vitro and in vivo evaluation. Mol Pharm 11 (4):1239–1249
12. Björnmalm M, Thurecht KJ, Michael M, Scott AM, Caruso F (2017) Bridging bio-nano science and cancer nanomedicine. ACS Nano 11 (10):9594–9613
13. O'Brien JR (1990) Shear-induced platelet aggregation. Lancet 335(8691):711–713
14. Gawaz M, Langer H, May AE (2005) Platelets in inflammation and atherogenesis. J Clin Invest 115(12):3378–3384
15. Alt K, Paterson BM, Ardipradja K, Schieber C, Buncic G, Lim B, Poniger SS, Jakoby B, Wang X, O'Keefe GJ, Tochon-Danguy HJ, Scott AM, Ackermann U, Peter K, Donnelly PS, Hagemeyer CE (2014) Single-chain antibody conjugated to a cage amine chelator and labeled with positron-emitting copper-64 for diagnostic imaging of activated platelets. Mol Pharm 11(8):2855–2863
16. Alt K, Paterson BM, Westein E, Rudd SE, Poniger SS, Jagdale S, Ardipradja K, Connell TU, Krippner GY, Nair AKN, Wang X, Tochon-Danguy HJ, Donnelly PS, Peter K, Hagemeyer CE (2015) A versatile approach for the site-specific modification of recombinant antibodies using a combination of enzyme-mediated bioconjugation and click chemistry. Angew Chem Int Ed 54 (26):7515–7519
17. Ardipradja K, Yeoh SD, Alt K, O'Keefe G, Rigopoulos A, Howells DW, Scott AM, Peter K, Ackerman U, Hagemeyer CE (2014) Detection of activated platelets in a mouse model of carotid artery thrombosis with 18F-labeled single-chain antibodies. Nucl Med Biol 41(3):229–237
18. Heidt T, Deininger F, Peter K, Goldschmidt J, Pethe A, Hagemeyer CE, Neudorfer I, Zirlik A, Weber WA, Bode C, Meyer PT, Behe M, von zur Muehlen C (2011) Activated platelets in carotid artery thrombosis in mice can be selectively targeted with a radiolabeled single-chain antibody. PLoS One 6(3):e18446
19. Paterson BM, Alt K, Jeffery CM, Price RI, Jagdale S, Rigby S, Williams CC, Peter K, Hagemeyer CE, Donnelly PS (2014) Enzyme-mediated site-specific bioconjugation of metal complexes to proteins: sortase-mediated coupling of copper-64 to a single-chain antibody. Angew Chem Int Ed 53 (24):6115–6119

Chapter 7

CRISPR/Cas9-Mediated Genetic Engineering of Hybridomas for Creation of Antibodies that Allow for Site-Specific Conjugation

Makan Khoshnejad, Jacob S. Brenner, Hamideh Parhiz, and Vladimir R. Muzykantov

Abstract

Covalent conjugation of chemical moieties to antibodies has numerous applications, including antibody--drug conjugates, antibody conjugation for diagnostics, and more. Most nonspecific chemical conjugation methods ligate onto any of a number of sites on the antibody, leading to multiple conjugated species, many of which perturb antibody function. To solve these problems, we used CRISPR/Cas9-edited hybridomas to introduce a Sortase tag (LPXTG) and a Flag tag at the 3′ end of the CH3 heavy chain region of a mouse monoclonal antibody. The Flag tag allows easy purification of the antibody, while the LPXTG is then acted on by the bacterial transpeptidase Sortase to site-specifically add on any of a number of chemical moieties that possess a triglycine repeat. This technique thus allows rapid production of an antibody onto which a wide array of chemical moieties can be site-specifically conjugated.

Key words CRISPR/Cas9, Antibody engineering, Hybridoma, Site-specific conjugation

1 Introduction

With increasing frequency, therapeutic monoclonal antibodies (mAbs) are made in the format of antibody–drug conjugates (ADCs), in which a mAb is covalently conjugated to a therapeutic drug [1]. ADCs provide numerous advantages over parent mAbs in terms of potency and clinical efficacy. Unfortunately, there are numerous challenges to creating ADCs, most notably the limited control over where the drug is conjugated to the mAb, which makes it difficult to regulate the number of attached drugs and often reduces the affinity and stability of the mAb [2, 3]. To overcome these challenges, a platform technology is needed to site-specifically conjugate drugs to antibodies.

To accomplish that task, we have combined CRISPR genomic DNA editing with Sortase-based protein conjugation to create a

Sam Massa and Nick Devoogdt (eds.), *Bioconjugation: Methods and Protocols*, Methods in Molecular Biology, vol. 2033, https://doi.org/10.1007/978-1-4939-9654-4_7,

technology that can site-specifically conjugate drugs to the C-terminus of mAbs [4]. CRISPR/Cas9 RNA-guided DNA nucleases have been used in numerous applications to modify genomic sequences in mammalian genomes. CRISPR/Cas9 is bacterial derived RNA-guided DNA endonuclease system that can be programmed to make sequence specific genomic editing. The Cas9 endonuclease through interaction with designed single guide RNA (sgRNA) can be programmed to target 20 nt sequences in the genomic DNA. Upon interaction with the target DNA strand, the Cas9 creates double-stranded DNA break (DSB) at the target site. This break activates nonhomologous end joining (NHEJ) or homology-directed (HDR) DNA repair pathways. For the purposes of site-specific incorporation of desired DNA sequence, homology-directed repair (HDR) plasmid is used where the DNA insert flanked by homologous arms complementary to the target DNA is site-specifically incorporated at the desired site through homologous recombination [5, 6]. Here, for the purposes of sortase-mediated conjugation, the sortase-recognition tag DNA sequence was incorporated into the C-terminal end of antibody. Sortase A is a transpeptidase enzyme from *Staphylococcus aureus* that can clip a Sortase-recognition-tag (LPXTG) from proteins, followed by conjugation of an oligoglycine peptide to the tagged protein [7]. Here we describe how to use CRISPR to add a Sortase-recognition-tag to the C-terminus of a mAb within a hybridoma cell line. We then describe how to conjugate to that sortase-recognition-tagged mAb a small molecule drug. This technology can be used to very easily site-specifically conjugate mAbs to drugs, making for rapid and reliable testing and design of ADCs. Further, the technology may allow for a broader range of types of ADCs, including not just small molecule drug cargoes but also large cargoes such as therapeutic proteins and nanoscale drug carriers such as liposomes.

2 Materials

2.1 Construction of the sgRNA-Expression Plasmid

1. Benchling CRISPR design tool (www.benchling.com).
2. pSpCas9(BB)-2A-GFP (PX458) plasmid (Addgene # 48138).
3. pUC19 (ThermoFisher Scientific).
4. 10× T4 DNA Ligase Reaction Buffer (New England Biolabs).
5. T4 polynucleotide kinase (New England Biolabs).
6. 10× Tango Buffer (ThermoFisher Scientific).
7. 1,4-Dithiothreitol (DTT).
8. 10 mM ATP solution (ThermoFisher Scientific).
9. FastDigest BbsI (ThermoFisher Scientific).

10. T7 DNA ligase (New England Biolabs).
11. One Shot™ TOP10 Chemically Competent *E. coli* (ThermoFisher Scientific).
12. LB Agar Plate with Ampicillin-100 (Sigma-Aldrich).
13. LB Broth (Sigma-Aldrich).
14. Ampicillin.
15. QIAprep Spin Miniprep Kit (Qiagen).
16. pSpCas9(BB)-2A-GFP Sequencing Primer: TTTATGGCGAGGCGGCGG.

2.2 Construction of the HDR Plasmid and Hybridoma Transfection

1. BamHI (New England Biolabs).
2. HindIII (New England Biolabs).
3. EcoRI (New England Biolabs).
4. Zero Blunt TOPO PCR Cloning Kit with TOPO Vector (ThermoFisher Scientific).
5. QIAquick Gel Extraction Kit (Qiagen).
6. QIAquick PCR Purification Kit (Qiagen).
7. T4 DNA Ligase (New England Biolabs).
8. One Shot™ TOP10 Chemically Competent E. coli (ThermoFisher Scientific).
9. LB Agar Plate with Ampicillin-100 (Sigma-Aldrich).
10. LB Broth (Sigma-Aldrich).
11. Ampicillin.
12. QIAprep Spin Miniprep Kit (Qiagen).
13. GeneJammer Transfection Reagent (Agilent Technologies).
14. Rat hybridoma YN1/1.7.4 (ATCC; Cat. No. CRL-1878; *see* **Note 1**).
15. RPMI 1640 Medium (ThermoFisher Scientific).
16. Fetal bovine serum, qualified, heat inactivated (ThermoFisher Scientific).
17. Penicillin–streptomycin (ThermoFisher Scientific).

2.3 Clonal Cell Isolation

1. 96 well plate U-bottom tissue culture plate.
2. RPMI 1640 Medium (ThermoFisher Scientific).
3. Fetal bovine serum, qualified, heat inactivated (ThermoFisher Scientific).
4. Penicillin–streptomycin (ThermoFisher Scientific).
5. 70 μm cell strainer (Fisher Scientific).

2.4 Gene Analysis

1. GeneJET Genomic DNA Purification Kit (ThermoFisher Scientific).
2. QIAquick Gel Extraction Kit (ThermoFisher Scientific).
3. Herculase II fusion polymerase (Agilent Technologies).
4. Primers for C-terminal end of the IgG2b constant:

   Forward primer (5′-ACAGAGCAGTTGACTGAGCAAACG-3′).

   Reverse primer (5′-AGCCCGTGACCTAGGCTCTATACC-3′).

2.5 Modified Antibody Production and Purification

1. RPMI 1640 Medium (ThermoFisher Scientific).
2. Fetal bovine serum, qualified, heat inactivated (ThermoFisher Scientific).
3. Penicillin–streptomycin (ThermoFisher Scientific).
4. PFHM-II Protein-Free Hybridoma Medium (ThermoFisher Scientific; Cat. No. 12040077).
5. Stericup Quick Release-GP Sterile Vacuum Filtration System (MilliporeSigma).
6. ANTI-FLAG® M2 Affinity Gel (Sigma-Aldrich; Cat. No. A2220).
7. Modified Lowry Protein Assay Kit (ThermoFisher Scientific).

2.6 Western Blot Analysis

1. 4–15% Mini-PROTEAN® TGX™ Precast Protein Gels (Bio-Rad Laboratories).
2. TBS-T buffer: 100 mM Tris–HCl, pH 7.5, 150 mM NaCl, 0.1% Tween 20.
3. Immobilon-P PVDF Membrane (EMD Millipore).
4. Anti-FLAG M2 antibody (Sigma-Aldrich; Cat. No. F3165).
5. Goat anti-mouse HRP secondary antibody (Abcam; Cat. No. ab6789).
6. Amersham ECL Western Blotting Detection Reagent (GE Life Sciences).

2.7 Sortase Conjugation Reaction

1. Sortase A enzyme (produced in-house, *see* **Note 2**).
2. GGGK-FITC peptide (Custom ordered from ThermoFisher Scientific).
3. Calcium chloride.
4. 10× Tris-buffered saline (Sigma-Aldrich; Cat. No. T5912-1L).
5. PBS buffer: pH 7.4, 138 mM NaCl, 2.7 mM KCl, 1.9 mM NaH_2PO_4, 8.1 mM Na_2HPO_4.

3 Methods

3.1 CRISPR sgRNA Design Through Benchling

1. Input target genomic DNA sequence into the Benchling CRISPR design tool. Here, the sequence for Rat IgG2b constant region from Ensemble rat genome build Rnor_6.0 using transcript Igh-6-201 ENSRNOT00000045874, which is confirmed by Sanger sequencing, is inserted into Benchling.
2. Select the sequence surrounding the stop codon at the 3′ end of the exon of the CH3 heavy chain and analyze through the CRISPR Benchling tool for 20 nt sgRNA targets sites containing a protospacer adjacent motif (PAM) sequence 5′-NGG-3′.
3. For the CRISPR guide design parameters select the following: single guide design, 20 nt guide length, Rnor_6.0 (*Rattus norvegicus*, *see* **Note 3**), and NGG as the PAM.
4. Select the sgRNAs closest to the stop codon (Fig. 1) and with the best On-Target/Off-Target scores (*see* **Note 4**). In this case, two sgRNAs, one upstream and one downstream of the

A

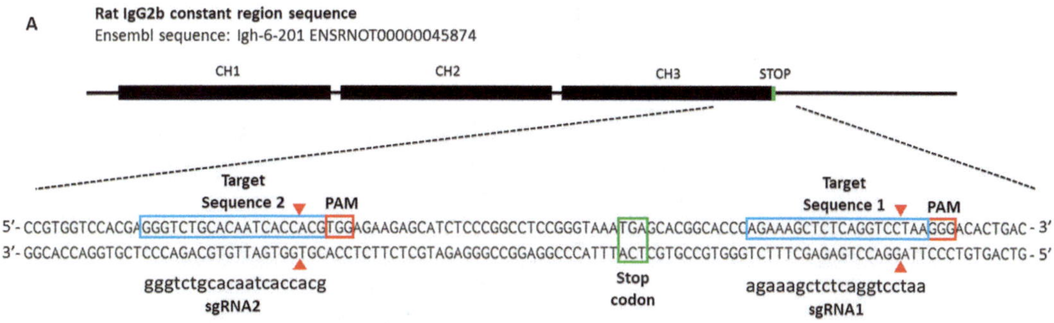

B

Strand	Sequence (5′ to 3′)	PAM	On-Target Score	Off-Target Score
▶ +	GGGTCTGCACAATCACCACG	TGG	85.7	45.3
-	TGGACCACGGAGCAGACGAA	GGG	82.3	42.0
▶ +	AGAAAGCTCTCAGGTCCTAA	GGG	80.6	38.1
-	GGGAGATGCTCTTCTCCACG	TGG	78.5	39.2
+	CGTCTGCTCCGTGGTCCACG	AGG	61.7	45.4
-	TGCCGTGCTCATTTACCCGG	AGG	60.6	45.7
-	TTATTTACACAAGGGAAGGG	TGG	54.7	32.1

C

Guide sequence oligos for cloning
sgRNA 1 - 27 bp downstream of end of stop sgRNA-1F-BbsI 5′-CACC GG agaaagctctcaggtcctaa-3′ sgRNA-1R-BbsI 5′-AAAC ttaggacctgagagctttct CC-3′
sgRNA 2 - 38 bp upstream of end of stop sgRNA-2F-BbsI 5′-CACC GGGTCTGCACAATCACCACG-3′ sgRNA-2R-BbsI 5′-AAAC CGTGGTGATTGTGCAGACCC-3′

Fig. 1 Schematic diagram of the Rat IgG2b constant region and sgRNA target selection. (**a**) Rat IgG2b constant region sequence from Ensemble rat genome build Rnor_6.0 using transcript Igh-6-201 ENSRNOT00000045874. Example of 20 nt sgRNA target sites (highlighted in blue) along with the 5′-NGG protospacer adjacent motif (PAM) (highlighted in red). (**b**) Sample results from the CRISPR Benchling tool for 20 nt sgRNA targets sites containing a protospacer adjacent motif (PAM) sequence 5′-NGG-3′. The red arrowheads indicate the two sgRNA sequences used for evaluation. (**c**) Guide sequences containing two guanines at their 5′ end and with CACC and CAAA overhangs for cloning into BbsI restriction enzyme site into the pSpCas9 plasmid

stop codon were selected. sgRNA 1 (agaaagctctcaggtcctaa) has an On-Target/Off-Target score of 80.6/38.1 and is located 27 bp downstream of the end of stop codon. sgRNA2 (gggtctgcacaatcaccacg) has an On-Target/Off-Target score of 85.7/45.3 and is located 38 bp upstream of the end of stop codon (*see* **Note 5**).

3.2 Generation of the sgRNA-Expression Construct

1. As shown in Fig. 2, design positive and negative strand sgRNA oligonucleotides by incorporating two guanines into 5′ end of the sgRNA positive strand (to enhance the U6 promoter transcription, *see* **Note 6**) and flank the sgRNA positive and negative strand with CACC and AAAC overhangs (for ligation into BbsI site in the pSpCas9 plasmid) respectively.
2. Order the designed ± strand DNA oligos.
3. Resuspend the sgRNA oligos to a final concentration of 100 μm. Phosphorylate and anneal the ± strands oligos using the method below:

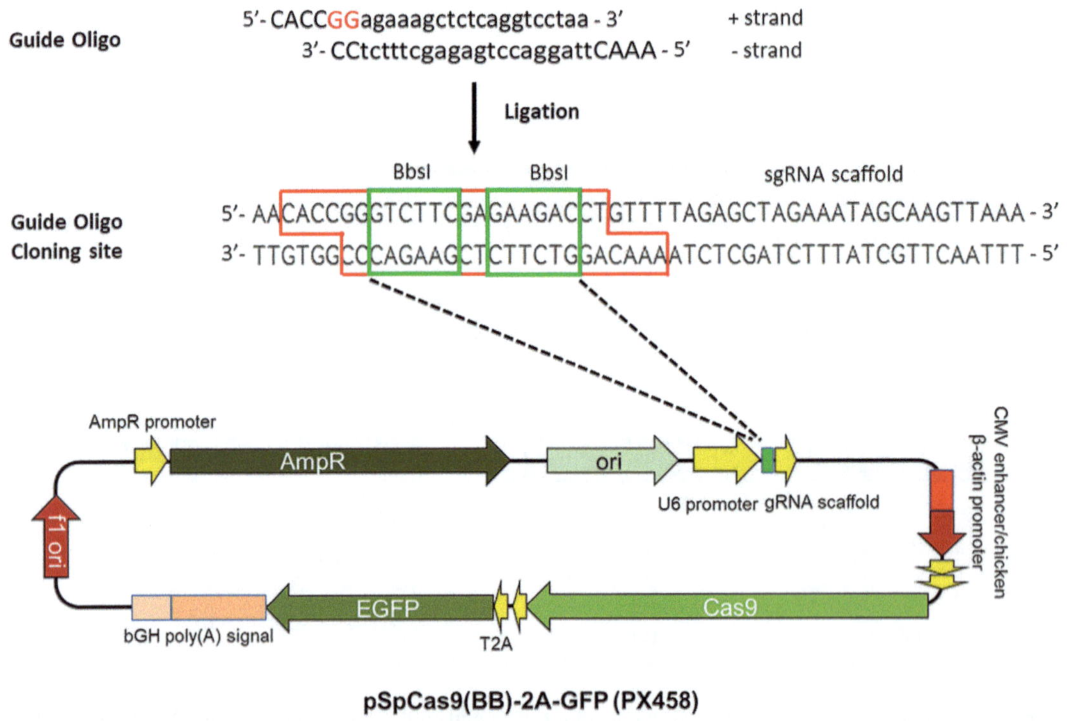

Fig. 2 Schematic diagram of the sgRNA cloning into pSpCas9 plasmid. Scarless cloning method similar to the Zhang protocol [6] is used. Two guanines are inserted at the 5′ end of the sgRNA to enhance the U6 transcription. The sgRNA ± strands are flanked with CACC and CAAA overhangs respectively, for ligation into BbsI site in the pSpCas9 plasmid

Reagent	Amount (μL)
+ Strand sgRNA	1
– Strand sgRNA	1
10× T4 ligation buffer	1
T4 PNK enzyme	1
ddH_2O	6
Total	*10*

4. Incubate the oligo phosphorylation and annealing reaction in a thermocycler using the following conditions: 30 min at 37 °C; 5 min at 95 °C; and ramping down to 25 °C at 5 °C per min.
5. Dilute the annealed oligos to a 1:200 dilution in ddH_2O.
6. Clone the annealed oligo into the pSpCas9(BB)-2A-GFP (PX458) plasmid as described below:

Reagent	Amount (μL)
pSpCas9 plasmid, 100 ng	X
Diluted annealed oligos	2
10× Tango buffer	2
10 mM DTT	1
10 mM ATP	1
FastDigest BbsI	1
T7 ligase	0.5
ddH_2O	Up to 20
Total	*20*

7. Incubate the reaction for 1 h with six cycles of 37 °C for 5 min; 21 °C for 5 min.
8. Transform the ligation reaction directly into One Shot Top10 competent *E. coli* using the manufacturer's transformation protocol.
9. Plate the cells onto LB agar plate with ampicillin (100 μg/mL) and incubate at 37 °C overnight.
10. Pick single colonies and incubate in 1–3 mL of LB Broth with ampicillin (100 μg/mL) in shaker at 37 °C overnight.
11. Harvest cells and extract DNA using QIAprep Spin Miniprep Kit according to the manufacturer's protocol.

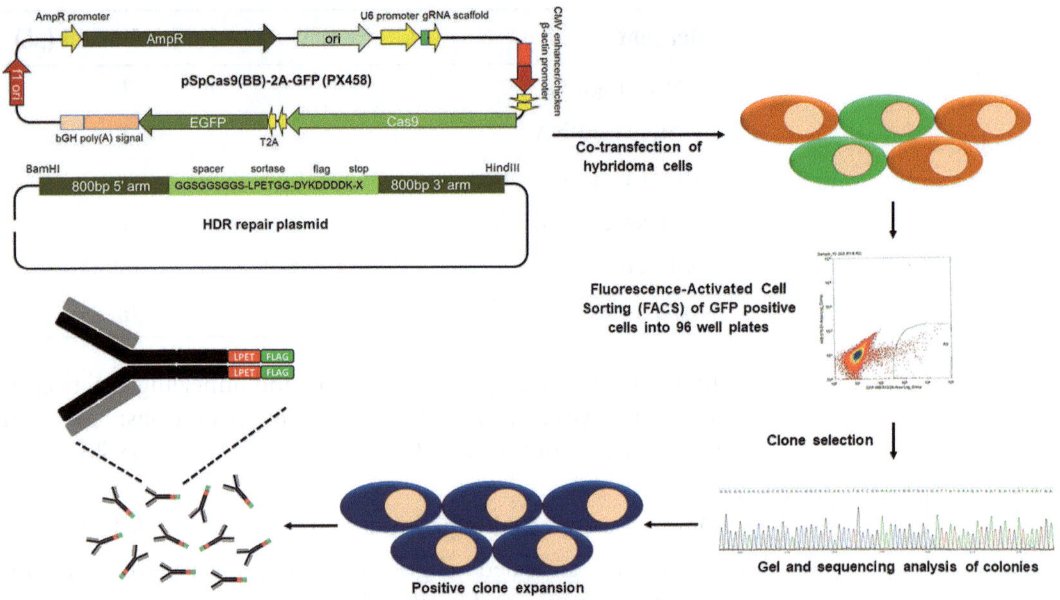

Fig. 3 Illustration of hybridoma cells cotransfection with genomic editing plasmids and clonal selection. Hybridoma cells are cotransfected with pSpCas9 and linearized HDR repair plasmids. After 48 h, single cell isolation of GFP-positive cells is performed using fluorescence-activated cell sorting (FACS). Positive clones are identified by gel and sequence analysis of the clones. The positive clones are scaled up for production of modified antibodies

12. Screen for insertion of sgRNA into the pSpCas9 plasmid by performing PCR using the pSpCas9(BB)-2A-GFP Sequencing Primer (TTTATGGCGAGGCGGCGG), and Sanger sequencing of the PCR product.

3.3 Generation of HDR Repair Plasmid and Transfection of Hybridoma Cells

1. Design the HDR repair plasmid, by selecting 800 bp complementary sequences upstream and downstream of the stop codon, as the left and right homology arms, each 800 bp (Fig. 3). In order to prevent recutting by the Cas9, mutate the PAM sequence of the sgRNA in the HDR homology arm (*see* **Note 7**).
2. Insert the DNA sequence for spacer-sortase tag-flag tag (GGSGGSGGS-LPETGG-DYKDDDDK) along with the new stop codon, in between the designed left and right homology arms of **step 1**.
3. Insert BamHI and HindIII restriction enzyme sites at ends of the HDR sequence for cloning into pUC19 vector.
4. Order a gBlock gene fragment containing the designed HDR sequence.
5. Clone the gBlock into a TOPO vector using Zero Blunt TOPO PCR cloning kit according to the manufacturer's protocol (*see* **Note 8**).

6. Perform restriction enzyme digest on 5 μg of the HDR-TOPO plasmid using BamHI and HindIII for 1 h at 37 °C.
7. Run the DNA digest reaction on a 1% (wt/vol) agarose gel.
8. Cut the band corresponding to the digested HDR and purify using QIAquick Gel Extraction Kit.
9. Digest 5 μg of the pUC19 vector using BamHI and HindIII for 1 h at 37 °C.
10. Purify the digest reaction using QIAQuick PCR Purification Kit.
11. Ligate the purified HDR DNA fragment into the digested pUC19 vector (around 100 ng total DNA in the reaction) using T4 DNA Ligase overnight at 16 °C.
12. Transform the ligation reaction directly into One Shot Top10 competent E. coli using the manufacturer's transformation protocol.
13. Plate the cells onto LB agar plate with ampicillin (100 μg/mL) and incubate at 37 °C overnight.
14. Pick single colonies and incubate in 1–3 mL of LB Broth with ampicillin (100 μg/mL) in shaker at 37 °C overnight.
15. Harvest the bacterial culture and extract DNA using QIAprep Spin Miniprep Kit according to the manufacturer's protocol.
16. Linearize the pUC19-HDR plasmid (5 μg DNA) using EcoRI enzyme located at the 5′ end of the upstream homology arm, for 1 h at 37 °C.
17. Purify the digested linearized pUC19-HDR plasmid using QIAQuick PCR Purification Kit.
18. Cotransfect of the pSpCas9 plasmid and the linearized HDR plasmid into the hybridoma cells cultured in 12 well plate, with GeneJammer or other transfection reagents according to the manufacturer's protocol (*see* **Note 9**), using premixed amounts of plasmids specified below.

Amounts of plasmids premixed for cotransfection	
pSpCas9 plasmid	500 ng
Linearized HDR plasmid	500 ng

19. Forty-eight hours posttransfection, evaluate cells for transfection efficiency and sort for GFP expression using clonal cell isolation (*see* Subheading 3.4).

3.4 Clonal Cell Isolation

1. Harvest the transfected hybridoma cells and resuspend a minimum of one million cells in a maximum of 500 μL RPMI media with 2% FBS.
2. Filter the cells through a 70 μm cell strainer.

3. Perform single cell isolation of GFP-positive cells using FACS cell sorter and deposit single cell isolates in 96 well U-bottom plate prefilled with 200 μL with RPMI media supplemented with 10% FBS. Cells transfected with GFP plasmid and untransfected cells can be used as positive and negative controls.
4. Incubate the cells for 5–7 days, followed by scaling up the cells into 24-well and eventually into 6-well plates according to standard cell culturing practices.
5. Perform genomic extraction using GeneJET Genomic DNA Purification Kit on the clonal cells, and screen clones using gel electrophoresis and Sanger sequencing (*see* Subheading 3.5).

3.5 Genomic DNA Analysis to Screen for Inserts

1. Perform PCR analysis of DNA extract from clonal cell isolate using Herculase II Fusion DNA Polymerase Kit to screen for the insert (*see* **Note 10**), with the following primers covering the C-terminal end of the IgG2b constant region.

Primers (PCR product size: 430 bp)	
Forward primer	5′-ACAGAGCAGTTGACTGAGCAAACG-3′
Reverse primer	5′-AGCCCGTGACCTAGGCTCTATACC-3′

2. Run the PCR amplicons on a 2% (wt/vol) agarose gel, stained with ethidium bromide.
3. Visualize the gel on an UV transilluminator and save gel images.
4. Cut the band corresponding to the PCR product containing the insert, and purify using QIAquick gel extraction kit.
5. Perform Sanger sequencing to confirm insert sequence.

3.6 Production of Modified Antibodies and Affinity Purification

1. The antibodies were produced according to standard hybridoma culturing technique used for monoclonal antibody production.
2. Culture the clonally selected CRISPR-modified hybridoma cells in RPMI media with 10% FBS and 1% penicillin–streptomycin at 37 °C, 5% CO_2, and 95% relative humidity for 2 weeks or longer to reach the desired cell culture scale.
3. Pellet the cells and resuspend in PFHM-II Protein-Free Hybridoma Medium. Incubate the cells for 7 days.
4. Pellet the cells, and filter the supernatant using Steritop bottle top filter unit with 0.22 μm pore size.
5. Purify the antibody from supernatant using an anti-FLAG (M2) affinity column, according to the manufacturer's protocol.
6. Determine antibody yield by standard protein quantification assay such as Modified Lowry Protein Assay.

3.7 Western Blot Analysis of Modified Antibodies

1. Run reducing and nonreducing mAb samples on sodium dodecyl sulfate–polyacrylamide gel electrophoresis (SDS-PAGE), 4–15% gradient gel (Mini-PROTEAN® TGX™ Gel) at 150 V for 45 min.
2. Transfer gel to PVDF membrane.
3. Block membrane for 1 h with 3% nonfat dry milk in TBS-T buffer.
4. Incubate with anti-FLAG M2 antibody at 1:2000 dilution in 3% nonfat dry milk in TBS-T buffer overnight at 4 °C, followed by the corresponding HRP-conjugated secondary antibody, goat anti-mouse IgG HRP, for 1 h at room temperature.
5. Perform chemiluminescence western blot analysis of the modified antibodies, using Amersham ECL Western Blotting Detection Reagent (GE Healthcare Bio-Sciences, Pittsburgh, PA).

3.8 Sortase-Mediated Conjugation to Modified Antibodies

1. Perform sortase-mediated conjugation reaction of modified antibody to peptide–fluorophore in Tris-buffered saline pH 7.4 at a total reaction volume of 50 μL, as specified below:

Reagent	Final concentration
Modified antibody	500 nM
GGGK-FITC peptide	5 μM
Sortase A enzyme	10 μM
Calcium	1 mM

2. Incubate reaction overnight at room temperature in the dark.
3. For sample purification and for quantitative analysis to determine percent of modified antibody labeling reaction, perform size-exclusion high-performance liquid chromatography (SE-HPLC) using a BioSep SEC-s3000 column with a fluorescence detector. The samples are run by isocratic method with 100% PBS pH 7.4 for 35 min at flow rate of 1 mL/min.

4 Notes

1. In this case, Rat hybridoma YN1/1.7.4 is used; however any other hybridoma cell line could be modified through this approach.
2. Sortase A is produced in *E. coli* as described in Warden-Rothman et al. [8]. Briefly, 59 amino acid truncated mutant of Staphylococcus aureus sortase A (Sa-SrtAΔ59) incorporated in a pRSET-A vector is transformed into competent *E. coli* Shuffle T7 cells. Standard IPTG Induction *E. Coli* Protein Expression method is used.

3. In this case, Rnor_6.0 (*Rattus norvegicus*) is used; however there are numerous other genome databases to choose from in the CRISPR Benchling tool, depending on specific source of hybridoma.
4. The Benchling On Target/Off Target from the CRISPR Benchling tool are on a scale of 0–100. Higher scores for both are ideal. Generally, it is good to keep the On target scores in between 60 and 100%, and preferably choosing the sgRNA with highest scores.
5. Here, two sgRNAs were evaluated; however, any number of sgRNAs can be selected either upstream or downstream of the target integration site, as well as be located on either the positive or the negative strand.
6. A guanine nucleotide is placed as the first base (in positive strand) of the U6 RNA polymerase III promoter transcript to enhance transcription of the sgRNA. Here, two guanine nucleotides were used to help increase the U6 promoter transcription efficiency. Incorporation of two guanine nucleotides is not necessary in the case of sgRNA2 as it already contains three guanines at the transcription start site.
7. To help prevent recutting of PAM sequences of the sgRNAs in the HDR, single nucleotide in the PAM site should be silently mutated. Here, for sgRNA1 target site, the PAM sequence is changed from ggg to gga, and for the sgRNA2 target site, the PAM sequence is changed from tgg to ttg.
8. Zero Blunt TOPO PCR cloning of the HDR gBlock into the TOPO vector is not mandatory. The gBlock can be directly cloned into the pUC19 vector if desired.
9. Many different transfection methods can be used for cotransfection of hybridoma cells. Hybridoma cells are generally difficult to transfect. Here, GeneJammer transfection method is used. Briefly, 500 ng of each plasmid (premixed) is added into a 50 μL Opti-MEM solution containing 1.5 μL GeneJammer reagent and incubated at room temperature for 30 min. Transfection efficiency of around 1% or less is achieved in the present method.
10. A high-fidelity polymerase such as Herculase II fusion polymerase should be used for genomic analysis of the inserts, in order to reduce errors in sequencing.

References

1. Sievers EL, Senter PD (2013) Antibody-drug conjugates in cancer therapy. Annu Rev Med 64:15–29. https://doi.org/10.1146/annurev-med-050311-201823
2. Kim EG, Kim KM (2015) Strategies and advancement in antibody-drug conjugate optimization for targeted cancer therapeutics. Biomol Ther 23:493–509. https://doi.org/10.4062/biomolther.2015.116
3. Chudasama V, Maruani A, Caddick S (2016) Recent advances in the construction of antibody–drug conjugates. Nat Chem 8:114–119. https://doi.org/10.1038/nchem.2415

4. Khoshnejad M, Brenner JS, Motley W et al (2018) Molecular engineering of antibodies for site-specific covalent conjugation using CRISPR/Cas9. Sci Rep 8:1–9. https://doi.org/10.1038/s41598-018-19784-2
5. Doudna JA, Charpentier E (2014) The new frontier of genome engineering with CRISPR-Cas9. Science 346:1258096–1258096. https://doi.org/10.1126/science.1258096
6. Ran FA, Hsu PDP, Wright J et al (2013) Genome engineering using the CRISPR-Cas9 system. Nat Protoc 8:2281–2308. https://doi.org/10.1038/nprot.2013.143. Genome
7. Antos JM, Ingram J, Fang T et al (2017) Site-specific protein labeling via sortase-mediated transpeptidation. Curr Protoc Protein Sci 89:15.3.1–15.3.19. https://doi.org/10.1002/cpps.38
8. Warden-Rothman R, Caturegli I, Popik V, Tsourkas A (2013) Sortase-tag expressed protein ligation: combining protein purification and site-specific bioconjugation into a single step. Anal Chem 85:11090–11097. https://doi.org/10.1021/ac402871k

[illegible] M, Brenner [illegible] (2018) [illegible] engineering of antibodies [illegible]

[illegible] et al (2017) [illegible] specific [illegible] labeling [illegible] mediated [illegible]

Chapter 8

Site-Specific Modification of Proteins via Trypsiligase

Sandra Liebscher and Frank Bordusa

Abstract

Site-specific incorporation of artificial functionalities into protein targets is an important tool in both basic and applied research and can be a major challenge to protein chemists. Chemical labeling methods often targeting multiple positions within a protein and therefore suffer from lack of specificity. Enzymatic protein modification is an attractive alternative due to the inherent regioselectivity and stereoselectivity of enzymes. In this contribution we describe the application of the highly specific trypsin variant named trypsiligase for the site-specific modification of virtual any target protein. We present two general routes of modification resulting in either *N*- or *C*-terminal functionalized protein products. Both reaction regimes proceed under mild and bioorthogonal conditions in a short period of time which result in homogeneously modified proteins bearing the artificial functionality exclusively at the desired position. We detail protocols for the expression and purification of trypsiligase as well as the construction of peptide or acyl donor ester probes used as substrates for the biocatalyst. In addition, we provide instructions how to perform the ultimate bioconjugation reactions and finally render assistance for the qualitative and quantitative analysis of the reaction course and outcome.

Key words Trypsin variant, Peptide ligation, Protein modification, Substrate-assisted catalysis, Substrate mimetic, Transpeptidation, Trypsiligase

1 Introduction

Trypsiligase represents a rationally designed, fourfold trypsin variant K60E, N143H, E151H, D189K derived from anionic rat trypsin II (Uniprot: P00763; Fig. 1). The native primary substrate specificity of wild-type (*wt*) trypsin for basic side chains is governed mainly by D189 at the base of the primary specificity pocket S_1 [1]. Substitution at this position in the trypsin variant D189K destructs the salt bridge responsible for the original Arg/Lys-specificity creating an enzyme species with only low-level esterase activity against P_1-tyrosine residues but negligible proteolytic activity [2]. The latter is partially restored by substrate affinity enhancement at secondary binding sites of the biocatalyst. In this context, the introduction of an acidic amino acid at the S_1' binding site in the trypsin variant K60E increases the specificity for P_1'-arginine by

Sam Massa and Nick Devoogdt (eds.), *Bioconjugation: Methods and Protocols*, Methods in Molecular Biology, vol. 2033, https://doi.org/10.1007/978-1-4939-9654-4_8,

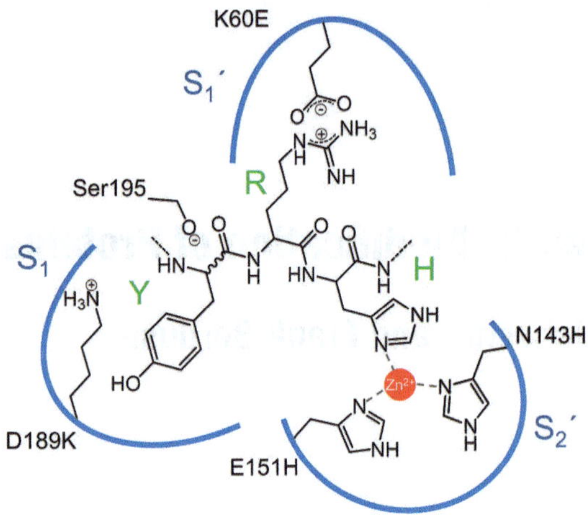

Fig. 1 Binding subsites of trypsiligase. Diagram of changes in the binding sites S_1, S_1', and S_2' that allow for recognition of the sequence YRH in specific substrates. Scissile bond cleavage (wavy line) is achieved by the reactive-site S195 hydroxyl of the catalytic triad in conjunction with the oxyanion hole (S195 and G193 backbone amides)

two orders of magnitude [3]. The replacements of N143 and E151 by histidine create a further unique recognition element in the enzyme's S_2' subsite based on an artificial metal binding site mediating a 350-fold increase in specificity toward P_2'-histidine substrates in the presence of zinc ions [4]. The amino acid exchanges in combination finally result in trypsiligase with its pronounced substrate specificity toward the tripeptide sequence YRH. Importantly, any variations even at only one position of the sequence lead to a significant decrease in catalytic efficiency due to the substrate-assisted mode of enzyme activation [5]. The latter relates to a disordered, zymogen-like structure of the apo-enzyme adopting an ordered, catalytically active conformation only upon binding to the whole YRH motif [5].

The small size and selective recognition of the YRH-motif makes the trypsiligase-catalyzed modification of proteins a valuable, minimally invasive improvement to most other described labeling procedures, which normally demand the installation of extended sequences up to polypeptide or even protein size. Despite its shortness, the recognition sequence shows a remarkably low natural currency in the human proteasome of less than 0.5% according to the SwissProt database.

N-terminal modifications of proteins via trypsiligase proceed preferably via the concept of substrate mimetics (such as peptidyl 4-guanidinophenyl esters; OGp), [6] whose ester leaving group mimics enzyme-specific amino acid side chains, allowing for the coupling of diverse and even nonpeptidic acyl moieties to the target protein (Fig. 2a). Based on the specificity and carbonyl activity,

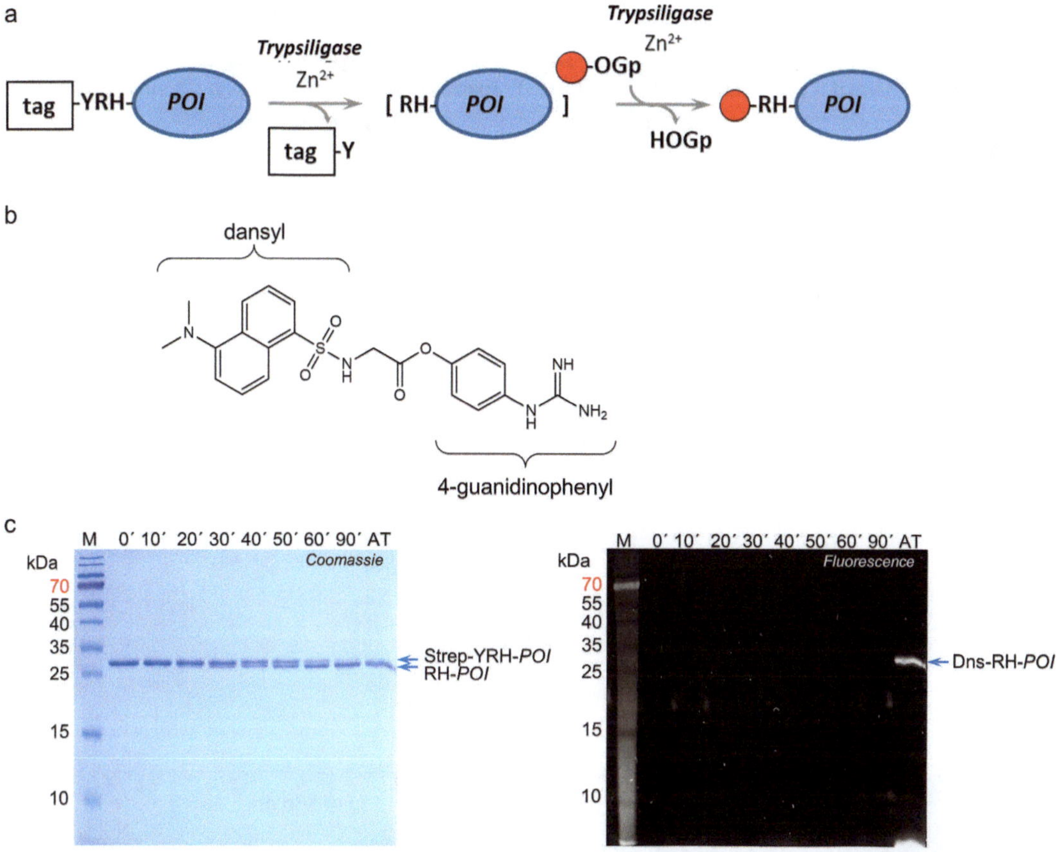

Fig. 2 Trypsiligase-catalyzed *N*-terminal protein modification. (**a**) POI is equipped with a tentative *N*-terminal (purification) tag and the recognition sequence YRH. Trypsiligase-catalyzed specific tag cleavage is followed by the selective bioconjugation reaction. The latter is initiated by the addition of the respective acyl 4-guanidinophenyl ester (OGp). (**b**) Structure of the substrate mimetic dansylglycine-4-guanidinophenyl ester. (**c**) SDS-PAGE analysis of the trypsiligase-catalyzed tag cleavage and subsequent acyl transfer of a model POI. Conditions: 100 μM StrepII-YRH-POI; 0.5 mM Dns-Gly-OGp; 10 μM trypsiligase; 100 mM HEPES–NaOH, pH 7.8; 0.1 mM $ZnCl_2$; 100 mM NaCl; 10 mM $CaCl_2$. After an initial time period of 90 min designated for the tag cleavage, the real bioconjugation reaction was ran for another 60 min. The 15% tris–glycine gel was visualized by in-gel fluorescent imaging using a simple transilluminator and Coomassie staining (*M* marker (PageRuler prestained), *AT* acyl transfer reaction)

OGp esters can acylate the enzyme already in the zymogen-like conformation resulting in a covalent acyl-enzyme intermediate with only low hydrolysis activity due to a dysfunctional oxyanion hole, that is at least partially formed in the presence of zinc and amino-terminal RH-containing peptides, leading to kinetic favoring of the aminolysis (peptide ligation) reaction. In a first step, trypsiligase removes the *N*-terminal affinity-fusion by cleaving the Tyr-Arg bond and finally, the modification agent (acyl OGp ester) is added in a one-pot-reaction initializing the trypsiligase-catalyzed acyl transfer reaction. With slight excess of the acyl donor component, the synthesis reaction usually proceeds quantitatively. The

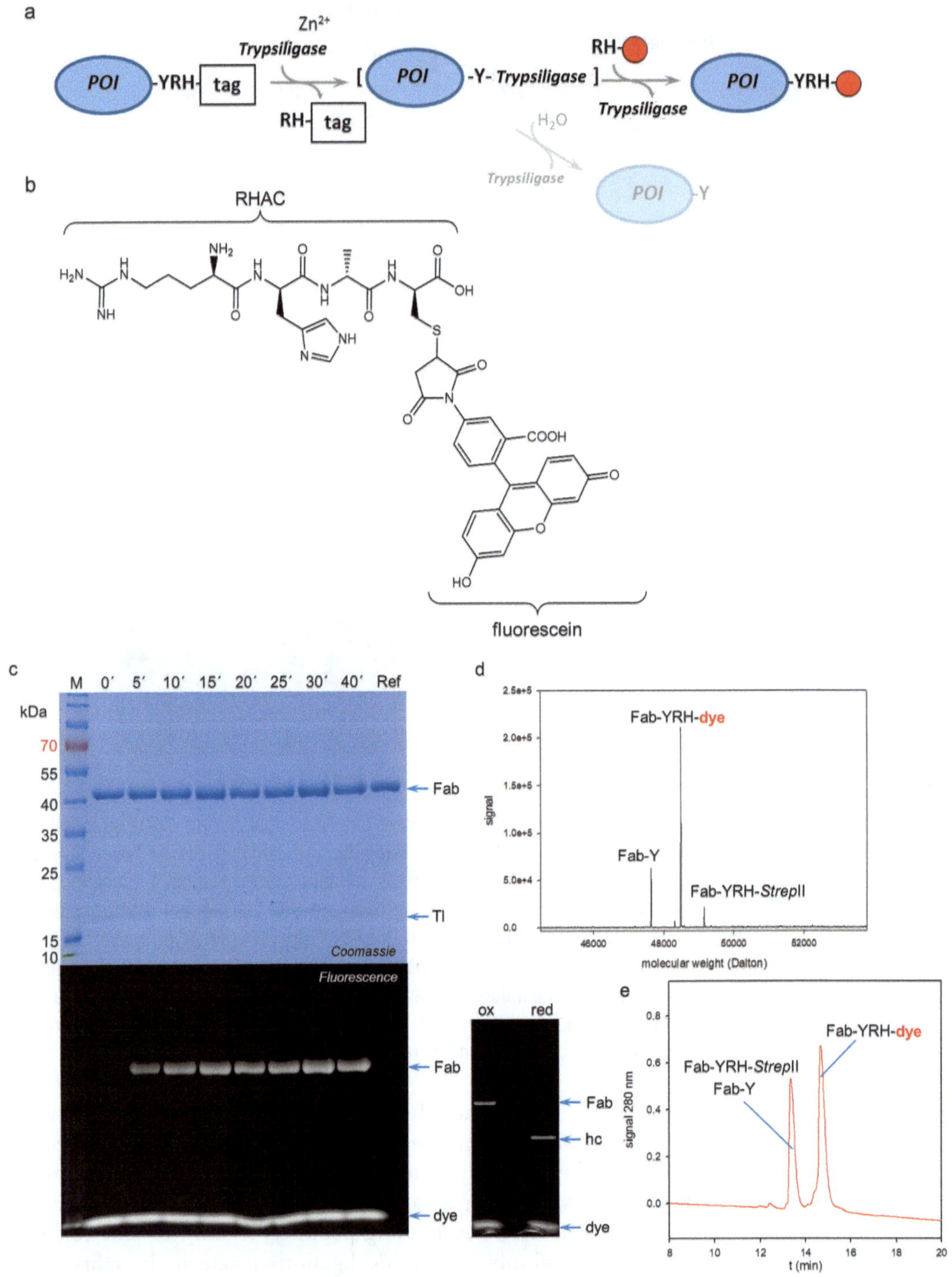

Fig. 3 Trypsiligase-catalyzed *C*-terminal protein modification. (**a**) POI's *C*-terminus is elongated by the recognition sequence YRH and a tentative (purification) tag. Trypsiligase-catalyzed modification reaction occurs via a transpeptidation mechanism. Cleavage of the tag initially results in a transient acyl—enzyme intermediate, where the POI is covalently linked to the enzyme. In the presence of Arg-His-containing acceptor peptides bearing the artificial functionality, the desired product is formed. Deacylation of the acyl-enzyme by

approach was successfully accomplished by a variety of protein targets in combination with a larger number of artificial functionalities (biotin, diverse fluorophores, polyethylene glycol, toxins, click labels, etc.) [5].

Contrary to the aforementioned *N*-terminal labeling procedure, *C*-terminal modifications are mechanistically comparable to transpeptidation reactions between YRH-bearing donor substrates and nucleophilic RH-acceptor peptides (RHX) (Fig. 3a). Such approaches are generally more challenging already from a thermodynamic point of view mainly with respect to the equilibrium constants which can be considered to be only around one for such reactions. To reach reasonable product yields, high ligation and low hydrolysis activity of the biocatalyst are in particular important. Trypsiligase fulfills both preconditions enabling the use of this approach also for selective *C*-terminal protein modifications even of protein targets of higher complexity, such as antibody fragments and full-length antibodies. The protein of interest (POI) (e.g., an antibody Fab fragment) is *C*-terminally elongated by the recognition sequence YRH and optionally by a further affinity label. Trypsiligase selectively cleaves the POI between the tyrosine and arginine moiety of the recognition sequence, thereby resulting in the formation of a covalent acyl–enzyme intermediate (and simultaneous liberation of optional affinity labels), which is then finally aminolyzed by the functionalized RH-acceptor leading to the modified POI. To this end, a large number of modification reagents such as diverse fluorophores, high molecular weight PEG and various toxins equipped with Arg-His-containing linkers were used as efficient nucleophiles resulting in covalent attachment to the Fab's *C*-terminus by a stable peptide bond as successfully

Fig. 3 (continued) competing water results in the hydrolysis product. (**b**) Structure of fluorescein-conjugated trypsiligase substrate. The peptide RHAC served as linker-scaffold for functionalization at the cysteine side chain. (**c**) Labeling reaction was quenched at distinct time points by the addition of sample buffer, products were separated via a nonreducing 15% tris–glycine gel and visualized by both in-gel fluorescent imaging using a simple transilluminator and Coomassie staining. Reduction of the interchain-disulfide bond results in a shift of the fluorescent band due to dissociation of the heavy and light chain. Conditions: 100 μM Fab-YRH-*Strep*II; 0.5 mM RHAK(CF); 10 μM trypsiligase; 100 mM HEPES–NaOH, pH 7.8; 0.1 mM $ZnCl_2$; 100 mM NaCl; 10 mM $CaCl_2$. *M* marker (PageRuler prestained), *R* reference, *Tl* trypsiligase, *hc* heavy chain, *ox* oxidized (without β-mercaptoethanol, *red* reduced (β-mercaptoethanol added). (**d**) LC-MS analysis of the reaction mixture at $t = 30$ min. Deconvoluted MS-data reveal three protein species representing the strep-tagged starting Fab, the hydrolysis product and the desired dye-conjugated product Fab-fragment. Fab-CF: $M_{calc} = 48{,}749$ Da; $M_{found} = 48{,}750$ Da, Fab-YRH-*Strep*II: $M_{calc} = 49{,}412$ Da; $M_{found} = 49{,}413$ Da, Fab-Y: $M_{calc} = 47{,}880$ Da; $M_{found} = 47{,}881$ Da. (**e**) Separation of the dye-conjugated Fab-fragments via HPLC-based hydrophobic interaction chromatography (HIC). The desired product eluates in a baseline-separated manner from the other protein species. HIC conditions: TSK-gel butyl-NPR; A = 1.5 M ammonium sulfate/0.05 M HEPES buffer, pH 7.5; B = 0.05 M HEPES buffer, pH 7.5; 0–2 min 100% A, 2–15 min 100% A to 100% B

demonstrated for the pharmaceutically interesting anti-Her2 and anti-TNFα Fab-fragments [7]. In addition to the direct introduction of the moiety of interest, trypsiligase also catalyzes the specific transfer of click-anchors to the POI offering a kit-of-parts system with maximum flexibility for site-specifically tagging proteins with a broad variety of artificial structures [8].

Both *N*- and *C*-terminal labeling reactions are ready to use and can be adapted for routine laboratory applications using recombinantly produced trypsiligase, a chemically synthesized RH-based nucleophile or OGp ester carrying a payload of choice, and the POI bearing the requisite YRH cleavage site.

Importantly, in contrast to *wt*-trypsin which is known for its autoproteolytic activity trypsiligase is a highly robust enzyme with a long-term stability and a wide tolerance for pH and additives. Once produced in the laboratory, the enzyme can be utilized for several years without any significant loss in activity. These characteristics together with the flexibility in application make trypsiligase a rather universal technology for *N*- as well as *C*-terminal modification of proteins in virtually any molecular- and cell-biological laboratory. The protocols presented for the expression and purification of active trypsiligase are inexpensive and can be performed in each appropriate laboratory. The same holds true for the synthesis of trypsiligase-compatible probes and, of course the real modification reactions themselves.

2 Materials

2.1 Substrate Synthesis

2.1.1 Ester Synthesis (for N-Terminal Protein Labeling)

1. Dichloromethane (DCM).
2. *N,N'*-bis-*tert*-butyloxycarbonyl-4-guanidinophenol (OGp $(Boc)_2$).
3. Dansylglycine.
4. Isobutyl chloroformate.
5. Boc deprotection cocktail: 50% (v/v) trifluoroacetic acid (TFA) in DCM.

2.1.2 Peptide Synthesis (for C-Terminal Protein Labeling)

1. 2-Chlorotrityl resin and a disposable fritted glass polypropylene peptide synthesis syringe.
2. Amino acid building blocks: Fmoc-Cys(Trt)-OH, Fmoc-Ala-OH, Fmoc-His(Trt)-OH, Fmoc-Arg(Pbf)-OH (Fmoc = fluorenylmethoxycarbonyl, Trt = trityl, Pbf = 2,2,4,6,7-pentamethyldihydrobenzofuran-5-sulfonyl).
3. Capping solution: 85% (v/v) dichloromethane (DCM), 10% (v/v) methanol, 5% (v/v) *N,N*-diisopropylethylamine (DIPEA).
4. *N,N*-Dimethylformamide (DMF).

5. DCM.
6. Amino acid coupling cocktail: 0.2 mmol amino acid building block, 0.2 mmol 2-(1*H*-benzotriazole-1-yl)-1,1,3,3-tetramethylaminium tetrafluoroborate (TBTU), 0.4 mmol DIPEA in 3 ml DMF. Prepare freshly before each single coupling step of the desired amino acid building block.
7. Fmoc deprotection cocktail: 20% (v/v) piperidine in DMF.
8. Cleavage cocktail: 94% (v/v) trifluoroacetic acid (TFA), 3% (v/v) triisopropylsilane (TIS).
9. Diethyl ether.
10. Methanol.
11. Speed vacuum concentrator system or freeze-dryer.
12. 20 mM sodium phosphate buffer, pH 7.2: 13.3 mM Na_2HPO_4, 6.7 mM $NaHPO_4$.
13. Maleimide coupling buffer (degassed): 20 mM sodium phosphate buffer, pH 7.2, 2 mM triscarboxyethyl phosphine (TCEP).
14. 0.2 M sodium hydroxide (NaOH).
15. Fluorescein maleimide.
16. Dimethyl sulfoxide (DMSO).

2.2 Substrate Purification

1. HPLC system with C18-reversed phase column or C18-based SPE cartridge.
2. 0.1% (v/v) TFA in degassed bidistilled water (solvent A).
3. 0.1% (v/v) TFA in acetonitrile (solvent B).

2.3 Trypsiligase Expression and Purification

1. *Pichia pastoris* (X-33) competent yeast cells.
2. pPICZα (Thermo Fisher Scientific, Waltham, USA).
3. YPDS + Zeocin plates: 1% (w/v) yeast extract, 2% (w/v) peptone, 2% (w/v) D-glucose, 1 M sorbitol, 1.5% (w/v) agar, 100 μg/ml Zeocin (allow the medium to cool to at least 55 °C before adding the Zeocin).
4. YNB medium with glucose: 1.34% (w/v) yeast nitrogen base (without amino acids); 2% (w/v) D-glucose.
5. YNB medium with methanol: 1.34% (w/v) yeast nitrogen base (without amino acids); 1% (v/v) methanol.
6. Strong cation exchange chromatography column (e.g., Toyopearl SP-650 M).
7. Binding buffer: 20 mM sodium acetate, pH 4.0.
8. Elution buffer: 100 mM hydroxyethyl piperazineethanesulfonic acid (HEPES), pH 7.8, 200 mM NaCl, 10 mM $CaCl_2$.

9. Fast liquid chromatography (FPLC) system equipped with a size exclusion chromatography (SEC) column (e.g., HiLoad Superdex 75 16/60).
10. Centrifugal Protein Concentrators 10 kDa MWCO (molecular weight cutoff).

2.4 Target Protein Preparation

1. pASK-IBA plasmid (IBA, Göttingen, Germany).
2. *E. coli* BL21 (DE3) competent cells.
3. LB media and agar plates: 10 g/l peptone, 5 g/l yeast extract, 20 g/l NaCl, supplemented with 20 g/l agar for plates.
4. 1 M Isopropyl β-D-thiogalactopyranoside (IPTG): Dissolve in water.
5. Strep-Tactin Sepharose column (e.g., StrepTrap HP).
6. Fast liquid chromatography (FPLC) system equipped with a size exclusion chromatography (SEC) column (e.g., HiLoad Superdex 75 16/60).
7. Binding buffer II: 100 mM tris(hydroxymethyl)aminomethane (Tris)–HCl, pH 8.0, 150 mM NaCl, 1 mM ethylenediamine-tetraacetic acid (EDTA).
8. Elution buffer II: 100 mM Tris–HCl, pH 8.0, 150 mM NaCl, 2.5 mM desthiobiotin.
9. Centrifugal Protein Concentrators 10 kDa MWCO.

2.5 Bioconjugation

1. Purified trypsiligase.
2. Purified YRH-containing target protein.
3. Purified labeling probe (OGp ester and RH-containing substrate, respectively) dissolved in water or DMSO (recommended for ester substrates) in a final concentration of 10–20 mM.
4. 10× Labeling buffer: 1 M HEPES–NaOH, pH 7.8, 1.5 M NaCl, 100 mM $CaCl_2$.
5. $ZnCl_2$: 10 mM $ZnCl_2$ in water.
6. Dialysis membrane (MWCO 10 kDa), desalting column (e.g., Sephadex G-25 PD-10), SEC column (e.g., HiLoad Superdex 75 16/60) or strong cation exchange chromatography column (e.g., Toyopearl SP-650 M), respectively.
7. Transilluminator.
8. InstantBlue staining solution (Expedeon, San Diego, USA).
9. 10 mM ethylenediaminetetraacetic acid (EDTA).

2.6 Hydrophobic Interaction Chromatography

1. TSKgel nonporous butyl-resin hydrophobic interaction chromatography column.
2. 50 mM HEPES–NaOH, 1.5 M ammonium sulfate, pH 7.5 (solvent A).

3. 50 mM HEPES–NaOH, pH 7.5 (solvent B).
4. 3 M ammonium sulfate.

3 Methods

3.1 Substrate Synthesis

Peptide and substrate ester syntheses are common procedures in appropriate chemical laboratories but might be the most challenging part for molecular biologists. An alternative option to guarantee fast success might be to employ one of the many commercial vendors offering professional synthesis services. Depending on the type of application, one could also consider whether instead of the commercial synthesis of the whole substrate that of a precursor may be sufficient. Proper selection of the precursor substrate might enable improved flexibility regarding the nature of the artificial functionality that can be coupled to the POI if one thinks on the additional use of click reactants. We recommend synthesis by hand, which is of course the most cost-efficient and flexible way of production.

3.1.1 Ester Synthesis (for N*-Terminal Protein Labeling)*

For *N*-terminal protein labeling we recommend the use of respective acyl-4-guanidinophenyl esters. Synthesis of OGp esters generally involves the initial preparation of *N,N′*-bis-*tert*-butyloxycarbonyl (Boc)-protected 4-guanidinophenol (OGp(Boc)$_2$) from 4-aminophenol and *N,N′*-bis-(Boc)-*S*-methylisothiourea in tetrahydrofuran as described previously [9]. According to this protocol, the synthesis is not sophisticated but somewhat time-consuming due to the need for vigorously stirring for 5 days (*see* **Note 1**). In case of shortage of time one can fall back on the use of phenyl esters, for which trypsiligase is only a little less specific but can be synthesized in a faster single step reaction [5].

The following protocol describes the synthesis of dansyl-glycyl-4-guanidinophenyl ester (Dns-Gly-OGp, Fig. 2b) but can be easily adapted to the corresponding phenyl ester by using phenol instead of OGp(Boc)$_2$ as the alcohol component. Individual ester synthesis protocols may have to be tailored to the particular acyl functionality designated for conjugation to the protein of interest (*see* **Note 2**).

1. Dissolve 1 eq. dansyl-glycine in DCM (start for example with 50 mg to usually obtain about 40 mg product after purification) in a round-bottom flask, add 1 eq. isobutyl chloroformate and incubate by agitation at −20 °C (for cooling bath add small amounts of dry ice to isopropyl alcohol until a temperature of −20 °C is reached).

2. Add 2 eq. DIPEA and 2 eq. $OGp(Boc)_2$ and slowly adjust temperature to room temperature (about 23 °C).
3. Stir at room temperature for 2 h.
4. Remove DCM by evaporation and dissolve residual substance in Boc deprotection cocktail.
5. Stir at room temperature for 1 h and remove DCM and TFA by evaporation. Ester synthesis is now completed and the reaction product can be purified.

3.1.2 Peptide Synthesis (for C-Terminal Protein Labeling)

For *C*-terminal labeling via trypsiligase-catalyzed transpeptidation we detail a protocol for the preparation of the fluorescent probe-containing peptide RHAC(fluorescein) (Fig. 3b). The synthesis scheme starts with the corresponding peptide precursor RHAC which can be synthesized on solid support by standard Fmoc-based peptide chemistry followed by coupling of fluorescein maleimide to the sulfhydryl moiety of the peptide's cysteine via a thiol-Michael addition [10, 11]. With the Cys-containing precursor peptide in hand, the synthesis protocol can be easily adapted to virtually any other thiol-reactive functionality of interest.

1. For 200 μmol synthesis scale, weigh out 100 mg 2-chlorotrityl resin (100–200 mesh) with a loading capacity of 1.0–1.6 mmol/g. Add the resin to a disposable fritted glass polypropylene peptide synthesis syringe.
2. Dissolve 0.2 mmol Fmoc-Cys(Trt)-OH in 3 ml DCM and add 0.4 mmol DIPEA. Attach an 18 G needle to the syringe and add the amino acid coupling cocktail to the beads. After removing the needle and replacing it with a plastic stopper, shake on a platform shaker for 2 h at room temperature to solvate the resin and couple the first amino acid.
3. Wash the beads three times with capping solution (3 ml) for 10 min each, followed by three washing steps with DCM (3 ml each) and three more washing steps with DMF (3 ml each).
4. Remove N^{α}-Fmoc-protection group by the addition of 3 ml deprotection cocktail for 7 min. Repeat this step two times and afterward wash the beads three times with DMF followed by the same procedure with DCM and DMF as well.
5. Add the amino acid coupling cocktail with Fmoc-alanine and shake for 1 h at room temperature.
6. Wash the beads three times with DMF, DCM and DMF in a similar way as described in **step 3**.
7. To remove the Fmoc-protection group of alanine repeat **step 4**.
8. Couple Fmoc-His(Trt)-OH by repeating **step 5** with the respective amino acid coupling cocktail.

9. After coupling of histidine repeat **steps 6** and 7 for washing and deprotection.
10. Finally couple Fmoc-Arg(Pbf)-OH and remove the terminal Fmoc-protection group by repeating **steps 5** and **6**. If the synthesis was running correctly, the side chain protected peptide Arg(Pbf)-His(Trt)-Ala-Cys(Trt) is now on the solid support.
11. Peptide liberation and side chain deprotection occur simultaneously in a one-pot reaction via freshly prepared cleavage cocktail. For this purpose, wash the resin three times with DCM and suspend it in 3 ml of cleavage cocktail. Shake for 1 h at room temperature. Collect the eluate in a round-bottom glass flask and repeat the cleavage procedure. Pool the eluates (about 6 ml in total).
12. Evaporate the TFA from the eluate using a vacuum pump and a cold trap until no more than 1 ml eluate is left (you can skip this step when you increase the amount of ether in the next step).
13. Chill 30 ml diethyl ether in a 50 ml Falcon tube with dry ice and add the TFA/peptide dropwise. Incubate on dry ice for further 30 min and centrifuge at 4000 × g for 15 min in a swinging bucket centrifuge. Decant the supernatant, dissolve the pellet in 1 ml methanol and repeat ether precipitation and centrifugation.
14. Finally, concentrate the peptide to dryness using a speed vacuum concentrator system or freeze-dryer. Mind for losses due to violently evaporating diethyl ether when applying vacuum; you may use sealing foil with pricked holes. Discard ether fractions not before you have confirmed that your peptide is in the pellet.
15. The cysteine-containing peptide can now be coupled to the maleimide-containing probe. For this purpose dissolve 40 mg peptide (2 eq.) in 1 ml maleimide-coupling buffer and check pH by transferring 1 μl on pH-paper. If necessary adjust the pH to 7.0–7.5 by the microliter-wise addition of 0.2 M sodium hydroxide (*see* **Note 3**). Dissolve 10 mg fluorescein maleimide (1 eq.) in 100 μl DMSO and protect stock solution from light by wrapping the tube in aluminum foil. Mix thiol-reactive probe and peptide solution and allow the reaction to proceed by agitation preferably in dark for at least 2 h or overnight at room temperature (*see* **Note 4**). Peptide synthesis and conjugation is now completed and the substrate can be purified.

3.2 Substrate Purification

Efficient bioconjugation via transpeptidation requires purification of the dye-labeled peptide from the unlabeled free peptide precursor which was used in twofold excess over the dye label in the synthesis procedure (*cf.* Subheading 3.1.1). Satisfying results will be usually obtained via preparative reversed-phase high-performance liquid chromatography (RP-HPLC) equipped with a C18 column (e.g., Merck Lichrosorb RP-18, 250 × 25 mm) and a linear H_2O–acetonitrile gradient for elution (0% solvent A to 30% solvent B in 45 min for RHAC(CF) using the aforementioned column). If supported by the equipment used, detection should be done at a wavelength of 220 nm combined with a second wavelength corresponding to the absorption maximum of the respective dye. Dual wavelength analysis easily enables to distinguish between the free precursor peptide and the dye-conjugated peptide fractions. Normally, the unconjugated peptide is significantly more hydrophilic than the conjugated one and will elute earlier from the RP-HPLC column. We recommend performing an analytical run before applying the whole peptide or ester fraction to the preparative column. Identity of the desired synthesis product should be validated via routine mass spectrometry. After RP-HPLC purification, respective substrate-containing fractions have to be pooled, subsequently evaporated from the solvent and finally lyophilized to dryness. If available, one can also use a speed vacuum concentrator system alternatively.

In case you don't have the possibility to isolate your substrates by RP-HPLC you can use disposable solid phase extraction (SPE) cartridges filled with C18 bonded silica sorbents. Suitable protocols for SPE have to be tailored to the manufacturer's information and the bed volume of the cartridge.

1. Add ten bed volumes of solvent B (0.1% (v/v) TFA in acetonitrile) to the C18 based SPE cartridge before equilibrating the cartridge with ten bed volumes of solvent A (0.1% (v/v) TFA in H_2O).
2. Fill up the volume of your crude peptide fraction (from synthesis step in Subheading 3.1) to two bed volumes with solvent A and gently pass the peptide solution through the resin. Collect the flow-through and wash the resin with five bed volumes of solvent A.
3. Prepare two bed volumes of different eluent solvents by mixing of solvent A and B to final acetonitrile concentrations of 5%, 10%, 20%, 30%, 50% and 100%. Carry out a stepwise elution with the former one and collect each eluate fraction.
4. To identify dye-containing peptide fractions you can use an UV-light source (e.g., transilluminator or hand lamp). To confirm identity of your dye-conjugated peptide mass spectrometry (MALDI or LC-MS) is the best choice. Alternatively one can test each fraction in a trypsiligase-catalyzed reaction.
5. Concentrate dye-containing peptide fractions to dryness using a freeze-dryer or a speed vacuum concentrator system and redissolve each fraction in water to a final concentration of

10 mM. The peptide can be stored in a refrigerator for several months. Test dye-containing peptide fractions as outlined in Subheading 3.5.

3.3 Trypsiligase Expression and Purification

Trypsiligase gene (derived from anionic rat trypsin II; Uniprot: P00763) has to be cloned into pPICZα for secretory expression in *P. pastoris* which is recommended in terms of protein yield and ease of purification [12]. Using this expression system, the biocatalyst becomes secreted into the culture medium and can be easily isolated by cation exchange and subsequent size exclusion chromatography (*see* **Note 5**). In contrast to native *wt*-proteases, which are usually toxic to the expression host and thus, need to be expressed as proteolytically inactive zymogens, trypsiligase is completely harmless and can be produced in its final active shape. Since trypsiligase's *N*-terminus is involved in the formation of a salt bridge with Asp194, which is essential for catalytic activity, it should be desisted from tentative modifications, for example, by *N*-terminal fusion to affinity tags.

1. Transform linearized plasmid DNA into *P. pastoris* X-33 and plate on Zeocin-containing YPDS plates.
2. Pick a single colony and grow at 30 °C and 130 rpm to saturation (48 h) in 50 ml glucose-containing YNB.
3. Centrifuge 20 ml of the preculture at 4000 × *g* for 5 min and discard supernatant.
4. Resuspend the pellet in 200 ml YNB supplemented with 1% (v/v) methanol in a 1 l flask and incubate for 5 days at 30 °C by continuously shaking at 200 rpm.
5. Add another 1% (v/v) methanol every 24 h to induce protein production.
6. Mix 20 μl of the supernatant with 10 μl sample buffer and verify protein expression by SDS-PAGE analysis (SDS-PAGE: sodium dodecyl sulfate–polyacrylamide gel electrophoresis).
7. Harvest cells by centrifugation at 4000 × *g* for 10 min and carefully decant the supernatant. Discard the pellet. If necessary, filtrate the supernatant with a single-use filter cartridge.
8. Add an equivalent of two volumes of binding buffer to the supernatant fraction and transfer the resulting solution to a strong cation exchange chromatography column equilibrated with binding buffer.
9. Wash the column with 5 column volumes binding buffer and finally elute trypsiligase with 2 column volumes elution buffer.
10. Trypsiligase may further be purified by an additional size exclusion chromatography step using elution buffer (e.g., over a Hi-Load Superdex 75 16/60 column via an appropriate FPLC system).

11. Concentrate the purified protein up to 100 μM (about 2.4 mg/ml) by using a centrifugal concentrator and finally determine the protein concentration photometrically at 280 nm (*see* **Note 6**).
12. Aliquots can be stored for several months at 4 °C in the fridge or for several years in a freezer without significant loss of enzymatic activity.

3.4 Target Protein Expression and Purification

Trypsiligase recognition of the target protein essentially needs the installation of the enzyme's recognition sequence YRH at either the *N*- or *C*-terminus of the protein substrate; which can be excellently combined with the installation of an additional affinity tag getting simultaneously cleaved during the bioconjugation reaction. The model IgG1 Fab fragment substrate described here carries a *Strep*-tag II fusion to simplify its purification, located downstream of the recognition sequence YRH at the very *C*-terminal end of the heavy chain (cloned into a pASK-IBA vector with ampicillin resistance cassette; *see* **Note 7**) [13]. The Fab substrate can be produced in *E. coli* and purified using this simplified protocol, which may have to be tailored to the specific protein to be labeled. The purification protocol outlined here corresponds to a standard affinity chromatography approach based on the Strep-tag system. Alternatively one may also apply Protein A as antibody related affinity label for chromatographic purification [14].

1. Transform plasmid into *E. coli* BL21(DE3) cells and plate on selective media.
2. Pick a single colony and grow to saturation (overnight) in 30 ml LB medium containing the appropriate antibiotic.
3. Inoculate 1 l of antibiotic-containing LB medium with 10 ml overnight culture and shake at 37 °C until an OD_{600} of 0.6 is reached.
4. Lower temperature to 18 °C, add IPTG to a final concentration of 0.5 mM and continue to grow cultures overnight.
5. Harvest bacteria by centrifugation at 6000 × *g* for 20 min.
6. Decant supernatant and resuspend pellet in 30 ml ice-cold binding buffer II.
7. Lyse bacteria using a sonifier (amplitude = 30–40%) with 6 pulses of 10 s with 1 min lapse between each pulse. Alternatively pass the bacteria through a prechilled French press cell twice at 1250 psi.
8. Clarify the lysate by centrifugation at 12,000 × *g* for 30 min. If necessary, resuspend the pellet, repeat **step** 7 and pool the supernatant fractions.
9. Equilibrate a Strep-Tactin Sepharose column with at least 5 column volumes binding buffer II.

10. Add the bacterial lysate to the column and wash with 5–10 column volumes binding buffer until the unbound proteins are stripped off and the baseline is constant.
11. Elute with five column volumes elution buffer II and analyze elution fractions by SDS-PAGE (*see* **Note 8**).
12. Concentrate Fab-containing fractions to less than 2 ml and perform size exclusion chromatography to remove tentative aggregates using elution buffer from Subheading 2.3 (e.g., over a Hi-Load Superdex 75 16/60 column using an appropriate FPLC system).
13. Concentrate the purified protein using a centrifugal concentrator and finally determine protein concentration photometrically at 280 nm.

3.5 Bioconjugation

Accessibility of the enzyme's recognition sequence depends on the nature of the target protein and may affect the reaction velocity and product yield. In the case of using POIs equipped with a purification tag in addition to the enzyme's recognition sequence the bioconjugations always involve the cleavage of the recognition sequence and liberation of the purification tag, which can be easily detected by respective shifts in the molecular weight. The cleavage reaction itself can take several minutes up to some hours depending on the enzyme and substrate concentration or the accessibility of the cleavage site. To determine suitable timeframes for the labeling reaction and for optimizing general reaction conditions we recommend initial hydrolysis studies for both the *N*- and *C*-terminal labeling procedure (*see* **Note 9**).

3.5.1 Initial Hydrolysis Studies

1. Mix labeling buffer (to a final concentration of 1×), substrate protein (100 μM), trypsiligase (10 μM), and $ZnCl_2$ (100 μM) together. For analytical purposes we recommend a final volume of 20–50 μl.
2. Incubate for 2 h at 30 °C (*see* **Note 10**).
3. Withdraw aliquots (1–2 μl) at distinct time points (e.g., 20, 30, 40, 60, 90, and 120 min) and quench them with SDS-PAGE sample buffer.
4. Analyze the progress of the reaction by SDS-PAGE (Fig. 2c) or by any other suitable method tailored to your POI (mass spectrometry, immunoblot analysis, etc.).

3.5.2 N-Terminal Bioconjugation

N-terminal protein labeling consists of the initial tag-cleavage (in the case of using POIs equipped with a purification tag in addition to the recognition sequence) generating the *N*-terminal RH-acceptor motif and subsequent transfer of the OGp ester's acyl moiety (synthesized according to Subheading 3.1.2) to the POI (Fig. 2a). After quantitative tag cleavage (proven by initial hydrolysis studies, *cf.* Subheading 3.5.1), the functionalized OGp ester (Dns-Gly-OGp) can be added to the reaction mixture.

1. Mix labeling buffer (to a final concentration of 1×), 100 μM substrate protein, 10 μM trypsiligase, and 100 μM $ZnCl_2$ together. For analytical purposes we recommend a final volume of 20–50 μl.
2. Incubate for 2 h at 30 °C (as long as determined in Subheading 3.5.1, respectively).
3. Add Dns-Gly-OGp in a final concentration of 0.2–0.5 mM (*see* **Note 11**) and incubate for 1 h at 30 °C.
4. Analyze the reaction mixture via SDS-PAGE (Fig. 2c) or by any other appropriate method tailored to the POI (mass spectrometry, spectroscopic or immunoblot analysis, etc.).
5. Typically, *N*-terminal labeling procedures via trypsiligase proceed quantitatively making the separation of conjugated and unconjugated POI unnecessary. Unconverted and/or hydrolyzed OGp ester can be easily removed via dialysis or a desalting column. Likewise, trypsiligase can be separated via ion exchange chromatography due to its low pI of 5.4 or alternatively, by size exclusion chromatography (molecular weight of trypsiligase: 23.8 kDa) which may remove the unconverted or hydrolyzed ester simultaneously.

3.5.3 C-Terminal Bioconjugation

For the *C*-terminal labeling approach, the functionalized peptide substrate (synthesized according to Subheading 3.1.1) is mixed with the POI (here the IgG1 Fab-fragment as prepared in Subheading 3.4) and the transamidation reaction is started by the addition of trypsiligase. Specific cleavage of the affinity tag as well as selective introduction of the fluorescent label proceed simultaneously in a one-pot reaction (Fig. 3a). Trypsiligase-catalyzed transpeptidation of the IgG1 Fab-fragment with fluorescein-conjugated RHAC-peptide reaches the maximum product yield at about 20–30 min reaction time. Accumulation of undesired hydrolysis product can be minimized by quenching the reaction at that time point (e.g., by the addition of 1 mM EDTA, shifting the pH to 5.0 or reduction of temperature). We recommend the addition of reducing agents such as 10 mM dithiothreitol in combination with 1 mM EDTA which was found to be effective also in terms of long-time stability of most target proteins. Due to the presence of functional disulfides, reducing agents should not be used; however, in the case of antibodies and antibody fragments. In such cases the addition of 1 mM EDTA is sufficient to quench the reaction completely.

For the separation of the desired transpeptidation product from the tentative hydrolysis product as well as to remove trypsiligase and conjugated acceptor peptide used in excess, we recommend the application of hydrophobic interaction chromatography (HIC) as outlined in Subheading 3.6. For individual reactions with individual POIs or acceptor peptide substrates we suggest initial transpeptidation experiments in analytical scale mainly to determine the time point of maximum product yield. Suitable analysis methods may need to be tailored to the individual POI and/or acceptor peptide functionality. If available, we generally recommend LC-MS analysis of the reaction mixture at least at the time of maximum product yield (Fig. 3d). After specifying the individual reaction in analytical reactions, scale up to preparative conditions was found to be unproblematic so far.

1. Mix labeling buffer (to a final concentration of 1×), 100 μM $ZnCl_2$, 100 μM POI, and 1 mM dye-conjugated acceptor peptide. For analytical purposes we recommend a final volume of 50 μl.
2. Add 10 μM trypsiligase.
3. Incubate the resulting reaction mixture at 30 °C (*see* **Note 10**).
4. Withdraw aliquots at distinct time points (e.g., 10, 20, 30, 40, 50, and 60 min) and quench them with SDS-PAGE sample buffer for analytical purposes (Fig. 3c).
5. Analyze the progress of reaction by SDS-PAGE and visualization by in-gel fluorescence and InstantBlue staining. Keep in mind that the addition of reducing agents to the SDS-PAGE sample buffer will separate the light and heavy chains of antibodies resulting in two distinct bands (of which only one is fluorescent). Estimate the time point of maximum product yield (in case of IgG1 Fab-fragments most intensive fluorescence protein band was obtained between 20 and 30 min).
6. Scale up to preparative conditions by repeating **steps 1–3** in the appropriate enlarged reaction volume (e.g., 1 ml).
7. Quench the enzymatic reaction by the addition of 1 mM EDTA and put the reaction vessel on ice. The fluorescent dye is now covalently linked to the POI which can be isolated from the reaction mixture as described in Subheading 3.6.

3.6 Hydrophobic Interaction Chromatography (HIC)

In contrast to the *N*-terminal labeling procedure, which results in a homogenous protein product, *C*-terminal bioconjugations usually lead the desired transpeptidation product as well as certain amounts of unwanted hydrolysis product due to the deacylation of the acyl-enzyme intermediate by competing water. If separation of both products is needed we recommend hydrophobic interaction chromatography using an HPLC or FPLC system (Fig. 3e). One option is the use of an analytical nonporous TSKgel-based butyl-column (dimension, e.g., 3.5 cm × 4.6 mm). An alternative approach might use a HiTrap butyl FF column of suitable dimension. In every case, perform initial reference runs including the tagged protein of interest, the hydrolyzed target protein and analytical quantities of the ultimate bioconjugation batch. If supported by the equipment used, combine 280 nm detection (analysis of activated aromatic amino acid side chains) with a second wavelength corresponding to the absorption maximum of the dye (490 nm for fluorescein), which enable discrimination of modified and unmodified POI. After preparative separation, the identity of the desired protein product has to be validated via MS analysis.

1. Equilibrate the TSKgel-based HIC-column with solvent A (flow rate: 1 ml/min).
2. Mix equal volumes of the conjugated Fab and 3 M ammonium sulfate (resulting in a final ammonium sulfate concentration of 1.5 M).
3. Inject 20 μl onto the column.
4. Apply the following gradient for elution: 0–2 min 100% A, 2–15 min 100% A to 100% B.
5. Collect each peak separately.
6. Analyze peaks by an appropriate method tailored to the POI (SDS-PAGE, mass spectrometry, spectroscopic or immunoblot analysis, etc.).

4 Notes

1. For research uses, moderate quantities of Boc-protected 4-guanidinophenol precursor may be available upon request (Frank Bordusa's Lab, Martin-Luther-University Halle-Wittenberg).
2. We recommend the use of X-glycyl- and X-alanyl-esters for trypsiligase-catalyzed *N*-terminal protein modification. Higher carboxyl activity of X-glycyl-esters results in high velocities of acyl-enzyme formation and therefore rapid product accumulation. The rather high reactivity may be associated with a reduced stability, particular in more alkaline buffer media resulting in a certain extent of spontaneous hydrolysis. To minimize such tentative events the ester substrates should be generally prepared as a water-free stock solution. In cases where higher stability of the ester is necessary one can additionally shift the pH of the bioconjugation buffer to 6.5 or/and exchange the ester's glycine by a less reactive alanine. The substrate mimetic concept allows even the transfer of completely nonpeptidic acyl moieties, which enables the direct esterification of such acyl functionalities without any amino acid or peptide spacers. Depending on the structure, esters of such nonpeptidic carboxylic acids may be rather stable against

chemical hydrolysis but may also require elongated incubation times and/or improved enzyme concentrations in the bioconjugation reaction. In the meantime, most fluorophores are commercially available as maleimides, which can be easily coupled to any thiol-containing carboxylic acid (cysteine, mercaptohexanoic acid, etc.) via thiol-Michael addition, resulting in a broad variety of suitable ester substrates.

3. Due to the TFA-cleavage step, especially basic amino acid moieties usually form TFA-salts, which can cause changes in the final pH of the buffered solution upon dissolving. Also, TFA-salts are relevant to the net weight and must be considered for the molecular weight of the respective peptide. Individual TFA contents of individual peptides can be determined via elementary analysis. As a rough estimation one might consider at least one equivalent TFA per positive charge of basic amino acid side chains and the *N*-terminus of the peptide, respectively).

4. Peptide-to-maleimide ratio usually depends on the individual conditions. We generally recommend an excess of peptide since the maleimide-conjugated fluorophore is often the more expensive part. The excess of starting peptide results in a product mixture of conjugated and unconjugated peptide, which needs of course a final separation step e.g. by classical RP-HPLC. If you do not have access to a preparative HPLC equipment, you may invert the ratio of precursor peptide and maleimide-functionalized probe (1:2) to realize complete conversion of the peptide. Please consider that the presence of nonconjugated peptide will compete for deacylation of the acyl-enzyme intermediate in the bioconjugation reaction resulting in transamidated but nonlabeled protein product.

5. Trypsiligase is produced in excellent product yields of 5–15 mg/l culture and is available upon request in moderate quantities (Frank Bordusa's Lab, Martin-Luther-University Halle-Wittenberg).

6. The molar extinction coefficient of trypsiligase at 280 nm is 34,670 M^{-1} cm^{-1} (determined empirically). Determination of individual protein concentrations by the classical Bradford assay also works well.

7. For Fab expression we recommend subcloning of the Fab fragment containing the heavy and light chain gene sequences of the corresponding antibody into a pASK vector for anhydrotetracycline-induced expression. For efficient disulfide bond formation, each protein chain needs to be equipped with an OmpA-leader sequence for periplasmic secretion [15]. IPTG-induced Fab-expression was found to be functional [16].

8. For SDS-PAGE analysis of Fab-fragments we recommend the use of sample buffer with and without reducing agent (e.g., β-mercaptoethanol) as well enabling the dissociation analysis of the heavy and light chain induced by reduction of the intramolecular disulfide bond.
9. Trypsiligase is a highly robust enzyme which is catalytically active in a wide range of reaction conditions tolerating for example 0.5 M guanidinium hydrochloride or up to 10% DMF in the reaction media. Although the pH-optimum of the enzyme was found at pH 7.8, it also tolerates pH-values in the range of pH 6.0–8.5. Trypsiligase recognition of YRH is however Zn^{2+}-induced wherefore we advise not to use phosphate-containing buffers which may result in the precipitation of zinc phosphate.

   Trypsiligase originally comes from anionic rat trypsin II which possesses six disulfide bonds resulting in a certain sensitivity toward reducing agents. Triscarboxyethyl phosphine (TCEP) up to concentrations of 2 mM does, however, not affect trypsiligase activity. In contrast, the use of dithiothreitol should be avoided.
10. Final reactant concentrations and reaction temperature depend on the nature of the individual POI and may need to be adapted. The trypsiligase-catalyzed modification reaction can proceed on ice as well as 37 °C with effect on the reaction rate. In case of sufficient accessibility of the recognition sequence complete conversion of the POI occurs after 2 h at 30 °C under the aforementioned conditions.
11. Due to spontaneous hydrolysis of esters, particularly under basic conditions, individual ester concentrations used may need to be adapted to realize intact ester concentrations sufficient for quantitative product yields.

References

1. Hartley BS, Shotton DM, Paul DB (1971) 10 pancreatic elastase. Enzymes 3:323–373. https://doi.org/10.1016/S1874-6047(08)60401-1
2. Graf L, Craik CS, Patthy A et al (1987) Selective alteration of substrate specificity by replacement of aspartic acid-189 with lysine in the binding pocket of trypsin. Biochemistry 26(9):2616–2623. https://doi.org/10.1021/bi00383a031
3. Kurth T, Grahn S, Thormann M et al (1998) Engineering the S1' subsite of trypsin: design of a protease which cleaves between dibasic residues. Biochemistry 37(33):11434–11440. https://doi.org/10.1021/bi980842z
4. Willett WS, Brinen LS, Fletterick RJ et al (1996) Delocalizing trypsin specificity with metal activation. Biochemistry 35(19):5992–5998. https://doi.org/10.1021/bi9530191
5. Liebscher S, Schoepfel M, Aumueller T et al (2014) N-terminal protein modification by substrate-activated reverse proteolysis. Angew Chem Int Ed 53(11):3024–3028. https://doi.org/10.1002/anie.201307736
6. Bordusa F (2002) Proteases in organic synthesis. Chem Rev 102(12):4817–4868. https://doi.org/10.1021/cr010164d
7. Liebscher S, Kornberger P, Fink G et al (2014) Derivatization of antibody fab fragments: a

designer enzyme for native protein modification. Chembiochem 15(8):1096–1100. https://doi.org/10.1002/cbic.201400059

8. Meyer C, Liebscher S, Bordusa F (2016) Selective coupling of click anchors to proteins via trypsiligase. Bioconjug Chem 27(1):47–53. https://doi.org/10.1021/acs.bioconjchem.5b00618
9. Sekizaki H, Itoh K, Toyota E et al (1996) Synthesis and triptic hydrolysis of p-guanidinophenyl esters derived from amino acids and peptides. Chem Pharm Bull 44(8):1577–1579. https://doi.org/10.1248/cpb.44.1577
10. Coin I, Beyermann M, Bienert M (2007) Solid-phase peptide synthesis: from standard procedures to the synthesis of difficult sequences. Nat Protoc 2:3247. https://doi.org/10.1038/nprot.2007.454
11. Hoyle CE, Lowe AB, Bowman CN (2010) Thiol-click chemistry: a multifaceted toolbox for small molecule and polymer synthesis. Chem Soc Rev 39(4):1355–1387. https://doi.org/10.1039/B901979K
12. Higgins DR, Cregg JM (1998) Introduction to Pichia pastoris. Methods Mol Biol 103:1–15. https://doi.org/10.1385/0-89603-421-6:1
13. Schmidt TG, Skerra A (2007) The Strep-tag system for one-step purification and high-affinity detection or capturing of proteins. Nat Protoc 2(6):1528–1535. https://doi.org/10.1038/nprot.2007.209
14. Bouvet J-P (1994) Immunoglobulin Fab fragment-binding proteins. Int J Immunopharmacol 16(5):419–424. https://doi.org/10.1016/0192-0561(94)90031-0
15. Skerra A (1994) Use of the tetracycline promoter for the tightly regulated production of a murine antibody fragment in *Escherichia coli*. Gene 151(1):131–135. https://doi.org/10.1016/0378-1119(94)90643-2
16. Skerra A (1993) Bacterial expression of immunoglobulin fragments. Curr Opin Immunol 5(2):256–262. https://doi.org/10.1016/0952-7915(93)90014-J

Chapter 9

Site-Selective Functionalization of Nanobodies Using Intein-Mediated Protein Ligation for Innovative Bioconjugation

Geert-Jan Graulus, Duy Tien Ta, Huong Tran, Rebekka Hansen, Brecht Billen, Erik Royackers, Jean-Paul Noben, Nick Devoogdt, Serge Muyldermans, Wanda Guedens, and Peter Adriaensens

Abstract

An expression strategy is presented in order to produce nanobodies modified with a clickable alkyne functionality at their C-terminus via the intein-mediated protein ligation (IPL) technique. The protocol focuses on the cytoplasmic expression and extraction of a nanobody–intein–chitin binding domain (CBD) fusion protein in *E. coli* SHuffle® T7 cells, in the commonly used Luria–Bertani (LB) medium. The combination of these factors results in a high yield and nearly complete alkynation of the nanobody at its C-terminus via IPL. The resulting alkynated nanobodies retain excellent binding capacity toward the nanobody targeted antigen. The presented protocol benefits from time- and cost-effectiveness and allows for a feasible upscaling of functionalized (here alkynated) nanobodies. The production of high quantities of site-specifically modified nanobodies paves the way to (1) novel biosurface applications that demand for homogeneously oriented nanobodies having their active site fully accessible for target (e.g., biomarker) binding, and (2) innovative applications such as localized drug delivery and image guided surgery by covalent "click" chemistry coupling of these alkynated nanobodies to a multitude of azide-containing counterparts as there are drug containing polymers and contrast labeling agents.

Key words Protein functionalization, Intein, Nanobody, Click chemistry

1 Introduction

When considering the coupling of a protein of interest (POI) to, for example, labeling moieties or material surfaces, the effectiveness of the resulting construct is strongly dependent on the applied conjugation methodology. Ideally, the conjugated protein should maintain its function (e.g., catalytic activity for enzymes, antigen binding potential for antibodies) making targeted, or site-specific, modification of proteins desirable. In addition, for most

Sam Massa and Nick Devoogdt (eds.), *Bioconjugation: Methods and Protocols*, Methods in Molecular Biology, vol. 2033, https://doi.org/10.1007/978-1-4939-9654-4_9,

applications the covalent coupling of the protein to the surface or other functional compounds is required.

Proteins are composed of various amino acids and thus generally contain multiple reactive functionalities (e.g., carboxylic acids, amines, thiols, alcohols) [1]. The possibility of possessing multiple copies of these functional groups and/or their presence in the binding pockets excludes many conventional chemical conjugation techniques (e.g., amide bond formation via carbodiimide chemistry) since such methods do not allow for the targeting of a specific region (i.e., a particular amino acid within the protein of interest) without altering the protein's conformation.

To circumvent this issue, bioorthogonal handles (i.e., not occurring naturally within a particular protein) can be introduced to subsequently allow for the selective coupling of the protein to a complementary functionality. In this respect the so-called click reactions have gained a lot of interest as such reactions are very efficient, highly specific and can be performed under the mild reaction conditions associated with protein modifications. The Cu-catalyzed Huisgen 1,3-dipolar cycloaddition (or Cu-catalyzed azide–alkyne cycloaddition, CuAAC) is a well-known example of such efficient coupling methodologies [2]. In this reaction an azide and an alkyne react to form a stable 1,2,3-triazole ring (*see* **Note 1**). The modification of the protein at a single site would thus enable the site-specific coupling of the POI to molecules and surfaces decorated with the complementary chemical functionality.

In the current chapter the intein-mediated protein ligation methodology is presented as a method to selectively introduce a particular chemical handle at the C-terminus of the POI [3, 4]. Inteins (also called protein introns) are self-cleavable peptide sequences ending in an N-terminal cysteine [5]. This cysteine facilitates an N-S shift at the C-terminus of the POI thereby forming a thioester linkage (Fig. 1). The formed thioester is prone to nucleophilic attack and can be cleaved via the addition of thiol-containing compounds (e.g., dithiothreitol, DTT; sodium 2-mercaptoethane sulfonate, MESNA) in what is called native chemical ligation [6, 7].

By fusing the POI with an intein sequence (*see* **Note 2**), the N-S shift can be exploited to make the POI susceptible to functionalized thiols. Within this protocol, the focus lies on a cysteine derivative containing an alkyne functionality (in short cysteine–alkyne linkers, Fig. 2a), which would enable the subsequent conjugation to various azide-functionalized compounds/surfaces [8–10]. These include azide-functionalized biotin for application in streptavidin-based assays (Fig. 2b), fluorescent markers for imaging applications (Fig. 2c), azide-pending polymers for controlled drug delivery applications (Fig. 2d) or azide-functionalized surfaces for biosensor development (Fig. 2e). The choice for introducing an

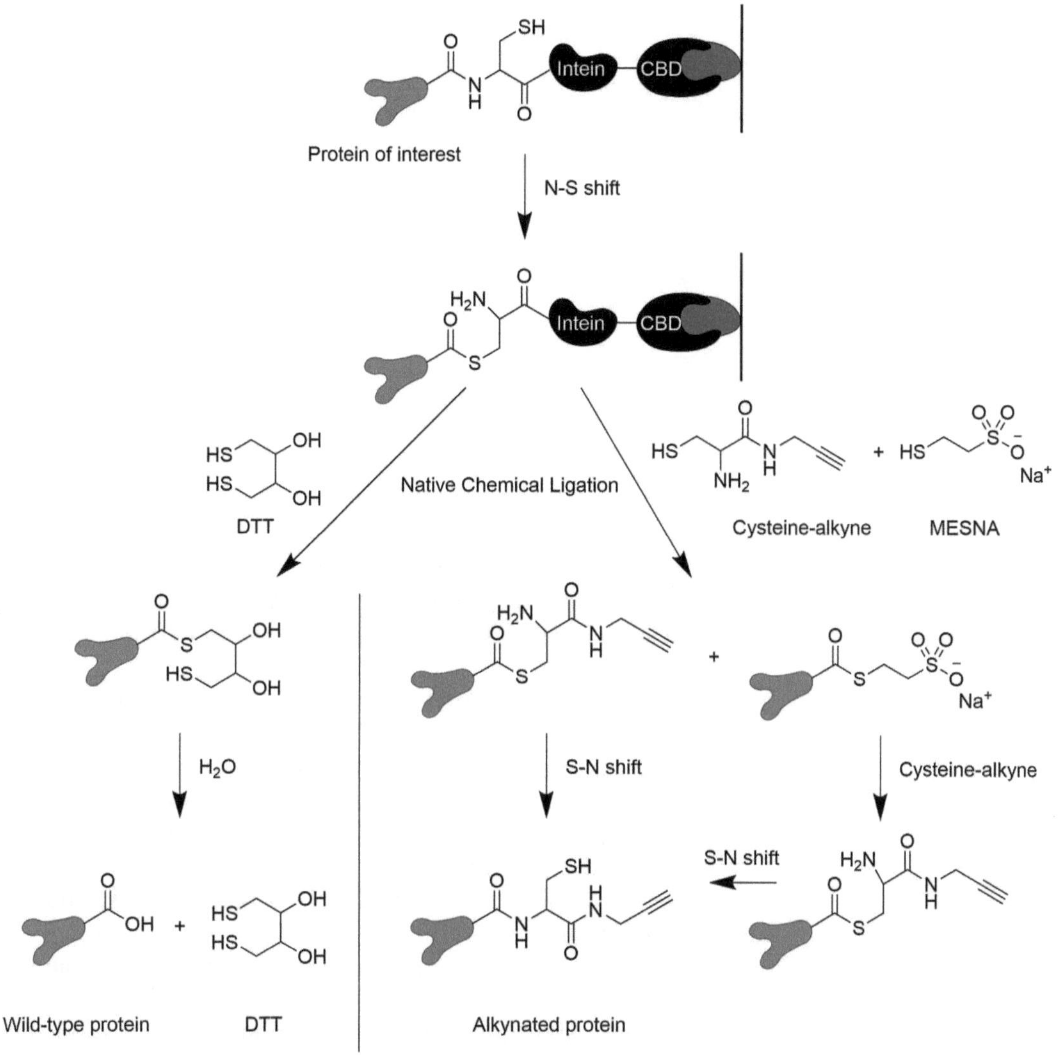

Fig. 1 Scheme showing the IPL mechanism. The protein of interest is C-terminally ligated to an intein–chitin binding domain (intein-CBD). The expressed fusion protein is purified via affinity chromatography on a chitin column and subsequently cleaved by addition of thiol-based nucleophiles. When cysteine–alkyne linkers are applied, on column cleavage of the POI occurs simultaneous to protein functionalization yielding alkynated proteins

alkyne functionality at the protein C-terminus rather than an azide is explained in **Note 3**.

In the current protocol the intein-mediated protein ligation technique is applied as a straightforward method to introduce an alkyne functionality at the C-terminus of a nanobody. Nanobodies are single domain antigen-binding proteins derived from camelid antibodies (or heavy chain antibodies) [11, 12]. Nanobodies exhibit similar binding capacities and affinities toward their antigen than conventional immunoglobulins, but are much smaller, more

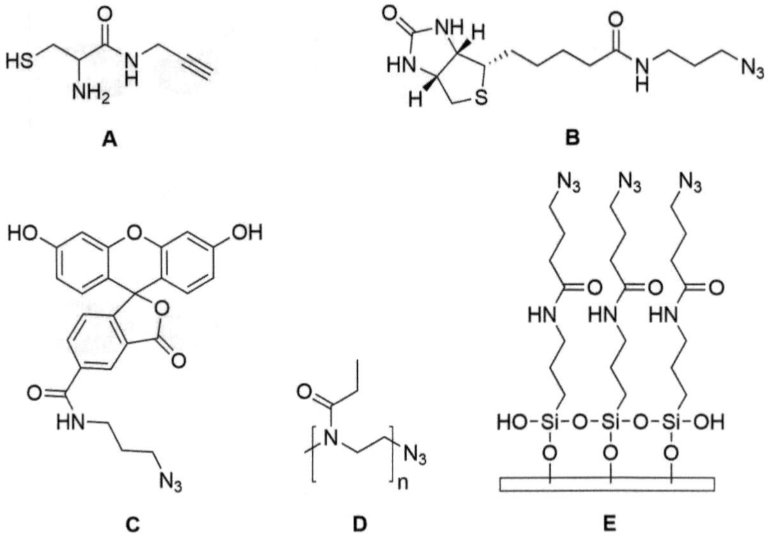

Fig. 2 Chemical structure of the (**a**) cysteine–alkyne, (**b**) azido-biotin, (**c**) 5-FAM-azide, (**d**) PEtOx-N_3 polymers, and (**e**) azide-functionalized surfaces to be coupled to the alkynated nanobodies by means of CuAAC "click" chemistry

thermostable and resistant to pH changes [13]. These properties, together with the fact that nanobodies are encoded by a single gene, make these proteins an interesting alternative for conventional antibodies in biosensing and targeted drug delivery applications [12, 14]. IPL is particularly interesting for functionalizing nanobodies as the C-terminus is located at the opposite site of the binding paratope, which contains the N-terminus [15]. Conjugating nanobodies at the C-terminus via the IPL methodology thus guarantees the target binding potential of the nanobody remaining unaffected.

The presented protocol describes the cloning of a nanobody in a commercially available plasmid, the production and purification of a nanobody–intein fusion construct out of the cytoplasm of fermented *E. coli* cells, the generation of alkynated nanobodies and the conjugation of azidified substrates to the alkyne-functionalized nanobody by CuAAC. Also methods are provided to evaluate the success of the conjugation reaction.

2 Materials

Unless stated otherwise, all reagents mentioned in this protocol are purchased from Sigma-Aldrich and used as received. Aqueous solutions and buffers are prepared using Milli-Q water ($\rho = 18.2\ M\Omega$).

2.1 Molecular Cloning of the Nanobodies

1. Plasmids containing nanobody sequences are needed. In our studies we have used pHEN6 plasmids containing coding for nanobodies followed by His_6 coding for purification purposes [16]. Dilute each template plasmid to 100 pg/mL in sterile water.
2. Forward primer (5′-GGTGGTCATATGCAGGTGCAGCTG CAGGAGTCT-3′, 10 μM concentration in sterile water).
3. Reverse primer (5′-GGTGGTCTCGAGTGAGGAGACGG TGACCTGGGTCCC-3′, 10 μM concentration in sterile water).
4. Thermocycling device.
5. 5 units/μL Taq polymerase (e.g., Dream Taq from Thermo Fisher Scientific).
6. Taq polymerase buffer (e.g., Dream Tag buffer from Thermo Fisher Scientific).
7. 10 mM dNTPs.
8. 10 units/μL restriction enzymes NdeI and XhoI (e.g., from Thermo Fisher Scientific).
9. QIAquick Gel Extraction Kit (Qiagen).
10. 1 U/μL T4 DNA Ligase.
11. 10× T4 DNA ligase buffer.
12. Double-digested (NdeI, XhoI) pMXB10 plasmid (New England Biolabs, *see* Fig. 3).

2.2 Transformation and Culture

1. Sterile Plates, flasks, glass tubes, pipette tips.
2. Glycerol stock *E. coli* BL21(DE3) at −80 °C (e.g., from Thermo Fisher Scientific).

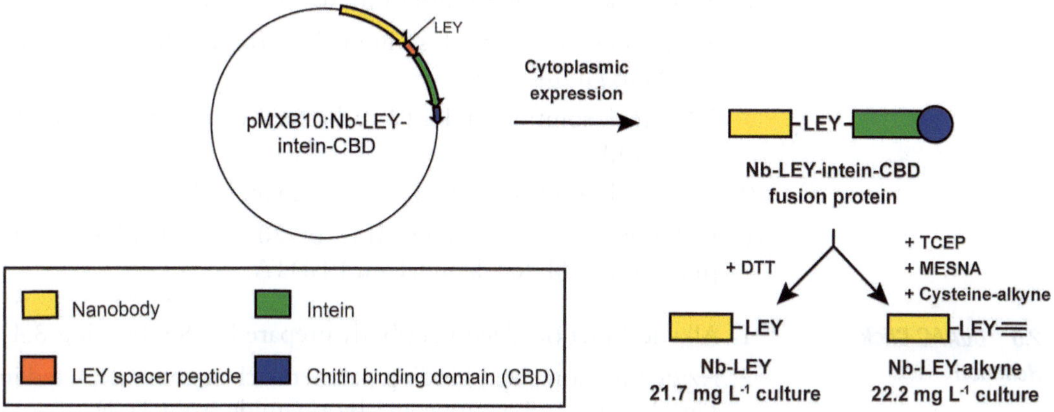

Fig. 3 Scheme for cloning and expression of the selected nanobody variant. The IPL-mediated production strategy of unmodified and alkynated nanobodies is shown with corresponding yields displayed at the bottom. Reproduced with permission from [10]

3. Luria–Bertani (LB) growth medium containing, per liter water, 10 g tryptone, 5 g yeast extract, and 10 g NaCl, adjusted to pH 7.0 with NaOH and autoclaved.
4. 200 μg/mL plasmid from Subheading 3.1 (*see* Fig. 3).
5. 100 mg/mL ampicillin stock solution.
6. 1 M isopropyl β-D-1-thiogalactopyranoside (IPTG) stock solution in Milli-Q.
7. Heating block or water bath (42 °C).
8. Incubator (37 °C).
9. LB^{AMP} (LB culture medium containing 100 μg/mL ampicillin).
10. LB^{AMP} agar plates (LB medium containing 100 μg/mL ampicillin and 1.5 w/v % agar).

2.3 Protein Extraction

1. Solid Phase Extraction (SPE) Cartridges.
2. Eppendorf vials.
3. Falcon tubes.
4. Micropipettes and tips.
5. Bacterial Protein Extraction Reagent (B-PER, Thermo Fisher Scientific) containing 1 U/mL DNaseI.

2.4 Protein Purification and IPL-Mediated Alkynation

1. Solid Phase Extraction Cartridges.
2. Eppendorf vials.
3. Falcon tubes.
4. Micropipettes and tips.
5. 50% (w/v) chitin slurry in 20% (v/v) ethanol (New England Biolabs).
6. 3 M stock solution of dithiothreitol (DTT) in Milli-Q.
7. 3 M stock solution of sodium 2-mercaptoethane sulfonate (MESNA) in Milli-Q.
8. 1 M stock solution of Tris(2-carboxyethyl)phosphine (TCEP) in Milli-Q.
9. 1 M stock solution cysteine–alkyne linker in Milli-Q.
10. Column buffer (CB) consisting of 20 mM HEPES–NaOH pH 8.5, 0.5 M NaCl, and 1 mM EDTA.

2.5 CuAAC Click Reaction

1. Alkyne-functionalized nanobody prepared in Subheading 3.4.
2. Azide functionalized binding partner such as those depicted in Fig. 3 (e.g. *N*-(3-azidopropyl)biotinamide from TCI).
3. CuAAC buffer: 1 mM Tris-(2-carboxyethyl)phosphine (TCEP), 0.1 mM Tris-(benzyltriazolylmethyl)amine (TBTA), and 1 mM $CuSO_4$ in PBS buffer pH 7.4.

2.6 Assessment of the Click Reaction Efficiency to Azidified Biotin by Western Blotting

1. 15% SDS-PAGE gel.
2. Amersham Hybond™-LFP PVDF membrane (GE Healthcare and Life Sciences).
3. 5% (w/v) bovine serum albumin (BSA).
4. Tris-buffered saline (50 mM Tris, 150 mM NaCl) containing 0.1% Tween 20 (TBST).
5. Streptavidin-alkaline phosphatase conjugate (2 mg/mL, e.g., from Thermo Fisher Scientific).
6. Nitro blue tetrazolium chloride/5-bromo-4-chloro-3-indolyl phosphate (NBT/BCIP) ready-to-use solution (e.g., from Sigma-Aldrich).

2.7 Electrospray Ionization Fourier Transform Mass Spectrometry

1. HPLC setup (Dionex 3000) equipped with autoinjector and coupled to the mass spectrometer (Orbitrap Velos Pro from Thermo Fisher Scientific).
2. Dionex Acclaim PolarAdvantage II C18 reversed-phase 2.0 × 10 mm guard column (particle diameter 5 μm, porosity 0.0120 μm).
3. Solvent A: 0.1% (v/v) formic acid in Milli-Q water.
4. Solvent B: HPLC grade acetonitrile.

3 Methods

3.1 Molecular Cloning of the Nanobodies in the pMXB10 Plasmid

This section describes the cloning of the nanobody (from a template plasmid) into the pMXB10 plasmid. As such the nanobody is cloned, via a linker peptide, in frame with an intein and a chitin binding domain (CBD) (*see* Fig. 3).

1. Thaw all reagents on ice.
2. Prepare a master mix containing 36.5 μL Milli-Q water, 5 μL Taq polymerase buffer, 5 μL dNTPs, 1 μL forward and 1 μL reverse primer, 1 μL nanobody-coding template plasmid, and 0.25 μL Taq polymerase.
3. Place the 96 well in a thermocycler, using the following temperature program:
   (a) 5 min at 95 °C.
   (b) 30 cycles of subsequently 30 s at 95 °C, 30 s at 55 °C, and 45 s at 72 °C.
   (c) A final elongation step of 10 min at 72 °C.
4. Digest the gene fragment using the appropriate restriction enzyme by mixing the reagents in the following ratios: 6.5 μL of Milli-Q water, 2 μL of 10× restriction buffer, 0.5 μL NdeI, 1 μL XhoI (10 units/μL), 10 μL PCR product.

5. Incubate this solution for 2 h at 37 °C and purify using the QIAquick gel Extraction kit.
6. Ligate the digested gene fragments into the pMXB10 expression vector by preparing a reaction mixture containing 100 ng double-digested and purified plasmid, 50 ng double-digested PCR-fragment (1–3 molar excess relative to the plasmid), 1 μL T4 DNA ligase, and 1 μL 10× T4 DNA Ligase buffer and sterile water for a total volume of 10 μL (*see* **Note 4**).
7. The solution is incubated at room temperature for 10 min followed by incubation at 4 °C until applied in transformation, which should take place on the same day.

3.2 Transformation and Culture

1. Take a vial of glycerol stock *E. coli* BL21 (DE3) and the plasmid solution (Fig. 3) and thaw on ice for 5 min.
2. Pipet 1 μL of the plasmid solution to the cells and do not mix (200 ng of plasmid per 50 μL of cell suspension).
3. Incubate the cells on ice for 30 min.
4. Give a heat shock by quickly transferring the cell vial to a heat block or water bath at 42 °C for 30 s.
5. Put the vial containing the cells immediately back on ice for 2 min.
6. Suspend the cells in 1 mL LB (without Amp) and incubate at 37 °C for 1 h while shaking.
7. Spread the cells in 2 LB^{Amp} plates (100 μg/mL, 1.5% Agar) at a ratio of 1:9.
8. Incubate the plates overnight at 37 °C (*see* **Note 5**).
9. Select a single colony and inoculate 3 mL of LB^{Amp} (100 μg/mL) in a glass tube (*see* **Note 6**).
10. Incubate the tube overnight at 37 °C.
11. Inoculate 300 mL liquid LB^{Amp} with the 3 mL of preculture (=1:100).
12. Incubate at 37 °C with shaking until an optical density (OD) of 0.5–0.6 is obtained (approx. 2.5–3 h).
13. Add IPTG to a final concentration of 1 mM to induce expression.
14. Incubate for 3 h at 37 °C.
15. Harvest the cells via centrifugation (5000 × *g* for 10 min at room temperature) (*see* **Note 7**).

3.3 Protein Extraction

1. Resuspend the cell pellet from 300 mL bacterial culture in 6 mL B-PER supplemented with DNAseI by pipetting up and down.
2. Incubate for 15 min at room temperature.

3. Centrifuge at 15,000 × *g*, 30 min, 4 °C.
4. Collect the supernatant for purification.

3.4 Protein Purification and IPL-Mediated Alkynation

1. Prewet the filter frits of 3 empty SPE columns with distilled water.
2. Add the appropriate amount of chitin slurry to each column (2.5 mL per column considering the 300 mL of growth medium) and let equilibrate by gravity.
3. Wash each column with 10 column volumes of Milli-Q water and subsequently 10 column volumes of column buffer (CB).
4. Divide the cell lysate in 3 portions and load each portion on a chitin column.
5. Wash each column with 20–30 column volumes of CB.
6. Quickly flush (<5 min) the columns with 1.5–3 mL of the freshly prepared cleavage buffers (*see* **Note 8**):
   (a) Column 1: column buffer (negative control).
   (b) Column 2: 30 mM DTT in CB.
   (c) Column 3: 30 mM cysteine–alkyne, 30 mM MESNA, and 1 mM TCEP in CB.
7. Close all SPE columns by putting a stopper on the column exit.
8. Add 1 column volume of the selected cleavage buffer to each column, mix gently and incubate overnight at 4 °C.
9. Elute the columns with 1.5 column volumes of column buffer while collecting the samples in falcon tubes.
10. The collected samples are finally purified via dialysis against PBS buffer (pH 7.4) with at least 3 buffer changes at 2 h intervals.

3.5 CuAAC Click Reaction with Alkynated Nanobody

1. Mix 10 μM purified alkyne-functionalized nanobody from Subheading 3.4 with 0.2 mM azide-functionalized substrate (*see* Fig. 2 and **Note 9**) in CuAAC buffer. As an easy quantifiable substrate we here use azide–biotin.
2. Shake the solution for 2 h at room temperature.

3.6 Assessment of the Click Reaction Efficiency

1. Perform electrophoresis on the nanobody–biotin conjugate using a 15% SDS-PAGE gel.
2. Transfer the proteins to an Amersham Hybond™-LFP PVDF membrane.
3. Block the blot with a 5% BSA solution for 1.5 h.
4. Incubate the membrane for 1 h in a sufficient amount of TBST containing 2 μg/mL streptavidin-alkaline phosphatase conjugate.
5. The proteins are finally visualized by adding 1 mL of the ready-to-use NBCP/BCP solution (approx. 10 min development) (*see* **Note 10**).

3.7 Electrospray Ionization Fourier Transform Mass Spectrometry

1. Trap and further desalt 30 μL aliquots of the dialyzed or desalted nanobodies and nanobody–conjugate solutions (20 μM) for 5 min on a C18 reversed-phase guard column at a flow rate of 0.5 mL/min using 0.1% (v/v) formic acid in Milli-Q (solvent A) as mobile phase.
2. Elute the sample in 50% (v/v) acetonitrile in solvent A at a flow rate of 0.1 mL/min into an electrospray ionization (ESI) source with sheath gas (8 arbitrary flow rate units). The source voltage and capillary temperature were + 4 kV and 275 °C, respectively.
3. Collect mass spectra in full scan mode in the mass range of 110–2000 Thomson at a resolution of 3×10^4 full width at half maximum with automatic gain control set to 1×10^6 ions in 100 ms and without microscan averaging.
4. Average the scans recorded in the 4–7 min time frame.
5. Deconvolute the average spectrum using Promass software for Xcalibur v2.8 (Novatia LLC) for molecular weight determination.
6. Compare the deconvoluted masses to the corresponding theoretical values derived from CLC Main Workbench 6 software (*see* Fig. 5 and **Note 11**).

4 Notes

1. The CuAAC reaction discussed in this chapter requires the presence of Cu(I) ions to catalyze the conjugation reactions. Since Cu^+ ions are known to complex to the formed 1,2,3-triazole ring and possibly to the protein, this may exclude the application of the CuAAC reaction for certain applications. For this reason, an alternative pathway has been proposed which relies on highly strained alkynes (e.g. cyclooctyne) [17, 18]. Due to the ring strain in this functional group, a strain promoted azide–alkyne click reaction (SPAAC) can be performed in absence of a metal catalyst, thereby further broadening the scope of the azide–alkyne click reaction.
2. The presented protocol lists the optimized conditions for one particular nanobody construct characterized by a LEY spacer between the nanobody and the intein domain. Especially the tyrosine adjacent to the intein domain has been reported to facilitate the thiol-initiated cleavage of the protein-of-interest from the intein. Other constructs exhibiting EFLEY or His_6 spacers have also been produced via the presented protocol. Translation of the IPL methodology to other proteins and/or substrates will probably require some optimization in terms of product yield and modification efficiency.

3. Cysteine-alkyne is preferred for the native chemical ligation over cysteine–azide based on the inherent reactivity of organic azides [19]. Azides are considered manageable when (1) the number of nitrogen atoms (N_N) in the compound does not exceed the number of carbon atoms (N_C) and (2) the equation $(N_C + N_O)/N_N \geq 3$ is fulfilled, with N_O the number of oxygen atoms in the molecule. Considering a cysteine analog carrying an azide functionality, the latter equation is not met unless very long spacers are introduced. Therefore, from a safety point of view, the azide functionality is generally introduced on the surface of a solid support or on the compound with the highest molecular weight (e.g., biotin, 5-carboxy-fluorescein (5-FAM), and poly(ethyl-2-oxazoline)).
4. It is advised to verify that the insert was ligated correctly into the vector via sequencing prior to applying the recombinant plasmid in the transformation of *E. coli*.
5. If it is not possible to immediately continue with the following step of the presented protocol (i.e., the preculture of a single colony), it is advised to store the plates at 4 °C to prevent the overgrowth of the bacteria.
6. The yield of the fusion protein can be further increased by applying the *E. coli* Shuffle® T7 strain and the nutrient-rich EnPresso® medium (supplemented with 100 ng/mL ampicillin) [8]. For EnPresso® the manufacturer's specifications were followed by dissolving the tablets in Milli-Q water, then 1/2000 of reagent A and bacterial preculture were added for growth at 30 °C overnight. The next day, 0.5 mM IPTG, reagent A (1/2000), and the booster tablet are added, followed by protein expression for 24 h. As can be seen in Fig. 4, yields in excess of 40 mg/mL could be reached by combining the Shuffle® T7 strain with the EnPresso® growth medium. However, it should be noted this increase in yield comes at a cost since EnPresso® is more expensive and the procedure is more time-consuming (48 h) than when using LB medium (3 h).
7. In case the harvested cells cannot be used in the subsequent protein extraction step, the cells should be stored at −20 °C in order to prevent protein degradation taking place prior to the protein extraction step.
8. It is advised to flush the column with cleavage buffer, to make sure the entire chitin column experiences the same cleavage conditions. If this step is omitted, adding cleavage buffer to the column will efficiently cleave off the nanobodies from the chitin beads near the top, while nanobodies near the column outlet are unaffected by the added thiols. Losses of protein during this step are limited since the intein-mediated cleavage does not

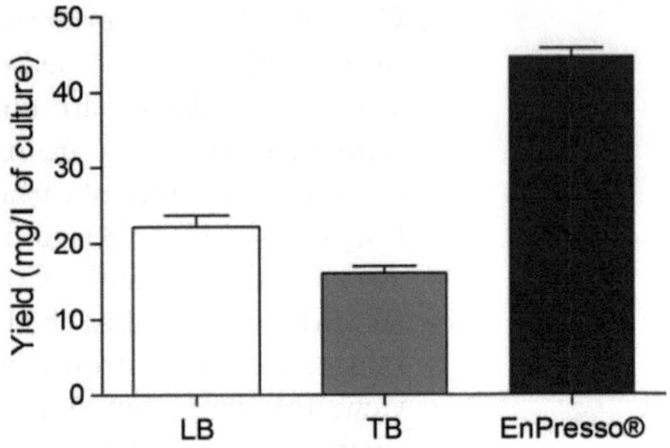

Fig. 4 Expression yield of the alkynated nanobody (NbVCAM1-LEY-alkyne) in the SHuffle® T7 *E. coli* strain using different media: Luria-Bertani (LB), terrific broth (TB) and EnPresso®. All cultures were grown in triplicate and the reported values correspond to the averages with their standard deviations. Reproduced with permission from [10]

take place instantaneously. However, the flushing step should not take more than 5 min.

9. Next to azido-biotin, the alkynated nanobodies have been successfully conjugated to 5-carboxyfluorescein-azide (5-FAM azide) and azidified poly(ethyl-2-oxazoline) (pEtOx-N_3). pEtOx-N_3 (M_n 5 and 10 kg/mol) was kindly provided by prof. Richard Hoogenboom (Ghent University). For these azides, slightly modified conditions were applied during the click reactions as shown in Table 1. In the case of 5-FAM, reaction efficiencies were higher when using sodium ascorbate as the reducing agent (5 mM) and Tris(3-hydroxypropyltriazolylmethyl)amine (THPTA) as copper ligand (0.1 mM). These reagents were also beneficial for the conjugation to pEtOx, but only ten equivalents of the azidified polymer were added to the reaction mixture since polymers reduce the solution's viscosity.
10. Although the methodology described in Subheading 3.6 enables the visualization of successful click reactions with azido-biotin as reagent, similar results can be obtained in a more straightforward manner when using alternative azide-pending compounds. For example, when azide-functionalized 5-carboxy fluorescein (5-FAM, Fig. 2) is used in click reactions, proteins that were successfully coupled to 5-FAM will fluoresce under direct UV illumination. In addition, coupling the nanobodies to high molecular weight azides (e.g., polymers carrying an azide end-group) will result in an apparent change in molecular weight compared to unmodified protein when the gel is stained (e.g., with the ready-to-use instant blue stain from Expedeon).

Table 1
Overview of CuAAC conditions applied for the bioconjugation of alkynated nanobodies with various azide-containing reaction partners

Reagents	Binding partner			
	Surface-bound azides [8]	Azido-biotin	5-FAM-azide	pEtOx-N_3
Alkynated nanobody	0.5 μM[a]	10 μM	10 μM	100 μM
Azide-functionalized compound	–	200 μM	200 μM	1000 μM
Catalyst	0.5 mM Cu $(CH_3CN)_4{}^+PF_6{}^-$	1 mM $CuSO_4$	1 mM $CuSO_4$	1 mM $CuSO_4$
Reducing agent	0.65 mM sodium ascorbate	1 mM TCEP	5 mM sodium ascorbate	5 mM sodium ascorbate
Ligand	0.5 mM TBTA	0.1 mM TBTA	0.1 mM THPTA	0.1 mM THPTA
Reaction time (h)	16	2	1.5	0.75
Reaction yield (%)	N.A.[b]	40	50–60	80

[a]Reaction performed in 95-5 mixture of acetate buffer (pH 4) and DMSO
[b]Successful deposition was observed, but this cannot be expressed in terms of yield

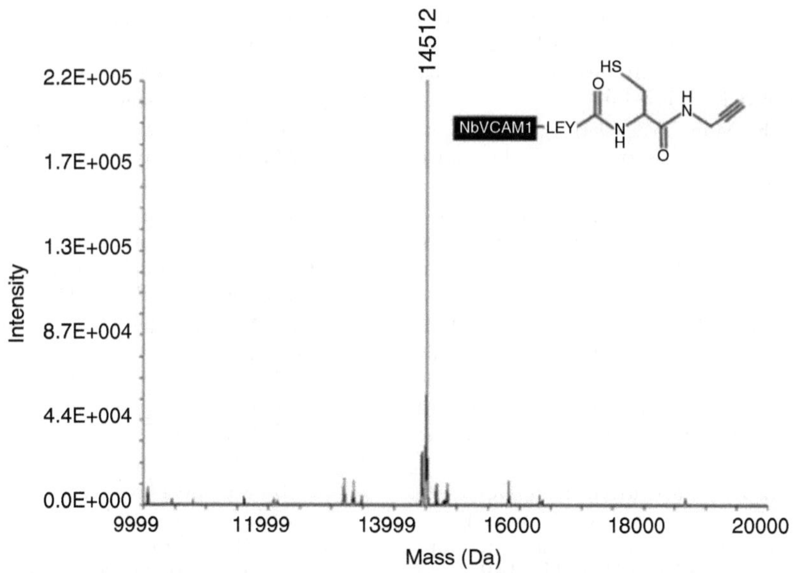

Fig. 5 ESI-FTMS spectra of desalted and buffer-exchanged NbVCAM1-LEY-alkyne (using amicon concentrator with N_2-aerated PBS buffer) showing quasi only monoalkynated NbVCAM1-LEY-alkyne. The representative structure of the corresponding nanobody species is displayed next to the corresponding mass peaks. Reproduced with permission from [10]

11. Although the above-mentioned evaluation of the click efficiency via SDS-PAGE is straightforward for high molecular weight compounds, this approach provides only qualitative information. Mass spectrometry was shown to be a semiquantitative method to determine the degree of alkynation of the nanobodies (Fig. 5). In addition, this technique clearly indicates that only a single alkyne functionality is introduced in the POI via IPL.

References

1. Steen Redeker E, Ta DT, Cortens D et al (2013) Protein engineering for directed immobilization. Bioconjug Chem 24:1761–1777
2. Kolb HC, Finn MG, Sharpless KB (2001) Click chemistry: diverse chemical function from a few good reactions. Angew Chem Int Ed Engl 40:2004–2021
3. Muir TW, Sondhi D, Cole PA (1998) Expressed protein ligation: a general method for protein engineering. Proc Natl Acad Sci 95:6705–6710
4. Ayers B, Blaschke UK, Camarero JA et al (1999) Introduction of unnatural amino acids into proteins using expressed protein ligation. Biopolymers 51:343–354
5. Shemella P, Pereira B, Zhang Y et al (2007) Mechanism for intein C-terminal cleavage: a proposal from quantum mechanical calculations. Biophys J 92:847–853
6. Dawson PE, Muir TW, Clark-Lewis I, Kent SBH (1994) Synthesis of proteins by native chemical ligation. Science 266:776–779
7. Muir TW, Dawson PE, Kent SBH (1997) [13] Protein synthesis by chemical ligation of unprotected peptides in aqueous solution. Methods Enzymol 289:266–298
8. Ta DT, Steen Redeker E, Billen B et al (2015) An efficient protocol towards site-specifically clickable nanobodies in high yield: cytoplasmic expression in *Escherichia coli* combined with intein-mediated protein ligation. Protein Eng Des Sel 28:351–363
9. Massa S, Xavier C, Muyldermans S, Devoogdt N (2016) Emerging site-specific bioconjugation strategies for radioimmunotracer development. Expert Opin Drug Deliv 13:1149–1163
10. Vranken T, Steen Redeker E, Miszta A et al (2017) In situ monitoring and optimization of CuAAC-mediated protein functionalization of biosurfaces. Sensors Actuators B Chem 238:992–1000
11. Massa S, Vikani N, Betti C et al (2016) Sortase A-mediated site-specific labeling of camelid single-domain antibody-fragments: a versatile strategy for multiple molecular imaging modalities. Contrast Media Mol Imaging 11:328–339
12. Muyldermans S (2001) Single domain camel antibodies: current status. Rev Mol Biotechnol 74:277–302
13. Muyldermans S, Atarhouch T, Saldanha J et al (1994) Sequence and structure of vh domain from naturally occurring camel heavy chain immunoglobulins lacking light chains. Protein Eng Des Sel 7:1129–1135
14. Huang L, Muyldermans S, Saerens D (2010) Nanobodies®: proficient tools in diagnostics. Expert Rev Mol Diagn 10:777–785
15. Kijanka M, Dorresteijn B, Oliveira S, van Bergen en Henegouwen PM (2015) Nanobody-based cancer therapy of solid tumors. Nanomedicine 10:161–174
16. Broisat A, Hernot S, Toczek J et al (2012) Nanobodies targeting mouse/human VCAM1 for the nuclear imaging of atherosclerotic lesions. Circ Res 110:927–937
17. Agard NJ, Prescher JA, Bertozzi CR (2004) A strain-promoted [3+2] azide-alkyne cycloaddition for covalent modification of biomolecules in living systems. J Am Chem Soc 126:15046–15047
18. Baskin JM, Prescher JA, Laughlin ST et al (2007) Copper-free click chemistry for dynamic in vivo imaging. Proc Natl Acad Sci U S A 104:16793–16797
19. Smith PAS (1965) The chemistry of open-chain organic nitrogen compounds : Smith, Peter Alan Somervail. W.A. Benjamin, New York

Chapter 10

Site-Specific Bioconjugation Using SMARTag® Technology: A Practical and Effective Chemoenzymatic Approach to Generate Antibody–Drug Conjugates

Junjie Liu, Robyn M. Barfield, and David Rabuka

Abstract

As a critical feature of the next generation of antibody–drug conjugates (ADCs), site-specific bioconjugation approaches can help to optimize stability, pharmacokinetics, efficacy, and safety as well as improve manufacturing consistency. The SMARTag® technology platform offers a practical and efficient chemoenzymatic solution for site-specific protein modifications. A bioorthogonal aldehyde handle is introduced through the oxidation of a cysteine residue, embedded in a specific peptide sequence (CxPxR), to the aldehyde-bearing formylglycine (fGly). This enzymatic modification is carried out by the formylglycine-generating enzyme (FGE). The broad recognition of this short sequence by FGE within the context of heterologous proteins allows for the introduction of fGly residues at chosen sites in proteins expressed in prokaryotic and eukaryotic systems. The protocol presented here describes the methods for expressing fGly-containing antibodies in eukaryotic cells and subsequent site-specific conjugation with a payload-linker using aldehyde-specific Hydrazino-*Iso*-Pictet–Spengler (HIPS) chemistry.

Key words Antibody–drug conjugate (ADC), Formylglycine (fGly), Formylglycine-generating enzyme (FGE), HIPS, Site-specific bioconjugation, SMARTag®

1 Introduction

Antibody–drug conjugates (ADCs) combine the antigen specificity and long half-life of an antibody with the biological activity of a small molecule to create a target-specific, potent agent with an improved therapeutic index [1]. Since the approval of Adcetris® and Kadcyla® earlier this decade, the ADC pipeline has expanded with over 60 ADCs currently in clinical trials for a variety of indications [2]. Conventional bioconjugation strategies use lysine side-chain amines or cysteine sulfhydryl groups that yield a mixture of regioisomers and stereoisomers with undefined stoichiometry [3–5]. This heterogeneity not only results in suboptimal pharmacological profiles but also confounds analytical characterization, impedes optimization, and causes batch-to-batch variability during

Sam Massa and Nick Devoogdt (eds.), *Bioconjugation: Methods and Protocols*, Methods in Molecular Biology, vol. 2033, https://doi.org/10.1007/978-1-4939-9654-4_10,

manufacturing [6, 7]. In recent years, significant progress has been made to develop site-specific bioconjugation strategies to produce homogeneous ADCs with greater manufacturing consistency and a larger therapeutic index. These site-specific approaches include engineering extra reactive cysteine residues [8], installation of biorthogonal handles through unnatural amino acids [9–11], selenocysteine [12] or unnatural sugars [13–17], and exploiting enzyme-assisted ligations [18–20].

The SMARTag® technology platform is a chemoenzymatic method for site-specific protein modification. Formylglycine-generating enzyme (FGE) is utilized for the conversion of cysteine (Cys) to an aldehyde-bearing amino acid, formylglycine (fGly). The fGly modified protein can then be chemically elaborated in a highly efficient and aldehyde-specific chemical transformation generating homogeneous and stable conjugates. In nature, FGE cotranslationally catalyzes the oxidation of Cys to fGly in Type I sulfatases. The minimal consensus sequence for Cys to fGly conversion by FGE is CxPxR (x is variable and usually S, T, A, or G) (Fig. 1a). Using standard molecular biology techniques, this consensus sequence can be inserted at a desired location within any protein of interest where it can be recognized and modified by FGE independent of context (Fig. 1b). FGE activity requires a copper cofactor. The addition of $CuSO_4$ to culture media significantly improves Cys to fGly conversion [21]. Under manufacturing conditions, SMARTag® antibodies have yielded titers up to 5 g/L with >95% conversion [21]. This converted protein contains the aldehyde functionality which can be used as a bioorthogonal handle for site-specific modification of proteins (Fig. 1b) [22].

Aldehyde chemical handles in proteins are typically modified using oxime or hydrazone ligations. The broad availability of oxyamine or hydrazide-bearing reagents and biocompatible reaction conditions make them useful tools for biological studies. However, the intrinsic instability of oxime or hydrazone makes them less desirable ligation tools for the preparation of ADCs which require long term in vivo stability [23]. To leverage SMARTag®'s bioorthogonal aldehyde handle for stable ADC linkage, the Pictet–Spengler reaction was studied and optimized to afford the Hydrazino-*Iso*-Pictet–Spengler (HIPS) ligation method. This bioconjugation proceeds efficiently at weak acidic conditions [24, 25]. An alkylhydrazine-functionalized indole reacts with an aldehyde to generate a hydrazonium ion intermediate followed by intramolecular alkylation with a nucleophilic indole (Fig. 1c). This constructed azacarboline linkage has been shown to have excellent in vivo stability [25].

The protocol presented in this chapter describes the methods for expression of an fGly-containing anti-HER2 antibody in CHO

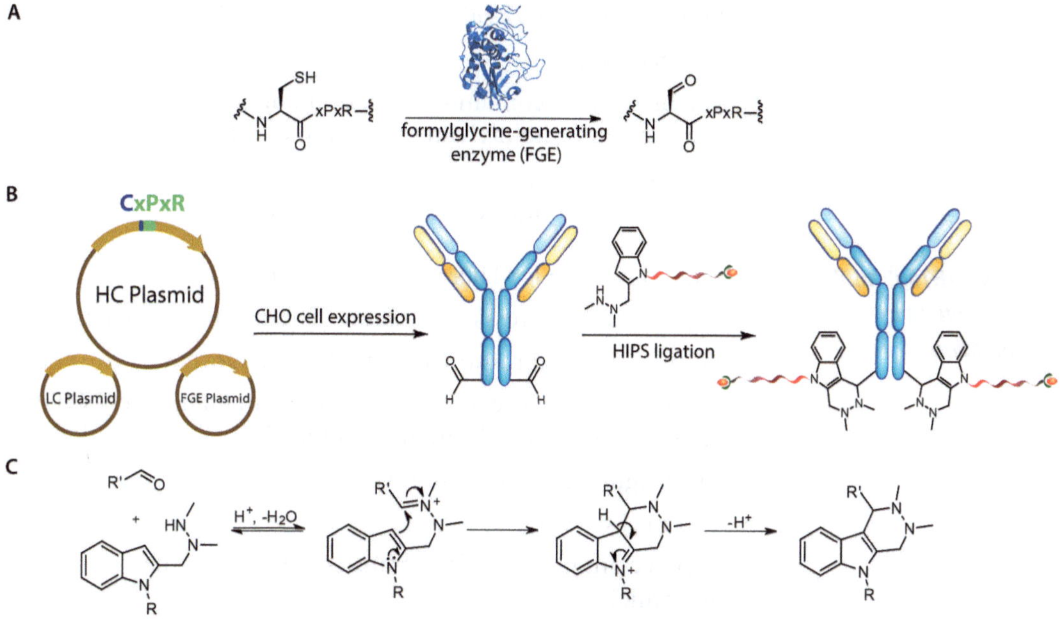

Fig. 1 SMARTag® technology for the production of site-specific ADCs. (**a**) FGE converts Cys in the CxPxR consensus sequence to an aldehyde-bearing fGly. (**b**) A reactive aldehyde functional group is engineered into an antibody by cloning CxPxR into the heavy or light chain expression vector and cotransfecting the heavy and light chain vectors along with an FGE expression vector (*see* **Note 3**). The resulting antibody contains an aldehyde handle at a specified location, which can be ligated to an aldehyde-specific payload-linker using HIPS chemistry to form a stable ADC. (**c**) Mechanism of HIPS ligation

cells and its purification, the key steps in the synthesis of a maytansine payload-linker for HIPS ligation, and subsequent site-specific bioconjugation, purification, and characterization of an anti-HER2-maytansine ADC.

2 Materials

All nongeneral reagents and materials are purchased from Millipore or Sigma and used as received, unless otherwise stated.

2.1 Generation of Heavy Chain Expression Vector Containing the LCTPSR Consensus Sequence (See Note 1)

1. 5 ng/μL IgG1 heavy chain expression vector.
2. 100 ng/μL forward and reverse primers with the desired LCTPSR consensus sequence insertions and substitutions (*see* **Note 2**).
3. 100 mM dNTP mix.
4. 2.5 U/μL Pfu Ultra high-fidelity DNA polymerase (Agilent Technologies).
5. 10× Pfu Ultra reaction buffer (Agilent Technologies).
6. 10 U/μL DpnI enzyme (New England BioLabs).

7. Supercompetent *E. coli* cells, such as XL1-Blue.
8. LB medium.
9. LB-agar plates containing selection antibiotic appropriate for heavy chain or light chain expression construct.
10. LB medium containing selection antibiotic appropriate for heavy chain or light chain expression construct.

2.2 SMARTag® mAb Production and Purification

1. ExpiCHO-S cells (Thermo Fisher Scientific).
2. ExpiCHO™ expression medium (Thermo Fisher Scientific).
3. Human FGE expression vector [22] (*see* **Note 3**).
4. IgG1 heavy chain expression vector containing the LCTPSR consensus sequence from Subheading 3.1.
5. IgG1 light chain expression vector.
6. 100 mM $CuSO_4$.
7. ExpiFectamine™ CHO transfection kit (Thermo Fisher Scientific).
8. OptiPRO™ SFM complexation reagent (Thermo Fisher Scientific).
9. ExpiFectamine™ CHO enhancer reagent (Thermo Fisher Scientific).
10. ExpiCHO™ Feed (Thermo Fisher Scientific).
11. 5 mL MabSelect SuRe column (GE Healthcare).
12. FPLC binding and wash buffer: 20 mM sodium citrate, pH 7.2, 150 mM NaCl.
13. FPLC elution buffer: 20 mM NaAc, pH 3.5, 50 mM NaCl.
14. Neutralization buffer: 250 mM triethanolamine.
15. mAb storage buffer: 20 mM sodium citrate, pH 5.5, 50 mM NaCl.
16. TFF Pellicon 3 Cassette with Ultracel PLCTK membrane, 88 cm^2, 30 kDa MWCO.
17. 15 mL 30 kDa MWCO centrifugal filter.

2.3 Synthesis of a Maytansine Payload-Linker (RED-106) for HIPS Ligation (See Note 4)

1. Acetic acid (HOAc; Acros).
2. Acetonitrile (CH_3CN), HPLC grade.
3. Acetonitrile with 0.1% (v/v) formic acid, LC-MS grade (LC-MS mobile phase B; Honeywell Burdick & Jackson).
4. 1,2-Dichloroethane.
5. Dichloromethane (DCM).
6. *N,N*-Diisopropylethylamine (DIPEA).
7. 4-(Dimethylamino)pyridine.

8. *N*,*N*-dimethylformamide (DMF).
9. 1,4-Dioxane.
10. Di-*tert*-butyl dicarbonate (Boc_2O; Chem-Impex International).
11. Ethyl acetate (EtOAc).
12. HATU (1-[Bis(dimethylamino)methylene]-1H-1,2,3-triazolo [4,5-b]pyridinium 3-oxid hexafluorophosphate; Matrix Scientific).
13. Hexanes.
14. H_2N-PEG_2-COO*t*-Bu (Broadpharm).
15. Methanol (MeOH).
16. 4 Å Molecular sieve, powder (Alfa Aesar).
17. Piperidine (redistilled).
18. Piperidone hydrochloride monohydrate.
19. Potassium carbonate (K_2CO_3).
20. Potassium permanganate ($KMnO_4$).
21. Sodium carbonate (Na_2CO_3).
22. Sodium chloride (NaCl; Thermo Fisher Scientific).
23. Sodium hydrogen carbonate ($NaHCO_3$; VWR).
24. Sodium hydroxide (NaOH).
25. Sodium sulfate, anhydrous (Na_2SO_4).
26. Sodium triacetoxyborohydride (STAB; Chem-Impex International).
27. Succinic anhydride (Alfa Aesar).
28. Tin(IV) tetrachloride ($SnCl_4$) solution, 1.0 M in DCM.
29. Water with 0.1% (v/v) formic acid, LC-MS grade (LC-MS mobile phase A; Honeywell Burdick & Jackson).
30. Agilent 6120 quadrupole mass spectrometer interfaced to an Agilent 1260 HPLC instrument equipped with an Agilent Poroshell 120, EC-C18, 2.7 μm, 3.0 × 50 mm analytical column and ChemStation software.
31. Biotage Isolera Prime flash chromatography system and proper SNAP cartridges and SNAP Samplet®.
32. Ventilated chemical enclosure for the handling of maytansine and its derivatives such as a Mettler Toledo enclosure by Labconco.

2.4 Bioconjugation and Conjugate Purification

1. 500 mM sodium citrate buffer, pH 5.5.
2. 20 mM sodium citrate buffer, pH 5.5, 50 mM NaCl.
3. 18 mg/mL mAb from Subheading 3.2.

4. 100 mM compound ***RED-106*** dissolved in *N,N*-dimethylacetamide (DMA) (*see* **Note 5**).
5. TFF Pellicon 3 Cassette with Ultracel PLCTK membrane, 88 cm^2, 30 kDa MWCO.
6. 5 mL HiTrap Phenyl HP FPLC Hydrophobic Interaction Chromatography (HIC) column (GE Healthcare).
7. FPLC sample diluent: 2 M $(NH_4)_2SO_4$, 25 mM sodium phosphate, pH 6.8.
8. FPLC mobile phase A: 1 M $(NH_4)_2SO_4$, 25 mM sodium phosphate, pH 6.8.
9. FPLC mobile phase B: 25% isopropanol, 18.75 mM sodium phosphate, pH 6.8.
10. 15 mL 30 kD MWCO centrifugal filter.

2.5 Conjugate Analysis and Characterization

1. TSK gel Butyl-NPR HIC column, 4.6 mm × 35 mm, 2.5 μm (Tosoh Bioscience).
2. HIC mobile phase A: 1.5 M $(NH_4)_2SO_4$, 25 mM sodium phosphate, pH 6.8.
3. HIC mobile phase B: 25% isopropanol, 18.75 mM sodium phosphate, pH 6.8.
4. ChemStation software (Agilent Technologies).
5. 0.5 mM DTT.
6. 1× PBS.
7. PLRP-S HPLC column, 1000 Å, 2.1 × 50 mm, 8 μm (Agilent Technologies).
8. PLRP mobile phase A: 0.1% trifluoroacetic acid in H_2O.
9. PLRP mobile phase B: 0.1% trifluoroacetic acid in CH_3CN.
10. G3000 SW XL SEC column, 7.8 × 300 mm (Tosoh Bioscience).
11. SEC buffer: 300 mM NaCl, 25 mM sodium phosphate, pH 6.8.

3 Methods

3.1 Generation of Heavy Chain Expression Vector Containing the LCTPSR Consensus Sequence

1. In a PCR tube, combine 5 μL 10× reaction buffer, 2 μL heavy chain expression vector, 1.25 μL forward primer, 1.25 μL reverse primer, 1 μL dNTP mix, and 39.5 μL H_2O. Then add 1 μL Pfu Ultra polymerase.
2. Incubate the sample in a thermal cycler for 2 min at 95 °C followed by 18 cycles of 30 s at 95 °C, 1 min at 52 °C, and 6 min at 68 °C.

3. Chill the sample on ice for 2 min, add 1 μL DpnI, mix, pulse-spin, and incubate at 37 °C for 1 h.
4. Add 1 μL of the DpnI-treated sample to 50 μL competent cells that have been thawed on ice. Gently mix and incubate on ice for 30 min.
5. Heat the transformation reaction at 42 °C for 45 s and then incubate on ice for 2 min.
6. Add 500 μL LB medium and shake in a 37 °C incubator for 30 min.
7. Plate 250 μL on two different LB-agar plates with antibiotic, and incubate overnight at 37 °C.
8. Inoculate single colonies in LB-medium containing appropriate selection antibiotic and shake in a 37 °C incubator for 8–16 h.
9. Isolate plasmid DNA using a Qiagen miniprep kit. Sequence the plasmid DNA to confirm insertion of the aldehyde-tag encoding sequence and to confirm there are no undesired mutations present.

3.2 SMARTag® mAb Production and Purification

1. Subculture ExpiCHO-S™ cells at 37 °C in an 8% CO_2 incubator to a density of 4–6 × 10^6 viable cells/mL at 125 rpm (*see* **Note 6**).
2. On day −1, dilute the cells to 3–4 × 10^6 and grow overnight at 37 °C in an 8% CO_2 incubator to a density of 6–10 × 10^6 viable cells/mL at 125 rpm.
3. On day 0, dilute the cells to 6 × 10^6 viable cells/mL with prewarmed ExpiCHO™ expression medium to a final volume of 200 mL in a 1 L flask. Add $CuSO_4$ to a final concentration of 100 μM.
4. Dilute 60 μg IgG1 heavy chain (containing the LCTPSR consensus sequence), 90 μg IgG1 light chain, and 50 μg FGE expression constructs with 7.9 mL cold OptiPRO™ SFM complexation reagent in a 50 mL Falcon tube. Invert the tube to mix.
5. Combine 640 μL ExpiFectamine™ CHO Reagent with 7.4 mL OptiPRO™ medium in a 50 mL Falcon tube. Invert the tube to mix.
6. Add the ExpiFectamine/OptiPRO™ mixture to the DNA/OptiPRO mixture, invert the tube to mix, and incubate at room temperature for 1–5 min.
7. Slowly transfer the DNA solution to a flask from **step 3** (*see* **Note 7**).
8. Incubate the cells at 37 °C overnight in an 8% CO_2 incubator with shaking at 125 rpm.

9. On day 1, add 1.2 mL ExpiFectamine™ CHO enhancer and 32 mL ExpiCHO™ feed to the flask and transfer the flask to 32 °C in a 5% CO_2 incubator with shaking at 125 rpm (*see* **Note 8**).
10. On day 5, add 32 mL ExpiCHO™ feed to the flask and continue incubating at 32 °C until day 10.
11. On day 10, transfer cultures to 50 mL Falcon tubes and spin for 15 min at 4000 × *g*. Filter supernatant through a 0.45 μm membrane.
12. Load the supernatant at 2 mL/min onto a 5 mL Mab Select SuRe chromatography column equilibrated in binding buffer. Wash with 10 column volumes (CVs) of wash buffer, and elute with 10 CVs of elution buffer into collection vials containing 10% neutralization buffer by volume (*see* **Note 9**).
13. Formulate the mAb in storage buffer by diafiltration using a 30 kDa TFF cassette and exchanging for eight diavolumes (*see* **Note 10**).
14. Collect mAb from TFF and concentrate in a 15 mL 30 kDa MWCO Amicon centrifugal filter to ~18 mg/mL (*see* **Note 11**).

3.3 Synthesis of a Maytansine Payload-Linker (Compound *RED-106*) for HIPS Ligation

Figure 2 provides a schematic, stepwise overview of the synthesis of compound ***RED-106***, starting from piperidone (compound *1*). Additional details can be found in Drake et al. [26].

3.3.1 Synthesis of Compound *2*

1. To a 100 mL round-bottom flask containing a magnetic stir bar, add piperidone hydrochloride monohydrate (compound *1*, 1.53 g, 10 mmol), Boc_2O (2.39 g, 11 mmol), sodium carbonate (1.22 g, 11.5 mmol), 1,4-dioxane (10 mL), and water (1 mL).
2. Stir the reaction mixture for 1 h at room temperature.
3. Dilute the mixture with water (100 mL). Extract the organic components with EtOAc (3 × 100 mL). Combine the organic layers, and then wash them with a saturated NaCl aqueous solution (20 mL). Dry the organic layer over anhydrous Na_2SO_4 (20 g). Remove the Na_2SO_4 by filtration, and concentrate the filtrate under reduced pressure.
4. Dry the resulting residue under high vacuum overnight. The expected yield for compound *2* is 87% (1.74 g) as a white solid. ^1H NMR ($CDCl_3$) δ 3.73 (t, 4H, J = 6.0), 2.46 (t, 4H, J = 6.0), 1.51 (s, 9H). MS (ESI) m/z: Calculated $[M+H]^+$ for $C_{10}H_{18}NO_3$ is 200.3, observed is 200.2.

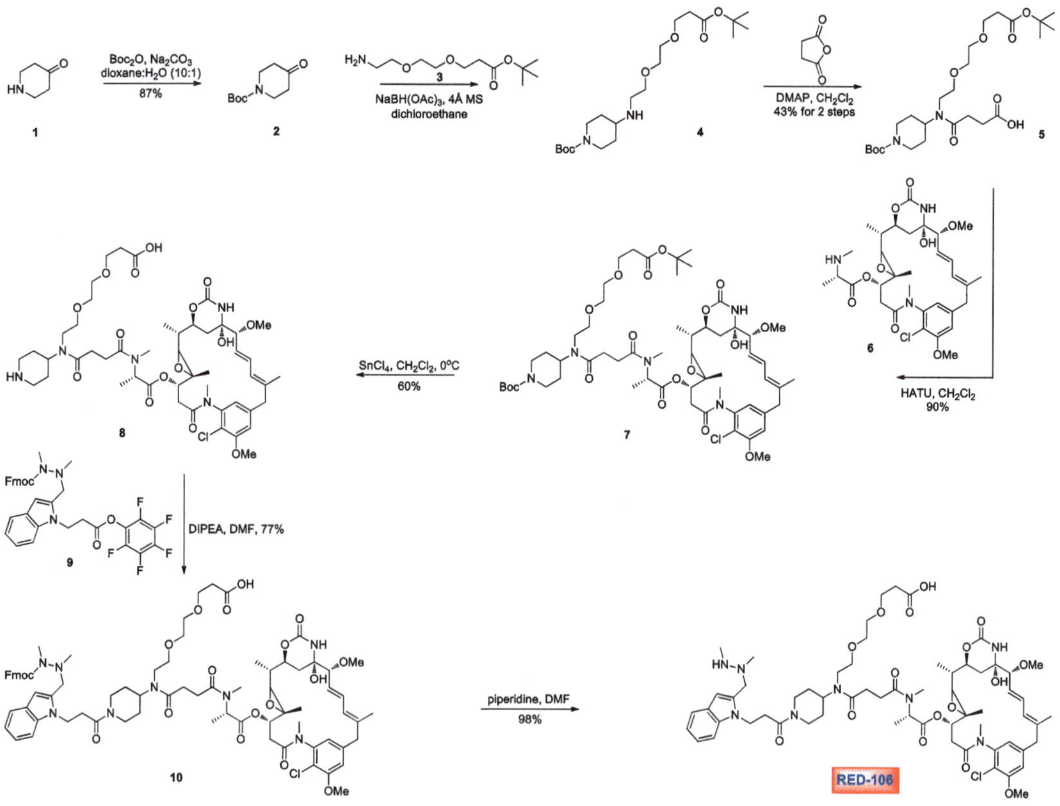

Fig. 2 Synthesis of a maytansine payload-linker ***RED-106*** for HIPS ligation

*3.3.2 Synthesis of Compound **4***

1. To a dried scintillation vial containing a magnetic stir bar, add compound *2* (399 mg, 2.0 mmol), H_2N-PEG_2-COO*t*-Bu (compound *3*, 550 mg, 2.4 mmol), 4 Å molecular sieves (activated powder, 200 mg), and 1,2-dichloroethane (5 mL).
2. Stir the mixture for 1 h at room temperature.
3. To the reaction mixture, add sodium triacetoxyborohydride (845 mg, 4.0 mmol).
4. Stir the mixture for 3 days at room temperature.
5. Partition the resulting mixture between EtOAc (100 mL) and saturated aqueous $NaHCO_3$ (100 mL). Wash the organic layer with a saturated NaCl aqueous solution (20 mL). Dry the organic layer over anhydrous Na_2SO_4 (20 g). Remove the Na_2SO_4 by filtration, and concentrate the filtrate under reduced pressure.
6. Dry the resulting residue under high vacuum overnight. Yield compound ***4*** as a viscous oil. MS (ESI) *m/z*: Calculated $[M+H]^+$ for $C_{21}H_{41}N_2O_6$ is 417.3, observed is 417.2.

3.3.3 Synthesis of Compound 5

1. To a dried scintillation vial containing a magnetic stir bar, add compound *4* (220 mg, 0.5 mmol), succinic anhydride (55 mg, 0.55 mmol), 4-(dimethylamino)pyridine (5 mg, 0.04 mmol), and CH_2Cl_2 (3 mL).
2. Stir the mixture for 24 h at room temperature.
3. Purify the reaction mixture by flash chromatography (elute 50–100% EtOAc/Hexanes). Combine the desired fractions (Rf ~ 0.5, EtOAc/$KMnO_4$) and remove the solvents under reduced pressure (*see* **Note 12**).
4. Dry the resulting residue under high vacuum overnight. The expected yield compound *5* is 43% (117 mg) for two steps as a colorless oil. MS (ESI) *m/z*: Calculated $[M+H]^+$ for $C_{25}H_{45}N_2O_9$ is 517.6, observed is 517.5.

3.3.4 Synthesis of Compound 7

1. To a dried scintillation vial containing a magnetic stir bar, add compound *5* (445 mg, 0.86 mmol), HATU (320 mg, 0.84 mmol), DIPEA (311 mg, 2.42 mmol), and DCM (6 mL).
2. Stir the reaction mixture for 5 min at room temperature.
3. Add the resulting solution to compound *6* (516 mg, 0.79 mmol) (*see* **Notes 13** and **14**).
4. Stir the reaction mixture for an additional 30 min at room temperature.
5. Purify the reaction mixture by flash chromatography (elute 3–10% MeOH/DCM). Combine the desired fractions and remove the solvents under reduced pressure.
6. Dry the resulting residue under high vacuum overnight. The expected yield for compound *7* is 90% (820 mg) as a light tan solid. MS (ESI) *m/z*: Calculated $[M+H]^+$ for $C_{57}H_{87}ClN_5O_{17}$ is 1148.6, observed is 1148.8.

3.3.5 Synthesis of Compound 8

1. To a dried scintillation vial containing a magnetic stir bar, add compound *7* (31 mg, 0.027 mmol) and DCM (1 mL).
2. Cool the solution to 0 °C.
3. Add $SnCl_4$ (1.0 M solution in DCM, 0.3 mL, 0.3 mmol).
4. Stir the reaction mixture for 1 h at 0 °C.
5. Purify the reaction mixture by C18 flash chromatography (elute 5–100% CH_3CN–water). Combine the desired fractions and remove the solvents under reduced pressure.
6. Dry the resulting residue under high vacuum overnight. The expected yield for compound *8* is 60% (16 mg) as a white solid. MS (ESI) *m/z*: Calculated $[M+H]^+$ for $C_{48}H_{71}ClN_5O_{15}$ is 992.5, observed is 992.6.

3.3.6 Synthesis of Compound *10*

1. To a dried scintillation vial containing a magnetic stir bar, add compound *8* (16 mg, 0.016 mmol), compound *9* (Fmoc-HIPS-OPFP, 13 mg, 0.02 mmol), DIPEA (8 μL, 0.05 mmol), and DMF (1 mL) (*see* **Note 15**).
2. Stir the solution for 18 h at room temperature.
3. Purify the reaction mixture by C18 flash chromatography (elute 5–100% CH_3CN–water). Combine the desired fractions and remove the solvents under reduced pressure.
4. Dry the resulting residue under high vacuum overnight. The expected yield for compound *10* is 77% (18 mg) as a white solid. MS (ESI) *m/z*: Calculated $[M+H]^+$ for $C_{77}H_{98}ClN_8O_{18}$ is 1457.7, observed is 1457.9.

3.3.7 Synthesis of Compound *RED-106*

1. To a dried scintillation vial containing a magnetic stir bar, add compound *10* (18 mg, 0.012 mmol), piperidine (20 μL, 0.02 mmol), and DMF (1 mL).
2. Stir the solution for 20 min at room temperature (*see* **Note 16**).
3. Purify the reaction mixture by C18 flash chromatography (elute 1–60% CH_3CN–water). Combine the desired fractions and remove the solvents under reduced pressure.
4. Dry the resulting residue under high vacuum overnight. The expected yield for compound ***RED-106*** is 98% (15 mg) as a white solid. MS (ESI) *m/z*: Calculated $[M+H]^+$ for $C_{62}H_{88}ClN_8O_{16}$ is 1235.6, observed is 1236.0 (*see* **Note 17**).

3.4 Bioconjugation and Conjugate Purification

1. Add 100 μL 500 mM sodium citrate, pH 5.5, 25 μL 100 mM payload-linker ***RED-106***, and 20 mM sodium citrate, pH 5.5, 50 mM NaCl to 15 mg of SMARTag® mAb to a final volume of 1000 μL (*see* **Note 18**).
2. Incubate the reaction mixture at 37 °C for 18–24 h (*see* **Note 19**).
3. Remove unconjugated ***RED-106*** by diafiltration using a 30 kDa TFF cassette and exchanging for 12 diavolumes into 20 mM sodium citrate, pH 5.5, 50 mM NaCl (*see* **Note 20**).
4. Collect ADC from the TFF and add FPLC sample diluent to the ADC sample to a final concentration of 0.8 M (*see* **Note 21**).
5. Load the diluted ADC at 4 mL/min onto a 5 mL HiTrap Phenyl HP HIC column equilibrated in mobile phase A, then wash with two column volumes of mobile phase A. An isocratic hold at 31% mobile phase B for 7 CVs is used to elute any DAR 0 material. This is followed by a 7 CV linear gradient of 41–95% mobile phase B to elute DAR 1 and DAR 2 species and then an isocratic hold at 95% mobile phase B to elute any remaining DAR 2 species. Fractions are collected into vials containing

33% 20 mM sodium citrate, pH 5.5, 50 mM NaCl by volume (*see* **Note 22**).

6. Pool fractions to give desired final DAR.
7. Formulate the mAb in 20 mM sodium citrate, pH 5.5, 50 mM NaCl by diafiltration using a 30 kDa TFF cassette and exchanging for eight diavolumes (*see* **Note 23**).
8. Collect ADC from the TFF and concentrate in a 15 mL 30 kDa MWCO Amicon centrifugal filter to desired final concentration.
9. Sterilize using a 0.2 μM filter.

3.5 Conjugate Analysis and Characterization

1. HIC analysis for quantifying DAR: Inject 20 μg of ADC, diluted 1:1 in HIC mobile phase A, onto a Butyl HIC column equilibrated with 88% mobile phase A and 12% mobile phase B (*see* **Note 24**).
2. Elute at 1.0 mL/min with a linear gradient from 12% to 42% mobile phase B for 5.5 min, followed by a linear gradient from 42% to 70% B for 2 min, followed by a linear gradient from 70% to 100% B for 1 min and a 1 min hold at 100% B. Monitor the protein at 280 nm UV absorbance.
3. Reverse phase analysis for quantifying DAR: Dilute 20 μg of ADC in PBS to a final volume of 90 μL. Add 10 μL DTT. Incubate at 37 °C for 30 min.
4. Load the entire sample onto a PLRP column equilibrated with 75% PLRP mobile phase A and 25% PLRP mobile phase B.
5. Elute at 2.0 mL/min with a linear gradient from 25% to 50% B for 8 min, followed by a linear gradient of 50% to 95% B for 0.1 min and a 1.9 min hold at 95% B. Monitor the protein at 280 nm UV absorbance.
6. Quantify % monomer by SEC analysis: Inject 20 μg of ADC onto an SEC column equilibrated with SEC buffer.
7. Elute at 1.0 mL/min for 20 min with an isocratic hold at 100% buffer *A. monitor* the protein at 280 nm UV absorbance.
8. Use ChemStation software to quantify ratios of different DAR species for HIC and PLRP and to quantify ADC monomer, high molecular weight, and low molecular weight species for SEC (Fig. 3).

4 Notes

1. While the minimal FGE recognition sequence across species is CxPxR, the optimal sequence for human FGE is LCTPSR.

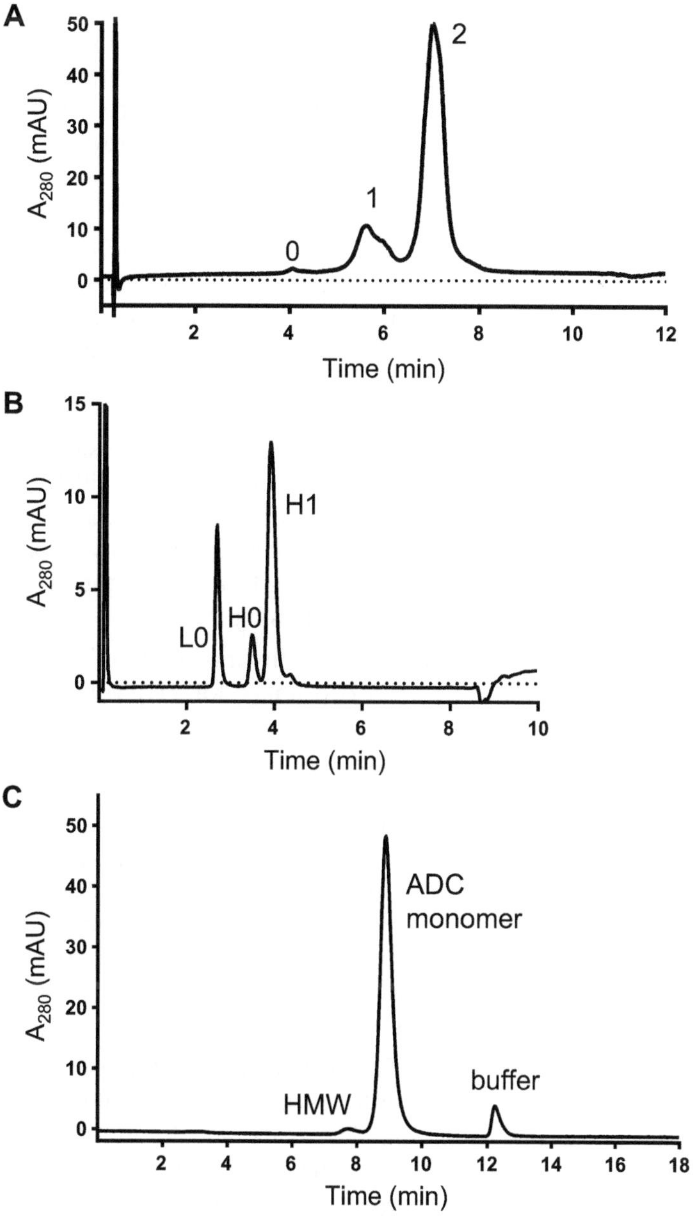

Fig. 3 Analysis of a SMARTag®-maytansine ***RED-106*** ADC. (**a**) Hydrophobic interaction chromatography shows separation between unconjugated mAb (0), mAb conjugated to one maytansine molecule (1), and mAb conjugated to two maytansine molecules (2). (**b**) Reverse phase chromatography of a DTT-reduced ADC is an orthogonal method to quantify DAR. This method shows separation between unconjugated light chain (L0), unconjugated heavy chain (H0), and heavy chain conjugated to one maytansine molecule (H1). (**c**) Size exclusion chromatography demonstrates a highly monomeric SMARTag® ADC with only a very small proportion of high molecular weight species (HMW)

2. Depending on the exact location and surrounding residues of the consensus sequence, LCTPSR can be introduced as a complete insertion or as a combination of insertions and substitutions. For the particular C-terminus location described here, primers are used to introduce LCTPSR such that the heavy chain constant region sequence is changed from SLSLSPGK* (where * indicates a stop codon) to SLSLSPGSLCTPSRGS*.
3. As an alternative to transfecting three separate vectors (expressing human FGE, IgG1 heavy chain, and IgG1 light chain), the human FGE DNA sequence can be incorporated into the light chain expression vector under a separate EF1 promoter. In this case, only two expression vectors (IgG1 heavy chain and IgG1 light chain/human FGE) are transfected at a heavy chain–light chain ratio of 2:3. Conversion of Cys to fGly is more efficient under these conditions, likely due to a greater transfection efficiency of FGE into cells where the heavy and light chains are being expressed. Conversion is even more efficient when transfecting heavy and light chain vectors into a cell line stably overexpressing FGE.
4. The alkylhydrazine-functionalized indole moiety in the payload-linker for HIPS ligation is designed to react rapidly with aldehydes. Therefore, even trace amounts of formaldehyde present in the payload-linker can quench the alkylhydrazine-functionalized indole moiety and form the unreactive [M+12] Da species. In addition, the alkylhydrazine-functionalized indole moiety is sensitive to oxidation, leading to an [M−2] Da species. To ensure the delivery of reactive payload-linker for HIPS bioconjugation, high quality solvents and reagents should be used. Furthermore, methanol, ethanol, and ethereal solvents should be avoided, if possible.
5. The SMARTag® technology is compatible with a variety of different payloads, including cytotoxins (such as maytansine), nucleic acids, and fluorophores. Depending on the payload-linker used for conjugation, different solvents may be used for optimal solubility.
6. A transient expression system for generating SMARTag® mAbs (where heavy chain, light chain, and FGE constructs are transfected) is outlined here for quicker production of aldehyde-tagged mAbs. SMARTag® mAbs can also be generated using cell lines stably expressing heavy chain, light chain, and FGE [22].
7. During **steps 7–12** in Subheading 3.2, avoid exposing the cell culture medium to white light.
8. The lower temperature (from 37 °C to 32 °C) as well as the feeding protocol help to improve titers of the mAb.

9. Depending on the scale and levels of mAb expression, the mAb can be purified in batch instead of by FPLC. Although Tris buffer is commonly used for the neutralization step, avoid using Tris because of its potential reactivity with the aldehyde on the fGly.
10. 20 mM Sodium citrate, pH 5.5, 50 mM NaCl works well as both a storage buffer and conjugation buffer for many SMAR-Tag® mAbs. Therefore, formulating the mAb in this buffer minimizes mAb manipulation before conjugation.
11. In order to minimize the payload:mAb equivalents used in the conjugation reaction, SMARTag® mAbs are conjugated at 15 mg/mL. Concentrating mAbs to 18 mg/mL supports conjugation reactions at 15 mg/mL. Small amounts of high molecular weight (HMW) species may form during concentration. This HMW species will dissociate upon conjugation, generating a highly monomeric final product.
12. Inclusion of acetic acid (0.2%) in the mobile phases improves chromatography.
13. Maytansine and all its derivatives should be handled as a class IV cytotoxic material.
14. The details for the preparation of compound *6* can be found in ref. 27.
15. The details for the preparation of compound *9* can be found in ref. 25.
16. Avoid extended reaction time.
17. Avoid keeping the product ***RED-106*** at room temperature for extended periods. Preferably, store ***RED-106*** at −80 °C.
18. The optimal pH range for HIPS ligation is 5.0–6.0. Alternative buffers, organic solvents and contents, surfactants, and excipients can be used but conjugation efficiency should be tested.
19. The protocol outlined here calls for 2.5 mM of payload-linker and a reaction time of 18–24 h. Alternatively, 1.7 mM or 0.85 mM payload-linker can be used for a reaction time of 48 or 72 h, respectively. The molarity of the payload-linker is more important than the number of payload:mAb equivalents. Rotation of the conjugation mixture is optional depending on reaction volume and payload.
20. Diafiltration is the most scalable method for the removal of free payload. For small-scale reactions and for reactions where the residual payload concentration is not a concern, alternative methods can be used (e.g., desalting and size exclusion chromatography).
21. The FPLC HIC purification step is optional. If the DAR after the free payload removal step meets specification, the HIC step

to enrich for higher DAR species can be eliminated. In that case, after the collection from the TFF, concentrate the ADC to the desired final concentration and sterilize using a 0.2 μM filter.

22. For different payload-linkers, the choice of HIC resin, gradient, and buffers will need to be adjusted to achieve optimal separation of different DAR species.
23. 20 mM Sodium citrate, pH 5.5, 50 mM NaCl works well as a storage buffer for many SMARTag® ADCs. Depending on the exact mAb/payload combination, other formulations could be ideal. SMARTag® ADCs can be stored for at least 1 month at 4 °C and at least 1 year at −80 °C.
24. HIC and PLRP methods to determine DAR rely on the increased hydrophobicity imparted by conjugation of hydrophobic payloads, such as maytansine. These methods may not separate differentially conjugated species for more hydrophilic payloads. In that case, DAR can be estimated by looking at the relative abundance of conjugated and unconjugated species by mass spectrometry.

References

1. Wu AM, Senter PD (2005) Arming antibodies: prospects and challenges for immunoconjugates. Nat Biotechnol 23(9):1137–1146
2. Beck A, Goetsch L, Dumontet C et al (2017) Strategies and challenges for the next generation of antibody-drug conjugates. Nat Rev Drug Discov 16(5):315–337
3. Wang L, Amphlett G, Blättler WA et al (2005) Structural characterization of the maytansinoid-monoclonal antibody immunoconjugate, huN901-DM1, by mass spectrometry. Protein Sci 14(9):2436–2446
4. Kim MT, Chen Y, Marhoul J et al (2014) Statistical modeling of the drug load distribution on trastuzumab emtansine (Kadcyla), a lysine-linked antibody drug conjugate. Bioconjug Chem 25(7):1223–1232
5. Bross PF, Beitz J, Chen G et al (2001) Approval summary: gemtuzumab ozogamicin in relapsed acute myeloid leukemia. Clin Cancer Res 7(6):1490–1496
6. Hamblett KJ, Senter PD, Chace DF et al (2004) Effects of drug loading on the antitumor activity of a monoclonal antibody drug conjugate. Clin Cancer Res 10 (20):7063–7070
7. Wakankar A, Chen Y, Gokarn Y et al (2014) Analytical methods for physicochemical characterization of antibody drug conjugates. MAbs 3(2):161–172
8. Junutula JR, Raab H, Clark S et al (2008) Site-specific conjugation of a cytotoxic drug to an antibody improves the therapeutic index. Nat Biotechnol 26(8):925–932
9. Axup JY, Bajjuri KM, Ritland M et al (2012) Synthesis of site-specific antibody-drug conjugates using unnatural amino acids. Proc Natl Acad Sci U S A 109(40):16101–16106
10. Zimmerman ES, Heibeck TH, Gill A et al (2014) Production of site-specific antibody-drug conjugates using optimized non-natural amino acids in a cell-free expression system. Bioconjug Chem 25(2):351–361
11. VanBrunt MP, Shanebeck K, Caldwell Z et al (2015) Genetically encoded azide containing amino acid in mammalian cells enables site-specific antibody-drug conjugates using click cycloaddition chemistry. Bioconjug Chem 26 (11):2249–2260
12. Hofer T, Skeffington LR, Chapman CM et al (2009) Molecularly defined antibody conjugation through a selenocysteine interface. Biochemistry 48(50):12047–12057
13. Li X, Fang T, Boons G-J (2014) The preparation of well-defined antibody–drug conjugates through glycan remodeling and strain promoted azide-alkyne cycloadditions. Angew Chem Int Ed Engl 53(28):7179–7182
14. Okeley NM, Toki BE, Zhang X et al (2013) Metabolic engineering of monoclonal antibody

carbohydrates for antibody-drug conjugation. Bioconjug Chem 24(10):1650–1655

15. Zhu Z, Ramakrishnan B, Li J et al (2014) Site-specific antibody-drug conjugation through an engineered glycotransferase and a chemically reactive sugar. MAbs 6(5):1190–1200
16. Tang F, Wang LX, Huang W (2017) Chemoenzymatic synthesis of glycoengineered IgG antibodies and glycosite-specific antibody-drug conjugates. Nat Protoc 12 (8):1702–1721
17. Zhou Q, Stefano JE, Manning C et al (2014) Site-specific antibody-drug conjugation through glycoengineering. Bioconjug Chem 25(3):510–520
18. Strop P, Liu SH, Dorywalska M et al (2013) Location matters: site of conjugation modulates stability and pharmacokinetics of antibody drug conjugates. Chem Biol 20(2):161–167
19. Dennler P, Chiotellis A, Fischer E et al (2014) Transglutaminase-based chemo-enzymatic conjugation approach yields homogeneous antibody-drug conjugates. Bioconjug Chem 25(3):569–578
20. Stefan N, Gébleux R, Waldmeier L et al (2017) Highly potent, anthracycline-based antibody-drug conjugates generated by enzymatic, site-specific conjugation. Mol Cancer Ther 16 (5):879–892
21. York D, Baker J, Holder PG et al (2016) Generating aldehyde-tagged antibodies with high titers and high formylglycine yields by supplementing culture media with copper(II). BMC Biotechnol 16:23. https://doi.org/10.1186/s12896-016-0254-0
22. Rabuka D, Rush JS, de Hart GW et al (2012) Site-specific chemical protein conjugation using genetically encoded aldehyde tags. Nat Protoc 7(6):1052–1067
23. Kalia J, Raines RT (2008) Hydrolytic stability of hydrazones and oximes. Angew Chem Int Ed Engl 47(39):7523–7526
24. Agarwal P, Kudirka R, Albers AE et al (2013) Hydrazino-Pictet-Spengler ligation as a biocompatible method for the generation of stable protein conjugates. Bioconjug Chem 24 (6):846–851
25. Drake PM, Albers AE, Baker J et al (2014) Aldehyde tag coupled with HIPS chemistry enables the production of ADCs conjugated site-specifically to different antibody regions with distinct in vivo efficacy and PK outcomes. Bioconjug Chem 25(7):1331–1341
26. Drake PM, Carlson A, McFarland JM et al (2018) CAT-02-106, a site-specifically conjugated anti-CD22 antibody bearing an MDR1-resistant maytansine payload yields excellent efficacy and safety in preclinical models. Mol Cancer Ther 17(1):161–168
27. Widdison WC, Wilhelm SD, Cavanagh EE et al (2006) Semisynthetic maytansine analogues for the targeted treatment of cancer. J Med Chem 49(14):4392–4408

Chapter 11

Protein Labeling and Bioconjugation Using *N*-Myristoyltransferase

Karin F. K. Ejendal, Julia G. Fraseur, and Tamara L. Kinzer-Ursem

Abstract

Methods that allow for labeling of proteins cotranslationally within protein expression systems have had wide-ranging applications in health, engineering, and medicine. Bioorthogonal chemistries that allow for conjugation of proteins or biomolecules of interest to substrates (fluorophores, gold nanoparticles, polymers, etc.) in living cells without prior enrichment or purification have likewise enabled advances in technology to study and engineer cellular and biomolecular systems. At the intersection of these, chemoenzymatic labeling of proteins at specific sites of interest and their subsequent selective bioconjugation to substrates without prior purification has dramatically streamlined workflows that allow proteins to reside in the native expression volumes as long as possible prior to conjugation, be readily isolated upon conjugation, and remain functionally active after conjugation. Here we present methods and protocols to express and label proteins of interest at the N-terminus with azide derivatives of myristic acid, a small, soluble, 14-carbon fatty acid, and conjugate the labeled protein to fluorophores and gold nanoparticle substrates. These methods can be extended to label proteins with other myristoyl derivatives and to conjugation to other solid or polymeric substrates of interest.

Key words *N*-myristoyltransferase (NMT), Myristic acid, 12-Azidododecanoic acid (12-ADA), Chemoenzymatic, Bioconjugation, Calmodulin, Bioorthogonal chemistry, Click chemistry, Azide–alkyne cycloaddition, CuAAC

1 Introduction

1.1 Cotranslational Labeling of Proteins

Methods that enable incorporation of reactive functional, groups onto proteins of interest for bioconjugation applications have gained much attention in recent decades. In particular, methods that allow for chemical functional groups to be incorporated onto proteins cotranslationally, while they are being synthesized in protein expression systems, have been especially promising. This allows proteins to be synthesized in their native or near-native environments and avoids multistep reactions and purification steps. Two general categories of cotranslational protein labeling methods have emerged: metabolic labeling and chemoenzymatic labeling. In metabolic labeling, nonnatural amino acids are incorporated into the

Sam Massa and Nick Devoogdt (eds.), *Bioconjugation: Methods and Protocols*, Methods in Molecular Biology, vol. 2033, https://doi.org/10.1007/978-1-4939-9654-4_11,

growing polypeptide chain as proteins are being translated at the ribosome. Metabolic labeling is designed to occur either at specific residues of interest (e.g., azidohomoalanine in the place of any methionine [1]) or at specific sites of interest (e.g., a specific amino acid residue at a known surface accessible site [2, 3]). *See* ref. [4] for review of both methods. In contrast, in chemoenzymatic labeling, enzymes are used to covalently modify a protein of interest with nonnatural analogs of posttranslational protein modifications (e.g., glycosylation [5–7], sortagging [8, 9], and fatty acylation [10–13] (including prenylation [14, 15], palmitoylation [16, 17], and myristoylation [18, 19])). For reviews of these and other chemoenzymatic methods *see* refs. [20, 21].

1.2 Chemoenzymatic Modifications of Proteins

To achieve selective chemoenzymatic modification of proteins for applications in bioconjugation, proteins of interest are either native targets or are engineered to include a target peptide sequence that is recognized by the enzyme that carries out the covalent modification. Ideally, the enzyme is able to recognize a variety of target peptide sequences and accept a variety of chemically functional labels for maximum flexibility in the design of the protein-conjugate system. *N*-myristoyltransferase (NMT) is a eukaryotic enzyme that cotranslationally or posttranslationally appends myristoyl-CoA, a 14-carbon chain saturated fatty acid (Myr, Fig. 1), to the N-terminal glycine residue of a six to ten amino acid residue recognition peptide sequence [22]. This is a multistep process. The first step is the removal of the N-terminal methionine by methionyl aminopeptidase (metAP) to expose the N-terminal glycine.

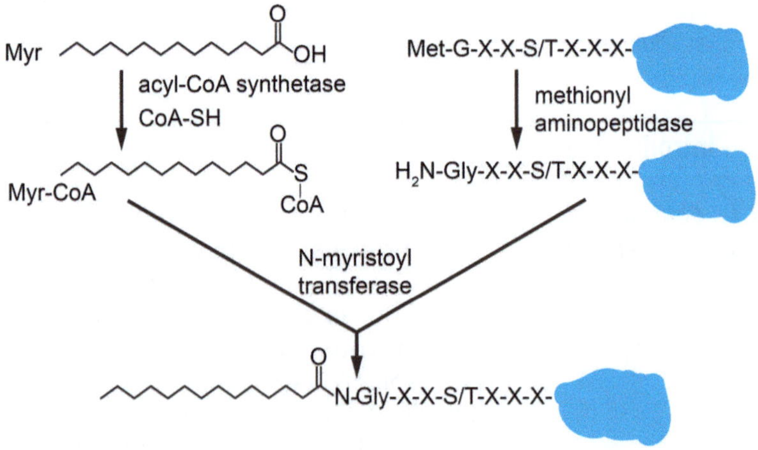

Fig. 1 Protein myristoylation. Diagram of the steps for the addition of myristic acid (Myr) to a protein by *N*-myristoyl transferase, NMT. Myristic acid (Myr) is activated by acyl-CoA synthetase to form Myr-CoA. A protein (in blue) carrying a recognition peptide sequence ((Met)GXXS/TXXX) is cleaved by methionyl aminopeptidase (MetAP) to expose an amino-terminal (NH_2) glycine (Gly). NMT catalyzes the covalent binding of Myr to the amino-terminal Gly, to which Myr-CoA is appended. This yields the myristoylated protein (bottom)

Myristoyl-CoA or CoA analogues of myristic acid are generated by acyl-CoA synthetase. Finally, the myristoyl-CoA or a functionalized analog, is appended via an (irreversible) amide linkage to the N-terminal glycine of the protein by the NMT enzyme (Fig. 1).

1.3 N-Myristoyltransferase Mediated Chemoenzymatic Modification of Proteins

NMT has been used in chemoenzymatic labeling of proteins for bioconjugation applications due to the convergence of several desirable characteristics: (1) specificity for the N-terminus of the protein, which is surface accessible in many proteins [23]; (2) an ability to accept many reactive analogs of Myr, including fatty acids modified with ketones, azides, alkynes, sulfur, and phenyl groups [24–26]; (3) recognition of short (6–10 amino acids) peptide sequences with flexibility in amino acid composition (general sequence ((M)GXXXS/TXXX) [27, 28]; and (4) orthogonality toward the proteome of prokaryotic expression systems in which only the protein of interest is labeled cotranslationally [29]. Taken together, we and others have shown that NMT can selectively label exogenously expressed NMT substrate proteins [30–32] and engineered, nonnatural substrate proteins [18, 19, 33] in bacteria with reactive myristic acid analogs. Facile conjugation of these proteins directly to substrates without prior purification further demonstrates the utility of NMT mediated chemoenzymatic labeling.

1.4 Bioorthogonal Chemical Functionality

The controlled conjugation of a single protein species to a substrate in the presence of cell lysates or protein expression volumes is challenging due to the diversity and abundance of endogenous reactive groups. However, a number of bioorthogonal chemistries have emerged that allow for facile and specific bioconjugation within these complex matrices. Key to these chemistries is that they do not interfere with the functional activity of the protein of interest, and do not cross react with natural biochemical functional groups of the cell. Due to the commercial availability of reagents, moderately fast kinetics, and well established protocols, azide–alkyne cycloaddition reactions, including both copper(I)-catalyzed azide–alkyne cycloaddition (CuAAC) and strain-promoted azide–alkyne cycloaddition (SPAAC), are widely used in bioconjugation applications [34, 35]. More recent additions to the "click chemistry" family of bioorthogonal chemistries include oxime or hydrazine ligation [15, 36], strain-promoted alkyne nitrone cycloaddition, tetrazine ligation [37, 38], and quadricyclane ligation [39, 40], among others. *See* also [20, 41, 42] for recent reviews of click chemistry in bioconjugation.

1.5 Combining Bioorthogonal Chemistries and Chemoenzymatic Labeling via NMT

One advantage of using *N*-myristoyltransferase as the protein-labeling mechanism is that the enzyme readily accepts myristic acid analogs as labeling partners. Gordon and colleagues have synthesized and described a number of analogs of myristic acid (e.g., ketones, azides, alkynes, sulfur and phenyl groups) that are appended to NMT peptide recognition sequences by the NMT

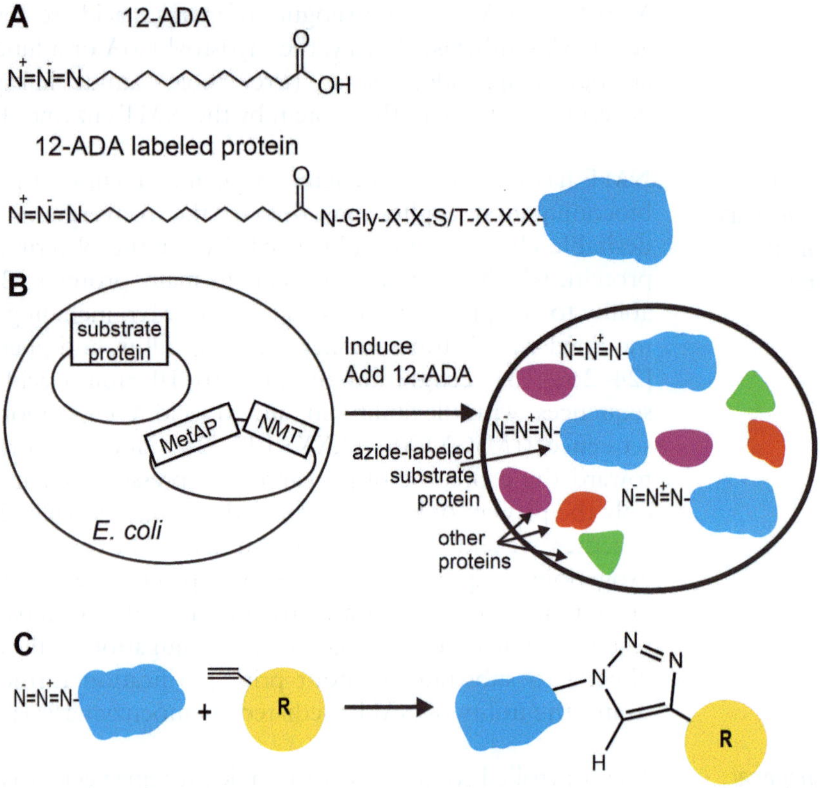

Fig. 2 Chemoenzymatic labeling with 12-ADA via NMT. (**a**) Structure of the azide-containing myristic acid analog 12-azidododecanoic acid (12-ADA) (top) and scheme of a 12-ADA labeled protein (in blue, below), (**b**) Scheme of protein expression and tagging with NMT. *E. coli* bacteria are transformed with plasmids encoding the ((Met)GXXS/TXXX)-tagged protein to be labeled and the modifying enzymes NMT and methionyl amino-peptidase (metAP). When protein expression is induced (typically with IPTG), 12-ADA is also added to the culture. After several hours of induction the substrate protein is site-specifically (at the N-terminus) and selectively labeled with 12-ADA. (**c**) When the cells are lysed the 12-ADA-labeled protein can be covalently conjugated (without prior protein purification) via click chemistry (CuAAC) to fluorophores, microarrays, polymers, or nanoparticles (R, gold circle). Here we describe bioconjugation to either fluorophores (Subheading 3.3) or gold nanoparticles (Subheading 3.4)

enzyme [24–26]. The wide array of NMT peptide recognition sequences found in native systems provides ample opportunity to optimize the properties of the recognition sequence for the protein or application of interest [27, 28]. NMT recognition sequences have also been successfully used in engineering proteins to be substrates of the myristoyl transferase reaction [18, 19, 33].

Here we outline the use of 12-azidododecanoic acid (12-ADA), an analog of myristic acid that is readily covalently attached to both native proteins [24, 30, 31] and proteins engineered to carry an NMT recognition sequence [18, 19, 33] via NMT (Fig. 2a). 12-ADA is commercially available and is relatively straight-forward to synthesize [18, 19, 24]. We have previously shown that engineered proteins can be selectively labeled by 12-ADA via coexpression with NMT in *E. coli* expression systems [18, 19] (Fig. 2b). This labeling is selective (no other proteins have the NMT recognition sequence), specific to the N-terminus of the protein, quantitative (at least 95% of the expressed protein of interest is labeled), and the protein remains soluble in the cytosol of the cells [18, 19, 43]. Moreover, we have shown that the calcium binding and regulatory protein calmodulin (CaM) retained wild-type level functionality after being labeled with 12-ADA [19, 33] and conjugated to gold nanoparticles [44]. The protocols we describe below take advantage of the ability of NMT to selectively label a single protein species of interest cotranslationally in *E. coli* expression systems. Due to the bioorthogonality of 12-ADA, bioconjugation of the protein of interest to fluorophores, microarrays, polymers, and nanoparticles directly from a cell lysate mixture is readily performed (Fig. 2c). Here we specifically focus our protocols on (1) chemoenzymatic labeling of a protein species (CaM) cotranslationally using NMT; (2) conjugating the 12-ADA labeled CaM (12-ADA-CaM) in lysate to fluorophores for imaging; and (3) conjugating the 12-ADA-CaM to functionalized gold nanoparticles from bacterial cell lysate.

2 Materials

Prepare all reagents and buffers in ultrapure water and use analytical grade reagents.

2.1 Bacterial Expression of Myristoylated Proteins

1. Chemically competent bacterial cells suitable for protein expression (e.g., BL21(DE3) (Invitrogen)).
2. Plasmids: the first one encoding the enzymes needed for chemoenzymatic tagging (NMT and aminopeptidase) and a kanamycin resistance gene; the second one encoding target protein (here: Calmodulin engineered to carry an NMT recognition sequence, RS1-CaM) and an ampicillin resistance gene. For cloning strategies, please refer to previous work [18, 19].
3. 50 mg/ml kanamycin (kan) stock. Store at −20 °C.
4. 100 mg/ml ampicillin (amp) stock. Store at −20 °C.
5. LB agar plates (solid medium for bacterial growth), containing both 50 μg/ml kan and 100 μg/ml amp. Store at +4 °C.
6. LB media (liquid broth, made according to standard protocols and autoclaved), supplemented with both 50 μg/ml kan and 100 μg/ml amp immediately prior to inoculation.
7. 1 M Isopropyl β-D-1-thiogalactopyranoside (IPTG) stock solution in water for induction of protein. Sterile filtered, use at a final concentration of 1 mM. Store at −20 °C.

8. 0.5 M myristic acid stock in DMSO, use at a final concentration of 500 μM. Store at −20 °C.
9. 0.5 M 12-ADA stock in DMSO, use at a final concentration of 500 μM. Store at −20 °C. 12-ADA is commercially available (e.g., Thermofisher) or it can be synthesized [18, 19, 24].
10. Phosphate buffered saline (PBS).
11. Temperature controlled incubator for plates.
12. Temperature controlled orbital shaker for cultures.
13. Spectrophotometer and cuvettes to measure optical density (OD) of *E. coli* cultures.
14. Centrifuge.
15. Low protein binding microfuge tubes (e.g., Eppendorf).
16. Glassware, pipettes, and laboratory consumables.

2.2 Cell Lysis

1. Lysis buffer: 50 mM Tris–HCl pH 7.5, 10 mM KCl, 0.5 mM EGTA, 0.5 mM EDTA, 1 mg/ml lysozyme, 10 units of DNase.
2. Protease inhibitor cocktail (PIC, e.g., Halt PIC without EDTA, Thermofisher).
3. Sonifier (e.g., Branson 450).
4. Method and reagent to measure protein concentration in sample that can accommodate detergent (e.g., Pierce 660 Protein assay reagent supplemented with Ionic Detergent Compatibility Reagent, IDCR) and a spectrophotometer that can read absorbance at 660 nm.

2.3 Reagents for Click Reactions

1. Click chemistry reaction buffer: 50 mM Tris–HCl pH 8.0, containing up to 1% SDS, supplemented with PIC.
2. 12-ADA labeled CaM protein, bacterial lysate, 1–10 mg/ml total protein, from Subheading 3.2.
3. 400 mM sodium ascorbate stock in water. Store at −20 °C.
4. 0.5 M iodoacetamide stock in water. Store at −20 °C.
5. 50 mM tris-hydroxypropyltriazolylmethylamine (THPTA) stock in water. Store at +4 °C.
6. 25 mM $CuSO_4$ stock in water. Store at +4 °C.
7. 0.8 mM alkyne–fluorophore (e.g., alkyne–Alexa 647 or alkyne–TAMRA) stock in DMSO. Protect from light. Store at −20 °C.
8. 100 mM aminoguanidine hydrochloride pH 7 stock in water. Store at +4 °C.

2.4 Gel Samples

1. Ice-cold acetone. Store at −20 °C.
2. Laemmli sample buffer with 5% v/v 2-mercaptoethanol.
3. Fume hood.

2.5 SDS-PAGE and Imaging

1. Sodium dodecyl sulfate polyacrylamide gel electrophoresis (SDS-PAGE) precast gradient gels or reagents for casting gels (e.g., Bio-Rad).
2. Electrophoresis apparatus and power supply for PAGE.
3. Protein ladder (e.g., Bio-Rad Kaleidoscope prestained standard).
4. Imager to detect fluorescence (e.g., gel imaging system from Azure Biosystems).
5. GelCode blue stain reagent (or similar, e.g., Coomassie) to stain protein gels.

2.6 Click Chemistry to Functionalize Gold Nanoparticles (AuNPs)

1. 100 nm NHS-activated gold nanoparticles kit (Cytodiagnostics, store at −20 °C) also containing protein resuspension buffer, reaction buffer, and quencher solution.
2. 0.5 M alkyne-PEG4-amine linker (Click Chemistry Tools) dissolved in DMSO. Store at −20 °C.
3. 20 mM HEPES buffer pH 7.4.
4. 2 mg/ml bovine serum albumin (BSA) prepared in 20 mM HEPES at pH 7.4.
5. 400 mM sodium ascorbate stock in water. Store at −20 °C.
6. 0.5 M iodoacetamide stock in water. Store at −20 °C.
7. 50 mM tris-hydroxypropyltriazolylmethylamine (THPTA) stock in water. Store at +4 °C.
8. 25 mM $CuSO_4$ stock in water. Store at +4 °C.
9. 100 mM aminoguanidine hydrochloride pH 7.0 stock in water. Store at +4 °C.
10. Tube-O-DIALYZER mini tubes, 1 k MWCO (G-Biosciences).
11. Beaker.
12. Centrifuge.
13. Rotator (at 4 °C and room temperature).
14. Dialyzing buffer: 20 mM HEPES buffer, pH 7.4.
15. Low protein binding microfuge tubes (e.g., Eppendorf).
16. Glassware, pipettes, and general lab consumables.

3 Methods

As introduced above, there are many possible applications for bioconjugation of proteins using incorporated clickable myristic acid analogs, hence, there are many variations to the basic protocol below. Here, we describe an approach for using an engineered protein, the Ca^{2+} binding protein calmodulin carrying an NMT

Table 1
Overview of different options for expression, conjugation, and analysis

NMT-site containing target protein	Engineered NMT recognition sequence, native myristoylated protein
Expression system	*Bacterial* (e.g., *E. coli*), cell-free expression systems (bacterial or insect cell extracts), mammalian or insect cell culture[a], in situ (tissue expression) [a]
Bioconjugation	Fluorophore, nanoparticles (e.g., AuNPs), resin (e.g., sepharose), solid surfaces (e.g., glass), magnetic beads
Analysis or application	SDS-PAGE (in-gel fluorescence), fluorescence correlation spectroscopy (FCS), enrichment of tagged protein or protein binding partners, mass spectrometry, nanoparticle delivery

Shown here is a selection of different possibilities for expression, conjugation, and analysis of myristoylated proteins via click chemistry. In this chapter, we describe the protocols for bacterial expression of an engineered target protein and bioconjugation to either a fluorophore or nanoparticles using copper(I)-catalyzed azide–alkyne cycloaddition
[a]Indicates protocols that will label all native or naturally myristoylated proteins. Will not be selective to one protein species of interest

recognition sequence, expressed in *E. coli.* We then proceed to describe two alternative paths: Subheading 3.3 details the expression, labeling, and conjugation of the labeled protein via click chemistry to a fluorophore followed by SDS-PAGE and analysis of fluorescence; whereas Subheading 3.4 describes conjugation of labeled protein to gold nanoparticles. These protocols can be further adapted to use a different expression system, protein of interest, or to suit a specific application (*see* Table 1).

3.1 Coexpression and Tagging Using the Biorthogonal Myristic Acid Analog 12-ADA

The coding sequence for the human calcineurin B subunit NMT recognition site (corresponding to the peptide recognition sequence MGNEASYPL) is cloned upstream of the CaM ATG start codon in the bacterial expression vector pET15b (CaM-pET15b), as described in detail [19]. The *N*-myristoyltransferase (NMT) and methionine aminopeptidase enzymes are both expressed from the same plasmid (pHV738, [45]).

1. Transform bacteria (e.g., BL1(DE3) chemically competent *E. coli* cells) according to the manufacturer's instructions with both the target plasmid (e.g., pET15b-RS1-CaM) and pHV738 plasmid and plate on LB plates containing both kan and amp to select for bacteria colonies that have both plasmids. Incubate overnight at 37 °C.
2. Pick a colony and inoculate 5–10 ml LB supplemented with kan and amp and grow overnight in an orbital shaker at 37 °C and 250 rpm.
3. Inoculate fresh LB (with kan and amp) with the overnight culture (e.g., 5–10 ml overnight culture into 500 ml sterile LB) and grow until OD600 reaches 0.8–1.

4. Induce protein expression with 1 mM IPTG, and add 12-ADA (or myristic acid as a negative control, *see* **Note 1**), at a final concentration of 500 μM; at the time of induction.
5. Incubate the culture for 3–4 h to allow for protein expression (*see* **Note 2**).
6. Harvest cells by centrifugation at 12,000 × *g* for 15 min at 4 °C.
7. Wash pellet once with ice-cold PBS (e.g., add 10 ml PBS, resuspend, centrifuge again, save pellet, discard wash). A large culture may be aliquoted into several pellets, so that each pellet is the result of 100–250 ml culture.
8. Save pellet at −80 °C until use.

3.2 Extraction of 12-ADA-Labeled CaM Protein

1. Resuspend the cell pellet in lysis buffer supplemented with PIC (*see* **Note 3**). Use 5 ml lysis buffer per gram of pellet. For example, it is expected that a pellet from a 100 ml culture weighs approximately 0.7 g, hence, will be resuspended in 3.5 ml lysis buffer.
2. Lyse cells by sonication (*see* **Note 4**).
3. To collect the soluble proteins, centrifuge the lysate at 12,500 × *g* for 20 min at 4 °C.
4. Save the supernatant and discard the pellet (*see* **Note 5**).
5. Measure the protein concentration in the clarified lysate. Dilute to 1–10 mg/ml, using lysis buffer.

3.3 Click Chemistry to Conjugate 12-ADA-CaM to Fluorophore

Fluorophore conjugation can be used to rapidly prove incorporation of the 12-ADA tag as an initial characterization or as a verification step prior to other applications such as conjugation to resin, nanoparticles, or solid substrates. *See* Fig. 3 for a representative example.

3.3.1 Click Reaction to Label Protein with Fluorescent Probe

1. Aliquot up to 120 μg protein from clarified lysate in a maximum volume of 60 μl in click chemistry reaction buffer (*see* **Notes 6**) into low protein binding microcentrifuge tubes. Add 10 μl of sodium ascorbate stock solution. Vortex, and incubate end-over-end for 5 min at room temperature.
2. Add 20 μl of iodoacetamide stock solution. Vortex and incubate end-over-end for 10 min at room temperature.
3. Add 1 μl of alkyne-labeled fluorophore stock solution, (e.g., alkyne–Alexa 647) (*see* **Note 7**).
4. Add 16 μl of $CuSO_4$ stock solution.
5. Add 40 μl of THPTA stock solution.
6. Add 40 μl of aminoguanidine stock solution.

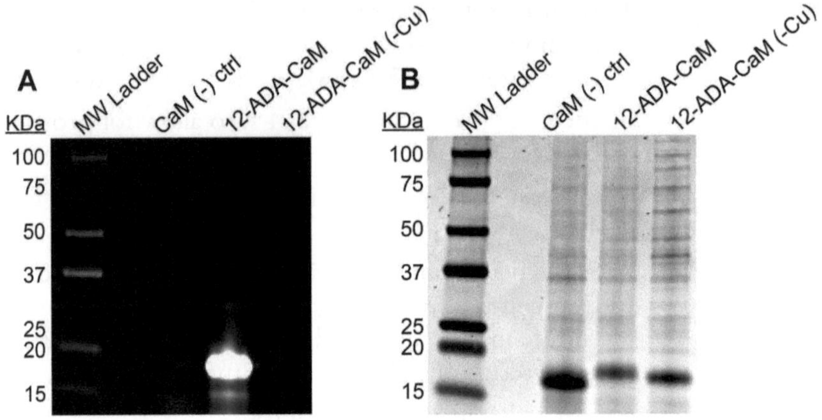

Fig. 3 Fluorescent conjugation of 12-ADA-CaM. *E. coli* cell lysate containing overexpressed and labeled 12-ADA-CaM was reacted with alkyne–Alexa 647 via CuAAC, separated using SDS-PAGE and analyzed via imaging. (**a**) Representative fluorescent image of SDS-PAGE gel (ex/em 650/665). Negative controls include cell lysate containing the wild-type CaM protein without 12-ADA added to the culture (CaM (−) ctrl) and cell lysate containing engineered, labeled 12-ADA-CaM but with no copper added in the CuAAC reaction (12-ADA-CaM (-Cu)). (**b**) The same SDS-PAGE gel as in (**a**) stained with Coomassie and imaged to show all the proteins contained in the samples

7. Vortex continuously or rotate end-over-end for 30 min at room temperature. Protect the samples from light.
8. To stop the reaction, proceed to precipitation in Subheading 3.3.2. Alternatively, the samples can be desalted to remove excess fluorophore prior to precipitation (*see* **Note 8**).

3.3.2 Preparation of Analytical Gel Samples

To stop the click reaction and prepare samples for electrophoresis, the proteins are precipitated. For 12-ADA-CaM, we use acetone-mediated precipitation, however, methanol–chloroform precipitation can also be used (*see* **Note 9**). When working with acetone, methanol, or chloroform, it is advisable to work in a fume hood, or well-ventilated area.

1. To the click reactions from Subheading 3.3.1, add 1 ml ice-cold acetone. Mix samples gently by inverting.
2. Incubate samples for a minimum of 60 min at −20 °C to precipitate the proteins.
3. Centrifuge for 20 min at 10,000 × *g* at 4 °C, then carefully remove and discard as much of the supernatant as possible while leaving the pellet containing the protein precipitate intact.
4. Wash by adding 1 ml cold acetone and centrifuge again for 10 min at 10,000 × *g*, at 4 °C.
5. Carefully remove supernatant without disturbing the pellet.

6. Cover the tube with a lint-free tissue and keep the tube cap open. Allow the pellet to air-dry for a minimum of 15 min (or overnight). Protect the samples from light.
7. Add 40 μl of 1× Laemmli sample buffer supplemented with 2-mercaptoethanol to protein pellets. Adjust volume based on pellet size.
8. Vortex continuously or rotate end-over-end for 5 min.
9. Place microcentrifuge tubes on at 95 °C heating block for 5 min.
10. Remove samples from heating block and vortex for 5 s. Let samples cool to room temperature and spin briefly to recover any droplets in the cap of the tube.
11. Determine the protein concentration of the samples (e.g., using a Pierce 660 nm protein assay supplemented with IDCR).
12. Adjust concentration with 1× Laemmli (supplemented with 5% 2-mercaptoethanol) for equal loading in gel (e.g., to 1 mg/ml).

3.3.3 SDS-PAGE and In-Gel Fluorescence Analysis

1. Load the samples on polyacrylamide gel for electrophoresis. For example, load 10 μl of each sample (1 mg/ml) into a 4–20% Mini-PROTEAN® TGX™ precast polyacrylamide gel. To assess protein size, include a protein standard (e.g., Kaleidoscope ladder, Bio-Rad). The standard may need to be diluted in 1× Laemmli 5% 2-mercaptoethanol to match the intensity of the fluorophore used in the experiment.
2. Run gels at 120 V for 70 min to allow for separation of proteins. The running time may be adjusted depending on size of protein of interest and the percentage of the gel.
3. Disassemble the gel cassette and place the gel carefully in the gel imager.
4. Image gel at appropriate wavelength (e.g., Alexa 647 at Ex 650 nm, Em 665 nm).
5. After imaging for fluorescence, stain the gel using GelCode Blue to visualize protein loading in the lanes.
6. Image stained gel.

3.4 Click Chemistry to Conjugate 12-ADA-CaM to Gold Nanoparticles

An alternative application for the clarified lysate containing 12-ADA labeled calmodulin (12-ADA-CaM) obtained in Subheading 3.2 is to functionalize the surface of gold nanoparticles, AuNP. There are several methods to append biomolecules to the surface of AuNPs. For example, the use of different linkers and chemistries, the specific application intended, or the starting material (e.g., purified protein or clarified lysate) will determine which method is the most suitable. We have found that for 12-ADA-CaM from lysate onto AuNPs, CuAAC conjugation is superior to SPAAC, due to the cyclooctyne causing AuNP aggregation. Here, we use NHS-coated 100 nm AuNPs and a heterobifunctional PEG4 linker, alkyne-PEG4-amine, to conjugate the 12-ADA-CaM to the surface of the AuNPs.

3.4.1 Conjugation of AuNPs with 12-ADA-CaM via Alkyne-PEG4-Amine Linker

1. In low protein binding 1.5 ml microcentrifuge tubes, combine 48 μl of alkyne-PEG4-amine linker with 60 μl AuNP reaction buffer (*see* **Note 10**).
2. Add 90 μl of mixture from **step 1** to a vial of prealiquoted NHS activated AuNPs and gently pipet the particles up and down until all clumps/aggregates are removed.
3. Rotate at room temperature for 4 h.
4. Quench by adding 10 μl of quencher solution and pipet up and down to stop the reaction.
5. Split the vial into three tubes (33.3 μl per tube, *see* **Note 11**).
6. Centrifuge vials at 200 × *g* for 20 min to settle the nanoparticles.
7. Remove the supernatant. Save the pellet. Record how much volume of supernatant was removed (*see* **Note 10**).
8. Add 10 μl of BSA solution to all vials and rotate tubes end-over-end for 30 min at room temperature.
9. Centrifuge tubes to settle nanoparticles at 200 × *g* for 20 min. Remove supernatant. Again record how much volume of supernatant was removed.
10. Add clarified lysate containing 12-ADA-CaM and click reagents to the tubes with the AuNPs. Include negative controls, such as non-azide-labeled protein (*see* **Note 1**). Hence, add 10 μl 12-ADA-CaM protein in lysate, 3.33 μl sodium ascorbate, 6.66 μl iodoacetamide, 13.33 μl aminoguanidine, 13.33 μl THPTA, and 5.33 μl $CuSO_4$.
11. Incubate on a rotator overnight at 4 °C to allow for bioconjugation of the 12-ADA-CaM to the linker-coated AuNPs.

3.4.2 Dialysis to Remove Nonconjugated Reagents

1. Fill beaker with dialyzing buffer (*see* **Note 10**).
2. Place conjugated AuNPs in a Tube-O-DIALYZER tube.
3. Dialyze overnight at 4 °C, stirring continuously.
4. Transfer dialyzed sample to low protein binding microfuge tubes. Bring sample to 100 μl using AuNP reaction buffer.
5. The AuNPs are now ready for the application of interest (*see* **Notes 12** and **13**).

4 Notes

1. Appropriate negative controls should be included. For example: (1) myristic acid labeled lysate of protein of interest, (2) lysate containing wild-type protein without the NMT recognition site with or without 12-ADA (e.g., Fig. 3), (3) 12-ADA labeled lysate with protein of interest, excluding copper in **step 6** of Subheading 3.3.1 (e.g., Fig. 3). If lysates from several cultures are used (e.g., 12-ADA vs. myristic acid), it is advantageous to normalize protein in the lysates to keep the volume between samples similar.
2. The parameters for protein expression given are based on expression of CaM, a small, soluble protein, where >95% of expression and labeling occurs within 4 h of IPTG induction [19, 33]. Depending on the properties of the target protein of interest, the expression time (from IPTG induction to harvest) may need to be optimized for the protein of interest.
3. Lysis buffer conditions may vary, depending on the requirements of the protein of interest. Here, we have given the lysis buffer for bacterial lysates expressing CaM. Of note, chelators in the lysis buffers can, at high concentrations, interfere with CuAAC click chemistry reactions, and are therefore not added to the reaction buffer. However, with copper in excess, as described here, the amount of chelator in the lysis buffer should not affect the reaction.
4. Settings for sonication will depend on the instrumentation, but a good starting point if using a Branson 450 sonifier is three times 1 min, output 4, and 50% duty cycles (1.5 min total sonication time).
5. After sonication and centrifugation: the supernatant is the clarified lysate where the soluble proteins should reside. Addition of myristic acid or 12-ADA does not change the solubility of calmodulin, but it is possible that the solubility of other target proteins may be affected.
6. For the click reactions with fluorophores we use a Tris buffer: 50 mM Tris–HCl, pH 8.0, supplemented with PIC. However, click chemistry reactions are relatively accommodating to different buffer systems such as HEPES and PBS, containing with low concentrations of detergents.
7. In the protocol above, we described bioconjugation of 12-ADA-labeled protein with an alkyne–Alexa 647 fluorophore using CuAAC, however, many alternative alkyne-tagged fluorophores exist (e.g., alkyne–TAMRA), as we have previously shown [19]. Another possibility is to switch the functionalities and label the protein with an alkyne-bearing myristic acid

analog (alkynyl myristic acid, commercially available from Click Chemistry Tools) and a fluorophore with an azide group. The reactions will be carried out in the same manner as described in the protocol above.

Another alternative is to use strain-promoted alkyne–azide cycloaddition (SPAAC), where the protein sample is labeled with an azide (e.g., 12-ADA) and is reacted with a fluorophore conjugated to a DBCO (DIBO) in the absence of copper, as well as in absence of sodium ascorbate, iodoacetamide, THPTA, and aminoguanidine. *See* [34, 46] for an overview of SPAAC.

8. Depending on the fluorophore and the protein of interest, it may be necessary to remove unbound, excess fluorophore directly after the click reaction and before protein precipitation (i.e., after Subheading 3.3.1) using desalting columns such as Zeba spin desalting columns (7 K MWCO, Thermofisher).
9. For CaM, especially purified CaM, acetone mediated precipitation is superior. However, for other proteins of interest, precipitation after click-conjugation can also be conducted with methanol–chloroform. Add 600 μl of methanol, 150 μl of chloroform, and 400 μl of water to each tube. Vortex between each addition. Centrifuge the tubes at top speed in a microfuge (~13,000 × *g*) for 10 min. Carefully remove upper aqueous phase without disturbing the lower phase or the layer in between (this is where the proteins reside). Add 450 μl methanol, centrifuge again for 5 min, and carefully remove the supernatant. The protein precipitate may appear like a thin disc or as specks at the bottom of the tubes. Be careful not to disturb these. Wash the pellet twice more by addition of 450 μl of methanol, centrifuge, and remove supernatant. Leave tubes open but cover with a lint free tissue and let samples dry. Then proceed to prepare the gel samples as for acetone precipitated samples, in Subheading 3.3.2, **step** 7).
10. Assays may need to be modified to fit the downstream application. For example, the AuNP reaction buffer that comes with the kit may not be compatible with the functional assay of the conjugated protein. To circumvent this, for our experiments, we substituted both the reaction buffer and the dialysis buffer with the HEPES based dialyzing buffer. Furthermore, depending on downstream applications such as protein activity, and also for traceability and reproducibility, it may be important to keep track of the exact volume is removed in the wash steps.
11. The amounts given here for the click reactions on the AuNPs are based on splitting one vial of AuNPs into three tubes, to conserve precious reagents. If one vial is used as is, without

aliquoting into three tubes, the click chemistry reagent volumes given here should also be tripled.

12. As a result of the tolerance of NMT for nonnatural myristic acid analogs, NMT-mediated functionalization of proteins and subsequent conjugation have a multitude of applications in biomedicine. For example, in addition to fluorophores and AuNPs, we have shown that the 12-ADA-CaM can be attached to sepharose resin via click chemistry, to be used as a protein purification affinity resin [33]. Furthermore, we have shown that a 12-ADA labeled protein (12-ADA-GFP) can be conjugated via click chemistry to an amine-functionalized glass surface directly from lysate [18], a procedure that may facilitate rapid characterization of proteins, such as protein–protein interactions, without the need for purification steps. Depending on the application and protein conjugated to the AuNPs, successful conjugation can be assessed by for example measuring particle size with dynamic light scatter (DLS) or similar particle sizing protocols; measuring protein content on the AuNPs via colorimetric assays (though we have found this difficult due to the properties of AuNPs); and spectroscopy by monitoring the maximum absorption wavelength of the particles before and after conjugation, or conducting an enzyme activity assay. We have shown that successful conjugation of the 12-ADA-CaM to the AuNPs is reflected by an increase in their size as measured with dynamic light scattering (DLS) and a shift in the maximum absorption wavelength [44].
13. For studies of individual proteins, a major advantage with expressing a target protein with an NMT recognition site in bacterial cells is that bacterial proteins are not naturally myristoylated, hence, the myristic acid analog would be exclusively added to the protein of interest. However, for studies of naturally myristoylated proteins in their native environment, or studies of the "myristome," other expression systems may be desirable [11–13, 17, 47–50]. The click chemistry protocol described would be the same, but instead of a single protein species such as CaM, a number of different labeled proteins would be observed on the gel or conjugated to the AuNPs. Furthermore, labeling of myristoylated proteins can be visualized in situ, for example for studies of changes in the distribution of myristoylated proteins [47].

Acknowledgments

The authors wish to thank Dr. Katherine N. Clayton who has been instrumental in developing the protocols described here.

References

1. Kiick KL, Saxon E, Tirrell DA, Bertozzi CR (2002) Incorporation of azides into recombinant proteins for chemoselective modification by the Staudinger ligation. Proc Natl Acad Sci U S A 99(1):19–24
2. Noren CJ, Anthony-Cahill SJ, Griffith MC, Schultz PG (1989) A general method for site-specific incorporation of unnatural amino acids into proteins. Science 244(4901):182–188
3. Bain JD, Diala ES, Glabe CG, Dix TA, Chamberlin AR (1989) Biosynthetic site-specific incorporation of a non-natural amino acid into a polypeptide. J Am Chem Soc 111 (20):8013–8014
4. Link AJ, Mock ML, Tirrell DA (2003) Non-canonical amino acids in protein engineering. Curr Opin Biotechnol 14(6):603–609
5. Palaniappan KK, Bertozzi CR (2016) Chemical glycoproteomics. Chem Rev 116 (23):14277–14306
6. Rexach JE, Clark PM, Hsieh-Wilson LC (2008) Chemical approaches to understanding O-GlcNAc glycosylation in the brain. Nat Chem Biol 4(2):97–106
7. Yang S, Rubin A, Eshghi ST, Zhang H (2016) Chemoenzymatic method for glycomics: isolation, identification, and quantitation. Proteomics 16(2):241–256
8. Ritzefeld M (2014) Sortagging: a robust and efficient chemoenzymatic ligation strategy. Chemistry 20(28):8516–8529
9. Popp MW, Antos JM, Grotenbreg GM, Spooner E, Ploegh HL (2007) Sortagging: a versatile method for protein labeling. Nat Chem Biol 3(11):707–708
10. Lanyon-Hogg T, Faronato M, Serwa RA, Tate EW (2017) Dynamic protein acylation: new substrates, mechanisms, and drug targets. Trends Biochem Sci 42(7):566–581
11. Hang HC, Wilson JP, Charron G (2011) Bioorthogonal chemical reporters for analyzing protein lipidation and lipid trafficking. Acc Chem Res 44(9):699–708
12. Thinon E, Hang HC (2015) Chemical reporters for exploring protein acylation. Biochem Soc Trans 43(2):253–261
13. Hannoush RN (2015) Synthetic protein lipidation. Curr Opin Chem Biol 28:39–46
14. Duckworth BP, Zhang Z, Hosokawa A, Distefano MD (2007) Selective labeling of proteins by using protein farnesyltransferase. Chembiochem 8(1):98–105
15. Mahmoodi MM, Rashidian M, Dozier JK, Distefano MD (2013) Chemoenzymatic site-specific reversible immobilization and labeling of proteins from crude cellular extract without prior purification using oxime and hydrazine ligation. Curr Protoc Chem Biol 5(2):89–109
16. Gao X, Hannoush RN (2018) A decade of click chemistry in protein palmitoylation: impact on discovery and new biology. Cell Chem Biol 25 (3):236–246
17. Hang HC, Linder ME (2011) Exploring protein lipidation with chemical biology. Chem Rev 111(10):6341–6358
18. Kulkarni C, Kinzer-Ursem TL, Tirrell DA (2013) Selective functionalization of the protein N terminus with N-myristoyl transferase for bioconjugation in cell lysate. Chembiochem 14(15):1958–1962
19. Kulkarni C, Lo M, Fraseur JG, Tirrell DA, Kinzer-Ursem TL (2015) Bioorthogonal chemoenzymatic functionalization of calmodulin for bioconjugation applications. Bioconjug Chem 26(10):2153–2160
20. Rashidian M, Dozier JK, Distefano MD (2013) Enzymatic labeling of proteins: techniques and approaches. Bioconjug Chem 24 (8):1277–1294
21. Rabuka D (2010) Chemoenzymatic methods for site-specific protein modification. Curr Opin Chem Biol 14(6):790–796
22. Farazi TA, Waksman G, Gordon JI (2001) The biology and enzymology of protein N-myristoylation. J Biol Chem 276 (43):39501–39504
23. Jacob E, Unger R (2007) A tale of two tails: why are terminal residues of proteins exposed? Bioinformatics 23(2):e225–e230
24. Devadas B, Lu T, Katoh A, Kishore NS, Wade AC, Mehta PP, Rudnick DA, Bryant ML, Adams SP, Li Q et al (1992) Substrate specificity of *Saccharomyces cerevisiae* myristoyl-CoA: protein N-myristoyltransferase. Analysis of fatty acid analogs containing carbonyl groups, nitrogen heteroatoms, and nitrogen heterocycles in an in vitro enzyme assay and subsequent identification of inhibitors of human immunodeficiency virus I replication. J Biol Chem 267(11):7224–7239
25. Kishore NS, Lu TB, Knoll LJ, Katoh A, Rudnick DA, Mehta PP, Devadas B, Huhn M, Atwood JL, Adams SP et al (1991) The substrate specificity of *Saccharomyces cerevisiae* myristoyl-CoA:protein N-myristoyltransferase. Analysis of myristic acid analogs containing oxygen, sulfur, double bonds, triple bonds, and/or an aromatic residue. J Biol Chem 266 (14):8835–8855

26. Lu T, Li Q, Katoh A, Hernandez J, Duffin K, Jackson-Machelski E, Knoll LJ, Gokel GW, Gordon JI (1994) The substrate specificity of *Saccharomyces cerevisiae* myristoyl-CoA: protein N-myristoyltransferase. Polar probes of the enzyme's myristoyl-CoA recognition site. J Biol Chem 269(7):5346–5357
27. Boutin JA (1997) Myristoylation. Cell Signal 9 (1):15–35
28. Johnson DR, Bhatnagar RS, Knoll LJ, Gordon JI (1994) Genetic and biochemical studies of protein N-myristoylation. Annu Rev Biochem 63:869–914
29. Duronio RJ, Jackson-Machelski E, Heuckeroth RO, Olins PO, Devine CS, Yonemoto W, Slice LW, Taylor SS, Gordon JI (1990) Protein N-myristoylation in *Escherichia coli*: reconstitution of a eukaryotic protein modification in bacteria. Proc Natl Acad Sci U S A 87 (4):1506–1510
30. Heal WP, Wickramasinghe SR, Bowyer PW, Holder AA, Smith DF, Leatherbarrow RJ, Tate EW (2008) Site-specific N-terminal labelling of proteins in vitro and in vivo using N-myristoyl transferase and bioorthogonal ligation chemistry. Chem Commun (4):480–482
31. Heal WP, Wickramasinghe SR, Leatherbarrow RJ, Tate EW (2008) N-Myristoyl transferase-mediated protein labelling in vivo. Org Biomol Chem 6(13):2308–2315
32. Heal WP, Wright MH, Thinon E, Tate EW (2012) Multifunctional protein labeling via enzymatic N-terminal tagging and elaboration by click chemistry. Nat Protoc 7(1):105–117
33. Fraseur JG, Kinzer-Ursem TL (2018) Next generation calmodulin affinity purification: clickable calmodulin facilitates improved protein purification. PLoS One 13(6):e0197120
34. Agard NJ, Prescher JA, Bertozzi CR (2004) A strain-promoted [3 + 2] azide-alkyne cycloaddition for covalent modification of biomolecules in living systems. J Am Chem Soc 126 (46):15046–15047
35. Presolski SI, Hong VP, Finn MG (2011) Copper-catalyzed azide-alkyne click chemistry for bioconjugation. Curr Protoc Chem Biol 3 (4):153–162
36. Rashidian M, Dozier JK, Lenevich S, Distefano MD (2010) Selective labeling of polypeptides using protein farnesyltransferase via rapid oxime ligation. Chem Commun 46 (47):8998–9000
37. Blackman ML, Royzen M, Fox JM (2008) Tetrazine ligation: fast bioconjugation based on inverse-electron-demand Diels-Alder reactivity. J Am Chem Soc 130(41):13518–13519
38. Oliveira BL, Guo Z, Bernardes GJL (2017) Inverse electron demand Diels-Alder reactions in chemical biology. Chem Soc Rev 46 (16):4895–4950
39. Sletten EM, Bertozzi CR (2011) A bioorthogonal quadricyclane ligation. J Am Chem Soc 133(44):17570–17573
40. Tomlin FM, Gordon CG, Han Y, Wu TS, Sletten EM, Bertozzi CR (2018) Site-specific incorporation of quadricyclane into a protein and photocleavage of the quadricyclane ligation adduct. Bioorg Med Chem
41. McKay CS, Finn MG (2014) Click chemistry in complex mixtures: bioorthogonal bioconjugation. Chem Biol 21(9):1075–1101
42. Sletten EM, Bertozzi CR (2009) Bioorthogonal chemistry: fishing for selectivity in a sea of functionality. Angew Chem Int Ed 48 (38):6974–6998
43. McCabe JB, Berthiaume LG, Pfeffer SR (1999) Functional roles for fatty acylated amino-terminal domains in subcellular localization. Mol Biol Cell 10(11):3771–3786
44. Clayton KN, Fraseur JG, Wereley ST, Kinzer-Ursem TL (2018) Functionalizing gold nanoparticles via click chemistry with highly active proteins directly from cell lysates. (in preparation)
45. Van Valkenburgh HA, Kahn RA (2002) Coexpression of proteins with methionine aminopeptidase and/or N-myristoyltransferase in *Escherichia coli* to increase acylation and homogeneity of protein preparations. Methods Enzymol 344:186–193
46. Agard NJ, Baskin JM, Prescher JA, Lo A, Bertozzi CR (2006) A comparative study of bioorthogonal reactions with azides. ACS Chem Biol 1(10):644–648
47. Witten AJ, Ejendal KFK, Gengelbach LM, Traore MA, Wang X, Umulis DM, Calve S, Kinzer-Ursem TL (2017) Fluorescent imaging of protein myristoylation during cellular differentiation and development. J Lipid Res 58 (10):2061–2070
48. Hannoush RN, Sun J (2010) The chemical toolbox for monitoring protein fatty acylation and prenylation. Nat Chem Biol 6(7):498–506
49. Tate EW, Kalesh KA, Lanyon-Hogg T, Storck EM, Thinon E (2015) Global profiling of protein lipidation using chemical proteomic technologies. Curr Opin Chem Biol 24:48–57
50. Thinon E, Serwa RA, Broncel M, Brannigan JA, Brassat U, Wright MH, Heal WP, Wilkinson AJ, Mann DJ, Tate EW (2014) Global profiling of co- and post-translationally N-myristoylated proteomes in human cells. Nat Commun 5:4919

Chapter 12

One-Step Fluorescent Protein Labeling by Tubulin Tyrosine Ligase

Dominik Schumacher, Heinrich Leonhardt, Christian P. R. Hackenberger, and Jonas Helma

Abstract

Tub-tag labeling, a novel chemoenzymatic protein functionalization method, facilitates one-step fluorescent labeling of functional biomolecules. The enzyme tubulin tyrosine ligase incorporates coumarin-amino acids to the terminal carboxylic acid of proteins containing a short peptidic recognition sequence called Tub-tag. Here we describe the one-step Tub-tag protein modification protocol in detail and explain its utilization to generate fluorescently labeled proteins for advanced applications in imaging and diagnostics.

Key words Chemoenzymatic labeling, Tubulin tyrosine ligase, Site-specific protein functionalization, Fluorescent amino acids, One-step functionalization of proteins

1 Introduction

Site-specific functionalization strategies for the labeling of proteins are on the rise since they enable the generation of uniform protein conjugates with defined activities [1–4]. Such conjugates are a prerequisite for the development of enhanced biologics and diagnostic tools for the detection and targeted treatment of severe illnesses [5–7]. We have recently expanded the toolbox of protein functionalization with Tub-tag labeling, a unique chemoenzymatic method based on small, amino acid-sized chemical reporters [8]. The enzyme tubulin tyrosine ligase (TTL) binds to a short, tubulin-derived peptide tag (the Tub-tag) fused to any protein of interest and catalyzes the C-terminal ligation of functional amino acid derivatives. For example, chemical reporter containing tyrosine derivatives facilitate protein functionalization by combining the TTL ligation with one of the many well established bioorthogonal reactions [9–11]. This versatile two-step strategy was applied for the functionalization of different biomolecules including antigen binding proteins with retained antigen binding properties.

Sam Massa and Nick Devoogdt (eds.), *Bioconjugation: Methods and Protocols*, Methods in Molecular Biology, vol. 2033, https://doi.org/10.1007/978-1-4939-9654-4_12,

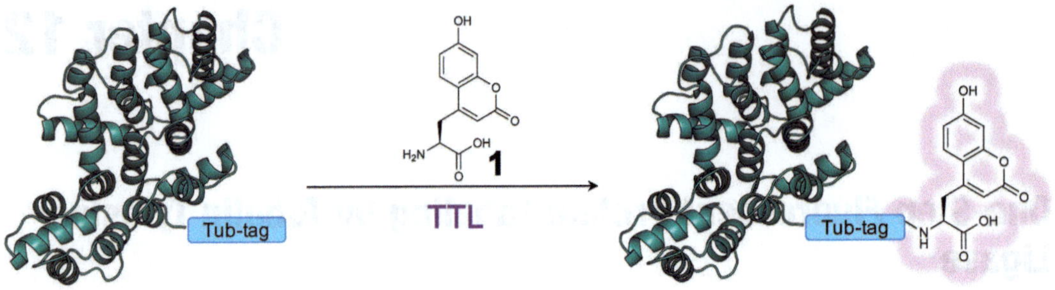

Fig. 1 One-step, site-specific protein labeling by tubulin tyrosine ligase (TTL). The functionalization of Annexin V is exemplarily illustrated (PDB Code: 1SAV) [15]. The enzyme ligates coumarin-derivative 1 to the terminal carboxylic acid of Tub-tagged protein facilitating fluorescent labeling for subsequent diagnostic and analytical applications

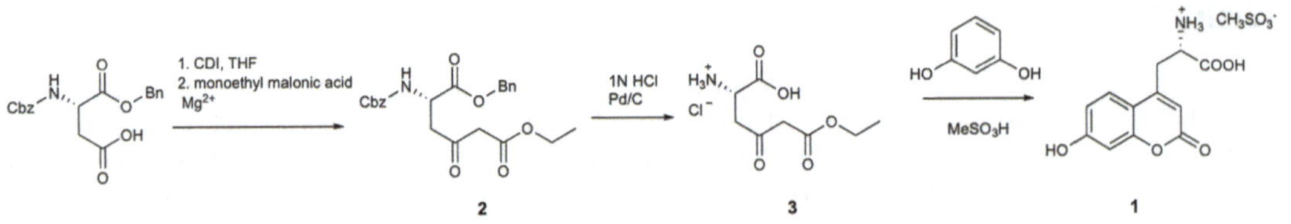

Fig. 2 Synthesis of coumarin-derivative **1** according to a published protocol [13, 14]

A Methods in Molecular Biology protocol has recently been published describing the synthesis of tyrosine derivatives, their ligation to proteins by the TTL and subsequent bioorthogonal reactions [12]. In this chapter, we expand this two-step Tub-tag protocol by a convenient one-step fluorescent labeling method based on coumarin-functionalized amino acids (Fig. 1) [13, 14]. We describe the expression of the TTL, the generation of functional Tub-tagged proteins, the synthesis of coumarin-amino acids (Fig. 2) and the ligation procedure by the TTL. Finally, we describe analytical methods for the characterization of fluorescently labeled proteins as well as labeling and microscopic detection of apoptotic cells by Annexin V-coumarin.

2 Materials

Prepare all solutions using ultrapure water (18.2 MΩ·cm at 25 °C), analytical grade reagents and solvents at ambient temperature (unless stated otherwise). Reagents and solvents are commercially available and do not require further purification.

2.1 DNA Cloning

1. Benchtop centrifuge.
2. Heating block.
3. Incubator shaker.

4. Incubator.
5. Microwave oven.
6. Petri dishes (*see* **Note 1**).
7. 1.5 mL sterile microcentrifuge tubes (autoclaved for at least 20 min at 120 °C and 2 bar).
8. Sterile PCR tubes (autoclaved for at least 20 min at 120 °C and 2 bar).
9. DNA purification kit.
10. DNA molecular-weight marker (e.g., SmartLadder, Eurogentec).
11. Midori Green Direct, SYBR Safe, Crystal Violet, or similar EtBr supplement.
12. Agarose.
13. 50× Rotiphorese buffer: 2 M Tris–HCl, 1 M acetic acid, 50 mM EDTA, pH 8.5 (*see* **Note 2**).
14. Phusion buffer (5×).
15. ddH_2O (preheated at 50 °C).
16. Phusion DNA Polymerase.
17. Restriction Enzymes: NdeI, NcoI, and XhoI.
18. 10× digestion buffer.
19. 10× T4 DNA Ligase Buffer (*see* **Note 3**).
20. T4 DNA Ligase.
21. dNTPs (10 mM).
22. Forward primer: Ubiquitin NdeI (5′-GGGGCCCATATGCAGATTTTTGTTAAAAC-3′).
23. Reverse primer (reverse complement): Ubiquitin Tub-tag XhoI

    (3′-GGGGCTCGAGTTATTCTTCGCCTTCTTCTTCG CCTTCGCCTTCCACGCTATCCACACCACCACGCAGA CG-5′).
24. Forward primer: Annexin V NcoI

    (5′-AATTCCATGGGACATCATCATCATCATCATAT GGCACAGGTTCTCAGA-3′).
25. Reverse primer (reverse complement): Annexin V Tub-tag XhoI

    (3′-AATTCTCGAGTTAGTCATCTTCTCCACAGAGCAGC -5′).
26. Ubiquitin template DNA.
27. Annexin V template DNA.
28. PCR-amplificant.

29. Target vector DNA for Ubiquitin (pET28a; 1 µg/µL).
30. Target vector DNA for Annexin V (pET22b; 1 µg/µL).
31. Kanamycin stock solution: Dissolve 30 mg of kanamycin sulfate in 1 mL water and sterile-filter with a 0.2 µm syringe filter.
32. Ampicillin stock solution: Dissolve 100 mg of D-α-aminobenzylpenicillin sodium salt in 1 mL water and sterile-filter with a 0.2 µm syringe filter.
33. LB-agar plates supplemented with ampicillin or kanamycin: Add 400 mL to a 1 L graduated cylinder. Weigh 10 g peptone, 5 g NaCl and 5 g yeast extract, 16 g agar and transfer to the cylinder. Add water to a volume of 1 L and mix thoroughly. Fill the solution in a 1 L Schott glass bottle and sterilize it for at least 20 min at 120 °C and 2 bar in an autoclave (*see* **Note 4**). Wait until the Schott bottle cooled down to ~40 °C and add 1 mL of ampicillin or kanamycin stock solution. Mix and pour agar directly in petri dishes. Wait until the agar is solid and store at 4 °C.
34. *E. coli* strain DH5α or similar, chemically competent (*see* **Note 5**).

2.2 Expression and Purification of the TTL and Tub-Tagged Proteins

1. Baffled Erlenmeyer flask, 1 L.
2. Incubator shaker.
3. High-speed centrifuge with cooling unit (~50,000 × *g*).
4. Centrifugation tubes.
5. Spectrophotometer to measure optical density (600 nm).
6. Ni-NTA column (*see* **Note 6**).
7. Protein desalting column (*see* **Note 7**).
8. Fast protein liquid chromatography system (FPLC system).
9. Ultrasonic cell disruptor.
10. *E. coli* strain JM109, BL21(DE3) or similar, chemically competent (*see* **Note 8**).
11. Ampicillin stock solution: Dissolve 100 mg of D-α-aminobenzylpenicillin sodium salt in 1 mL water and sterile-filter with a 0.2 µm syringe filter.
12. Kanamycin stock solution: Dissolve 30 mg of Kanamycin sulfate in 1 mL water and sterile-filter with a 0.2 µm syringe filter.
13. LB medium: Add about 400 mL to a 1 L graduated cylinder. Weigh 10 g peptone, 5 g NaCl, and 5 g yeast extract and transfer to the cylinder. Add water to a volume of 1 L and mix thoroughly. Fill the medium in a 1 L Schott glass bottle and sterilize it for at least 20 min at 120 °C and 2 bar in an autoclave (*see* **Note 4**). Store at 4 °C.

14. IPTG stock solution: 1 M isopropyl-β-D-thiogalactopyranoside (IPTG). Dissolve 238 mg of IPTG in 1 mL water and sterile-filter with a 0.2 μm syringe filter. Store at 4 °C.
15. PMSF stock solution: 0.2 M phenylmethanesulfonyl fluoride (PMSF). Dissolve 34.8 mg PMSF in 1 mL isopropanol. Store at 4 °C.
16. TTL IMAC wash buffer: 50 mM Tris–HCl, pH 7.4, 0.5 M NaCl, 10 mM DTT. Add about 400 mL to a 1 L graduated cylinder. Weigh 6.06 g Tris, 29.22 g NaCl, and 1.5 g dithiothreitol (DTT) and transfer to the cylinder. Add water to a volume of 900 mL and mix thoroughly. Adjust the pH to 7.4 with HCl (*see* **Note 9**). Adjust to 1 L with water. Store at 4 °C.
17. TTL IMAC resuspension buffer: TTL IMAC wash buffer supplemented with 100 μg/mL lysozyme, 25 μg/mL DNAse, and 100 μL of PMSF stock solution.
18. TTL IMAC elution buffer: 50 mM Tris–HCl, pH 7.4, 0.5 M NaCl, 500 mM imidazole, 10 mM DTT. Add about 400 mL to a 1 L graduated cylinder. Weigh 6.06 g Tris, 29.22 g NaCl, 34 g imidazole, and 1.5 g dithiothreitol (DTT) and transfer to the cylinder. Add water to a volume of 900 mL and mix thoroughly. Adjust the pH to 7.4 with HCl (*see* **Note 9**). Adjust to 1 L with water. Store at 4 °C.
19. Protein IMAC buffer: 50 mM Tris–HCl, pH 7.4, 0.5 M NaCl. Add about 400 mL to a 1 L graduated cylinder. Weigh 6.06 g Tris and 29.22 g NaCl and transfer to the cylinder. Add water to a volume of 900 mL and mix thoroughly. Adjust the pH to 7.4 with HCl (*see* **Note 9**). Adjust to 1 L with water. Store at 4 °C.
20. Protein IMAC resuspension buffer: Protein IMAC buffer supplemented with 100 μg/mL lysozyme, 25 μg/mL DNAse, and 100 μL of PMSF stock solution.
21. Protein IMAC elution buffer: 50 mM Tris–HCl, pH 7.4, 0.5 M NaCl, 500 mM imidazole. Add about 400 mL to a 1 L graduated cylinder. Weigh 6.06 g Tris, 29.22 g NaCl and 34 g imidazole and transfer to the cylinder. Add water to a volume of 900 mL and mix thoroughly. Adjust the pH to 7.4 with HCl (*see* **Note 9**). Adjust to 1 L with water. Store at 4 °C.
22. Protein storage buffer: 20 mM MES–KOH, 100 mM KCl, 10 mM $MgCl_2$, pH 7.0. Add about 400 mL to a 1 L graduated cylinder. Weigh 3.90 g MES, 7.45 g KCl, and 0.95 g $MgCl_2$ and transfer to the cylinder. Add water to a volume of 900 mL and mix thoroughly. Adjust the pH to 7.0 with 1 M KOH. Adjust to 1 L with water. Store at 4 °C.
23. LB-agar plates supplemented with ampicillin or kanamycin: Add about 400 mL to a 1 L graduated cylinder. Weigh 10 g

peptone, 5 g NaCl and 5 g yeast extract, 16 g agar and transfer to the cylinder. Add water to a volume of 1 L and mix thoroughly. Fill the solution in a 1 L Schott glass bottle and sterilize it for at least 20 min at 120 °C and 2 bar in an autoclave (*see* **Note 4**). Wait until the Schott bottle cooled down to ~40 °C and add 1 mL of ampicillin or kanamycin stock solution. Mix and pour agar plates directly. Wait until the agar is solid and store at 4 °C.

2.3 Synthesis of Coumarin-Derivative 1

1. 100 mL round-bottom flask with conical ground joint (NS) 14/23.
2. 25 mL round-bottom flask with conical ground joint (NS) 14/23.
3. Rotary evaporator.
4. High vacuum pump.
5. Fluorescence spectrometer.
6. NMR spectrometer.
7. Mass spectrometer.
8. Chromatography column with fused-in frit (porosity 0) and PTFE stopcock.
9. Silica gel (60 Å, ~0.035–0.070 mm).
10. Preparative HPLC system.
11. Preparative C18 column like Macherey-Nagel Nucleodur C18 HTec Spum column.
12. Round-bottom flask freeze-dryer for lyophilizing.
13. Centrifuge.
14. 50 mL falcon tube.
15. Cbz-(*L*)-Asp-OBn (*see* **Note 10**).
16. Monoethyl malonic acid magnesium salt (*see* **Note 11**).
17. 3-Hydroxyphenol (*see* **Note 12**).
18. 99% methanesulfonic acid (*see* **Note 13**).
19. Tetrahydrofuran.
20. Ice–water bath.
21. Ice-cold ether.
22. Conc. HCl.
23. Ethyl acetate (EtOAc).
24. 1 N HCl.
25. Hexane.
26. Ethanol (EtOH).
27. 95% Ethanol.

28. Saturated NaCl solution.
29. 10% $NaHCO_3$: Dissolve 10 g of $NaHCO_3$ in 1 L water.
30. H_2O.
31. 4:1 H_2O:$KHSO_4$ (1 N): Dissolve 136.17 g of $KHSO_4$ in 1 L water to prepare a 1 N $KHSO_4$ solution. Mix 800 mL water and 200 mL 1 M $KHSO_4$ solution.
32. Dry $MgSO_4$.
33. 10% Palladium (Pd) on charcoal.
34. Trifluoracetic acid (TFA).
35. Acetonitrile (ACN).

2.4 Ligation of Coumarin 1 to Proteins by the TTL

1. Dry block incubator.
2. Dialysis tube with a 1 kDa cutoff.
3. 1.5 mL microcentrifuge tube.
4. Ligation buffer: 20 mM MES–KOH, 100 mM KCl, 10 mM $MgCl_2$, pH 7.0. Add about 400 mL to a 1 L graduated cylinder. Weigh 3.90 g MES, 7.45 g KCl, and 0.95 g $MgCl_2$ and transfer to the cylinder. Add water to a volume of 900 mL and mix thoroughly. Adjust the pH to 7.0 with 1 M KOH. Adjust to 1 L with water. Store at 4 °C.
5. 2 mg/mL TTL stock solution (*see* Subheading 3.2).
6. 4 mg/mL Protein-Tub-tag stock solution (*see* Subheading 3.3).
7. 100 mM adenosine triphosphate (ATP) stock solution: Weigh 50.72 mg ATP and dissolve in 1 mL water. Store at −20 °C (*see* **Note 14**).
8. 11 mM Coumarin-derivative stock solution: Dissolve the corresponding amount of the respective tyrosine derivative in ligation buffer. Store at 4 °C.
9. 2 M dithiothreitol (DTT) stock solution: Weigh 308.5 mg DTT and dissolve in 1 mL water. Store at −20 °C (*see* **Note 15**).
10. 1 N KOH.
11. 0.1 N KOH.
12. Protein storage buffer: 20 mM MES–KOH, 100 mM KCl, 10 mM $MgCl_2$, pH 7.0. Add about 400 mL to a 1 L graduated cylinder. Weigh 3.90 g MES, 7.45 g KCl, and 0.95 g $MgCl_2$ and transfer to the cylinder. Add water to a volume of 900 mL and mix thoroughly. Adjust the pH to 7.0 with 1 M KOH. Adjust to 1 L with water. Store at 4 °C.

2.5 Analysis of Successful Ligation by SDS Polyacrylamide Gel and In-Gel Fluorescence

1. SDS-PAGE chamber.
2. Benchtop centrifuge.
3. Dry block incubator.
4. 1.5 mL microcentrifuge tube.
5. Imager with possibility to perform fluorescence analysis.
6. Microwave (*see* **Note 16**).
7. Prestained molecular weight standard.
8. 30% acrylamide solution with 0.8% bisacrylamide in water (ratio of 37.5:1).
9. 10% ammonium persulfate solution in water. Leave one 1 mL aliquot at 4 °C. Store remaining aliquots at −20 °C.
10. 10% sodium dodecyl sulfate (SDS) solution in water (*see* **Note 17**).
11. *N,N,N,N'*-tetramethylethylenediamine (TEMED): Store at 4 °C.
12. Resolving gel buffer: 1.5 M Tris–HCl, pH 8.8. Add about 400 mL water to a 1 L graduated cylinder. Weigh 181.7 g Tris and transfer to the cylinder. Add water to a volume of 900 mL and mix thoroughly. Adjust the pH to 8.8 with HCl (*see* **Note 9**). Adjust to 1 L with water. Store at 4 °C.
13. Stacking gel buffer: 0.5 M Tris–HCl, pH 6.8. Add about 400 mL to a 1 L graduated cylinder. Weigh 60.6 g Tris and transfer to the cylinder. Add water to a volume of 900 mL and mix thoroughly. Adjust the pH to 6.8 with HCl (*see* **Note 9**). Adjust to 1 L with water. Store at 4 °C.
14. SDS running buffer (10×): 0.25 M Tris, 1.92 M glycine, and 1% SDS (*see* **Note 18**).
15. Sample buffer (5×): 0.3 M Tris–HCl, pH 6.8, 10% SDS, 25% ß-mercaptoethanol, 0.1% bromophenol blue, and 45% glycerol. Leave one 1 mL aliquot at 4 °C. Store remaining aliquots at −20 °C.
16. Staining solution: Add about 400 mL water to a 1 L graduated cylinder and dissolve 2.5 g Coomassie Brilliant Blue, 300 mL MeOH and 100 mL acetic acid. Mix thoroughly and adjust to 1 L with water.
17. Destaining solution: Add about 400 mL water to a 1 L graduated cylinder and combine 300 mL MeOH and 100 mL acetic acid. Mix thoroughly and adjust to 1 L with water.

2.6 Tryptic Digest and Tandem Mass Spectrometry (MS/MS)

1. Benchtop centrifuge.
2. Vacuum centrifuge.
3. Acetonitrile.

4. Ultrasonic bath.
5. Tandem mass spectrometer.
6. 1.5 mL microcentrifuge tube.
7. Equilibration and digestion buffer: 50 mM $(NH_4)HCO_3$, ~pH 7.8. Dissolve 1.975 g $(NH_4)HCO_3$ in 500 mL water. There is no need to further adjust the pH (*see* **Note 19**).
8. Wash and dehydration buffer: 50 mM ammonium bicarbonate/acetonitrile 1:1. Add 100 mL acetonitrile to 100 mL of equilibration and digestion buffer.
9. Trypsin stock solution: Dissolve 20 μg trypsin in 100 μL 1 mM HCl and store aliquots of 0.4 μg trypsin at −20 °C (*see* **Note 20**).
10. Stop solution: 0.5% trifluoroacetic acid in acetonitrile. Add 500 μL of trifluoracetic acid to 90 mL acetonitrile. Mix and add acetonitrile to a volume of 100 mL.
11. Reducing buffer: 10 mM dithiothreitol in equilibration and digestion buffer. Dissolve 15.4 mg DTT in 9 mL of equilibration and digestion buffer. Adjust pH to 7.8 with 1 N HCl. Add water to a volume of 10 mL. Store aliquots of 20 μL at −20 °C (*see* **Note 20**).
12. Alkylation buffer: 55 mM iodoacetamide in equilibration and digestion buffer. Dissolve 102 mg iodoacetamide in 9 mL of equilibration and digestion buffer. Adjust pH to 7.8 with 1 N HCl. Add water to a volume of 10 mL. Store aliquots of 20 μL at −20 °C (*see* **Note 20**).
13. MSMS solution: Water, 5% acetonitrile and 0.1% trifluoracetic acid. Add 5 mL acetonitrile and 100 μL trifluoracetic acid to 90 mL water. Mix and add water to reach a volume of 100 mL.

2.7 Detection of Apoptotic Cells by Annexin V-Coumarin 1 and Confocal or Widefield Microscopy

1. Confocal or widefield microscope (e.g., Operetta high-content imaging platform (Perkin Elmer, USA)).
2. Eukaryotic cells (e.g., HEK or HeLa cells).
3. Staurosporine from *Streptomyces* sp. (*see* **Note 21**).
4. Cell culture multiwell plates (*see* **Note 22**).
5. Cell culture CO_2 incubator.
6. Phosphate-buffered saline buffer (PBS) (*see* **Note 23**).
7. 3.7% formaldehyde in PBS.
8. PBS–Tween (PBS-T): Add 50 μL of Tween 20 to 100 mL of PBS solution (*see* **Notes 24** and **25**).
9. Propidium iodide (1.0 mg/mL in water) (*see* **Note 26**).
10. 0.5% Triton X-100 solution: Add 500 μL of Triton-X100 to 99.5 mL ddH_2O.

3 Methods

All procedures should be performed at ambient temperature, unless stated otherwise.

3.1 DNA Cloning

1. The cloning procedure is exemplified for ubiquitin and annexin V. Experimental steps can be transferred to any protein of interest. Amplify and equip the ubiquitin and annexin V sequence with a N-terminal H_6-Tag and C-terminal Tub-tag using the respective forward and reverse primers via a standard PCR protocol (*see* **Note 27**).
2. Cast a 1% agarose gel by adding 1 g of agarose in 100 mL 1× Rotiphorese buffer in a 250 mL Erlenmeyer flask and heat until boiling using a microwave oven. Let the solution cool down for 5 min and transfer to the casting apparatus of your DNA electrophoresis system.
3. Add a EtBr substituent like Midori Green Direct according to the manufacturer's manual. Load the PCR reaction mixture on the agarose-gel next to 5 μL of SmartLadder and perform electrophoresis at a constant voltage of 100 V for 30–40 min.
4. Excise designated PCR product under 365 nm UV light from the gel and transfer into a 1.5 mL microcentrifuge tube (*see* **Note 28**).
5. Purify the PCR product using standard DNA purification columns according to the manufacturer's protocol and resolve DNA fragment in ddH_2O.
6. Digest both PCR product and vector backbone (pET22b, NcoI, XhoI in case of Annexin V; pET28a, NdeI and XhoI in case of Ubiquitin) according to the manufacturer's protocol to provide sticky ends for plasmid ligation (*see* **Note 29**).
7. Purify the digested DNA fragments using standard DNA purification columns according to the manufacturer's protocol and resolve them in ddH_2O at 10–20 ng/μL (PCR fragment) and 50–100 ng/μL (vector backbone).
8. Ligate the protein fragment into the bacterial expression vector (Annexin V into pET22b, Ubiquitin into pET28a) at molar ratio of 3:1 using T4 DNA Ligase for 18 h at 4 °C, then inactivate T4 DNA Ligase at 65 °C.
9. Thaw a 50 μL aliquot of chemically competent *E. coli* DH5α or similar on ice.
10. Add 1 μL of ligation product and incubate for 30 min on ice. Heat-shock the cells for 45 s at 42 °C, add 500 μL of pre-warmed LB medium (37 °C) and incubate for 1 h at 37 °C with shaking (180 rpm).

11. Streak 100 μL of the transformed cells on an LB agar plate supplemented with ampicillin (Annexin V) or kanamycin (Ubiquitin) (*see* **Notes 30** and **31**).
12. Incubate the agar plate overnight at 37 °C in an incubator.
13. Pick 20 colonies and inoculate 5 mL of LB medium supplemented with 5 μL of ampicillin stock solution for Annexin V or kanamycin stock solution for Ubiquitin with a single colony picked from the agar plate and incubate at 37 °C, 180 rpm overnight in an incubator shaker.
14. Isolate plasmid DNA using DNA Mini-Prep Kit according to the manufacturer's protocol.
15. Verify the correct insert via analytical restriction digest and DNA sequencing.

3.2 Expression and Purification of the TTL

1. Thaw a 50 μL aliquot of chemically competent *E. coli* BL21 (DE3) on ice.
2. Add 1 μL of plasmid encoding the TTL and incubate for 30 min on ice. Heat shock the cells at 42 °C for 45 s, add 500 μL of prewarmed LB medium and incubate for 1 h at 37 °C with shaking (180 rpm).
3. Streak 100 μL of the transformed cells on an LB agar plate supplemented with ampicillin (*see* **Note 30**).
4. Incubate the agar plate overnight at 37 °C in an incubator shaker.
5. Inoculate 5 mL of LB medium supplemented with 5 μL of ampicillin stock solution with a single colony picked from the agar plate and incubate at 37 °C, 180 rpm overnight (preculture) in an incubator shaker.
6. Inoculate 250 mL of LB medium supplemented with 250 μL of ampicillin stock solution in a 1 L baffled Erlenmeyer flask with 2 mL of preculture and incubate at 37 °C, 180 rpm.
7. Use a spectrophotometer (600 nm) to measure the optical density of the cell suspension. As soon as an optical density of ~0.7 is reached, induce the expression of the TTL by the addition of 125 μL IPTG stock solution.
8. Incubate the cells for 18 h at 18 °C, 180 rpm in an incubator shaker.
9. Harvest the cells by centrifugation (18,000 × *g*, 4 °C, 30 min) and suspend in 10 mL TTL IMAC resuspension buffer.
10. Perform ultrasonic disruption of the cells via three consecutive 90-s cycles of short ultrasonic pulses using an ultrasonic cell disruptor. Keep the cells on ice during the whole process and make sure to allow the solution to cool down between single cycles (*see* **Note 32**).

11. Remove cell debris by centrifugation (49,000 × *g*, 4 °C, 30 min) and directly apply the lysate to an Ni-NTA column using a FPLC system (*see* **Note 33**).
12. Wash the column with 10 column volumes of TTL IMAC wash buffer and elute the immobilized H_6-TTL by slowly increasing the amount of TTL IMAC elution buffer to 100% within 2 column volumes.
13. Analyze the purity of eluting protein fractions by SDS-PAGE (*see* Subheading 3.6), combine pure fractions and desalt the purified protein with a protein desalting column and protein storage buffer.
14. Shock-freeze 20 μL TTL aliquots at ~2 mg/mL using liquid nitrogen and store at −80 °C until further use.

3.3 Expression and Purification of Tub-Tagged Annexin V and Ubiquitin

1. Thaw a 50 μL aliquot of chemically competent *E. coli* BL21 (DE3) on ice.
2. Add 1 μL of plasmid encoding H_6-Annexin V-Tub-tag or H_6-Ubiquitin-Tub-tag and incubate for 30 min on ice. Heat shock the cells at 42 °C for 45 s, add 500 μL of prewarmed LB medium and incubate for 1 h at 37 °C with shaking (180 rpm).
3. Streak 100 μL of the transformed cells on an LB agar plate supplemented with the respective antibiotic (ampicillin for annexin V; kanamycin for ubiquitin) (*see* **Notes 30** and **31**).
4. Incubate the agar plate overnight at 37 °C in an incubator.
5. Inoculate 5 mL of LB medium supplemented with 5 μL of antibiotic stock solution with a single colony picked from the agar plate and incubate at 37 °C, 180 rpm overnight (preculture) in an incubator shaker.
6. Inoculate 250 mL of LB medium supplemented with 250 μL of antibiotic stock solution in a 1 L baffled Erlenmeyer flask with 2 mL of preculture and incubate at 37 °C, 180 rpm.
7. Use a spectrophotometer to measure the optical density of the cell suspension. As soon as an optical density of ~0.7 is reached, induce the expression of the Tub-tagged protein by the addition of 125 μL IPTG stock solution.
8. Incubate the cells for 5 h at 37 °C, 180 rpm in an incubator shaker (*see* **Note 34**).
9. Harvest the cells by centrifugation (18,000 × *g*, 4 °C, 30 min) and suspend in 10 mL Protein IMAC resuspension buffer.
10. Perform ultrasonic disruption of the cells via three consecutive 90 s cycles of short ultrasonic pulses using an ultrasonic cell disruptor. Keep the cells on ice during the whole process and make sure to allow the solution to cool down between single cycles (*see* **Note 32**).

11. Remove cell debris by centrifugation (49,000 × *g*, 4 °C, 30 min) and directly apply the lysate to an Ni-NTA column using a FPLC system (*see* **Note 33**).
12. Wash the column with 10 column volumes of protein IMAC buffer and elute the immobilized H_6-POI-Tub-tag by slowly increasing the amount of protein IMAC elution buffer to 100% within 2 column volumes.
13. Analyze the purity of eluting protein fractions by SDS-PAGE (*see* Subheading 3.6), combine pure fractions and desalt the purified protein with a protein desalting column and protein storage buffer.
14. In case of unsatisfactory purity after His-tag purification, perform a following size exclusion chromatography with protein storage buffer. Analyze the purity of eluting protein fractions by SDS-PAGE (*see* Subheading 3.6) and combine pure fractions.
15. Shock-freeze 20 μL protein aliquots at ~4 mg/mL using liquid nitrogen and store at −80 °C until further use.

3.4 Synthesis of Coumarin-Derivative 1

1. A scheme summarizing the synthesis of coumarin-derivative 1, for TTL-ligation to Tub-tagged proteins, is shown in Fig. 2.
2. Weigh 2 g of Cbz-(*L*)-Asp-OBn and transfer to a 100 mL round-bottom flask with conical ground joint (NS) 14/23 equipped with a magnetic stirring bar. Mount the round-bottom flask using a stand clamp over a magnetic stirrer.
3. Completely dissolve Cbz-(*L*)-Asp-OBn in 20 mL of THF.
4. While stirring, slowly add 1 g of carbonyldiimidazole. Close the reaction flask with a septum.
5. Keep stirring for 2 h.
6. Place an ice–water bath on the magnetic stirrer and mount the round-bottom flask half covered with ice using a stand clamp.
7. Cool the solution to 0 °C.
8. While stirring, slowly add 470 mg of monoethyl malonic acid magnesium salt. Close the reaction flask with a septum.
9. Keep stirring for an additional 12 h and allow the ice to thaw at ambient temperature.
10. Add 15 mL of ether.
11. Place a fresh ice–water bath on the magnetic stirrer and mount the round-bottom flask half covered with ice using a stand clamp.
12. Cool the solution to 0 °C.
13. Slowly acidify the solution with a few mL conc. HCl.

14. After removal of the aqueous layer (*see* **Note 35**), wash the organic layer with 30 mL of 10% $NaHCO_3$ solution in a separation funnel to remove side-products.
15. After removal of the aqueous layer (*see* **Note 35**), wash the organic layer with 30 mL of 4:1 H_2O–$KHSO_4$ (1 N) solution followed by H_2O and saturated NaCl solution in a separation funnel to remove side-products.
16. Discard the aqueous layer (*see* **Note 35**).
17. Add anhydrous $MgSO_4$ to the organic layer to remove residual water. Filter off the $MgSO_4$ and remove THF under reduced pressure with a rotary evaporator set to 40 °C and 330 mbar. Further decrease the pressure once no additional solvent is evaporating (*see* **Note 36**).
18. Purify compound **2** (Fig. 2) by flash column chromatography. For this, dissolve silica gel in a 1:1 mixture of EtOAc and hexane and transfer to a chromatography column with fused-in frit (porosity 0) and PTFE stopcock. Carefully place the impure compound **2** on top of the silica gel and cover with a small layer of sand. Perform chromatographic purification using a 1:1 mixture of EtOAc and hexane and collect the eluting fractions with test tubes and check for the elution of the product by TLC.
19. Weigh 1.5 g of compound 2 and transfer to a 100 mL round-bottom flask with conical ground joint (NS) 14/23 equipped with a magnetic stirring bar. Mount the round-bottom flask using a stand clamp over a magnetic stirrer.
20. Completely dissolve compound **2** in 10 mL of a 1:1 solution of EtOAc:95% EtOH.
21. Add 4 mL of 1 N HCl and 75 mg of 10% Pd on charcoal.
22. Perform hydrogenolysis under atmospheric pressure at ambient temperature while stirring.
23. Filter off the Pd, wash with 95% EtOH and remove EtOAc and EtOH under reduced pressure with a rotary evaporator set to 40 °C and 200 mbar. Further decrease the pressure once no additional solvent is evaporating to 150 mbar. Once no additional solvent is evaporating further decrease the pressure (*see* **Note 37**).
24. Completely dissolve the residue in water and remove solvent using a round-bottom flask freeze-dryer for lyophilization to give compound **3** (Fig. 2).
25. Compound **3** can be used without further purification.
26. Weigh 200 mg of compound **3** and transfer to a 25 mL round-bottom flask with conical ground joint (NS) 14/23 equipped with a magnetic stirring bar. Add 161 mg 3-hydroxyphenol.

Place an ice–water bath on a magnetic stirrer and mount the round-bottom flask half covered with ice using a stand clamp.

27. Cool solution to 0 °C.
28. While stirring, slowly add 1.59 mL of 99% methanesulfonic acid.
29. Allow the ice to thaw at ambient temperature and keep stirring for additional 2 h.
30. Add 10 mL of ice-cold ether and transfer solution to a 50 mL falcon tube.
31. Centrifuge the mixture for 20 min at 4000 × *g* and carefully remove the ether.
32. Dissolve the residue in 3 mL of water.
33. Purify coumarin-derivative **1** (Fig. 2) by preparative HPLC with a C18 column (like Macherey-Nagel Nucleodur C18 HTec Spum column) using the following gradient: A = H_2O + 0.1% TFA, B = ACN + 0.1% TFA) flow rate 32 mL/min, 10% B 0–5 min; 10–100% B 5–35 min; 100% B 35–40 min. Collect fractions and analyze by MS and NMR analysis (*see* **Note 38**). Combine pure fractions and remove solvent using a round-bottom flask freeze-dryer for lyophilization. Analyze absorbance and emission spectra using a fluorescence spectrometer (Fig. 3). Store coumarin-derivative **1** (yellow powder) at −20 °C until further use.

3.5 TTL Reaction

1. Add 100 μL of ligation buffer to a 1.5 mL microcentrifuge tube.
2. Add 5 μL of ATP stock solution (final concentration 2.5 mM) and 13.64 μL of coumarin-derivate stock solution (final concentration 1 mM).
3. Add 0.38 μL of DTT stock solution (final concentration 5 mM).

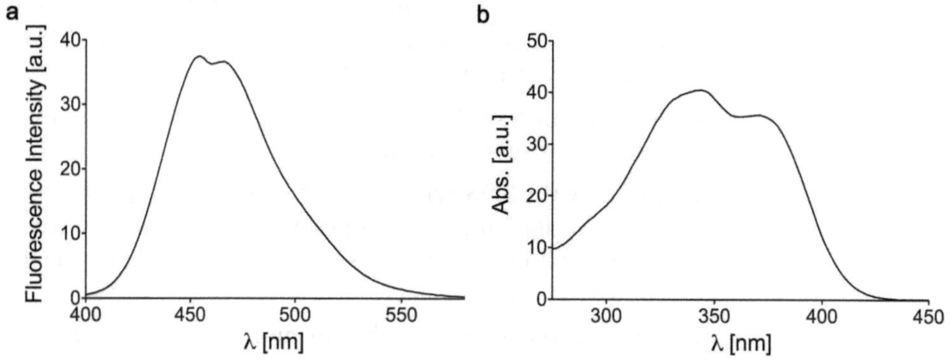

Fig. 3 Fluorescent properties of coumarin-derivative **1**. (**a**) Emission spectrum. (**b**) Absorption spectrum

4. Adjust pH to 7.0 using 1 N and 0.1 N KOH (*see* **Note 39**).
5. Add ligation buffer to 134.72 μL.
6. Add 11.68 μL of protein-Tub-tag stock solution to a final concentration of 10 μM and 3.6 μL of TTL stock solution to a final concentration of 1 μM.
7. Incubate the protein mixture at 37 °C for 1–3 h in a dry block incubator (*see* **Note 40**).
8. Remove excess of reducing agent and tyrosine derivative using dialysis at 4 °C for 3 h (cutoff 1 kDa) against protein storage buffer. The success of ligation can be analyzed by SDS-PAGE, tryptic digest and MSMS analysis (*see* Subheadings 3.6 and 3.7). The fluorescent protein can be stored at 4 °C.

3.6 Analysis of Successful Ligation by SDS–Polyacrylamide Gel and In-Gel Fluorescence

1. Mix 1.7 mL of resolving gel buffer, 5 mL of acrylamide mixture and 3.1 mL of water in a 15 mL falcon tube. Add 100 μL of SDS solution, 100 μL of ammonium persulfate solution and 10 μL of TEMED and cast gel within a ~7.25 cm × 10 cm × 1.5 mm gel cassette. Allow space for stacking gel and gently overlay with isopropanol (*see* **Note 41**). Polymerize the gel at ambient temperature for 20 min.
2. Carefully remove isopropanol after polymerization.
3. Prepare the stacking gel by mixing 1.25 mL of stacking gel buffer, 0.65 mL of acrylamide mixture and 3 mL of water in a 15 mL falcon tube. Add 50 μL of SDS solution, 50 μL of ammonium persulfate solution, and 5 μL of TEMED. Transfer to the resolving gel and immediately insert a 10-well gel comb (*see* **Note 42**).
4. Add 1.5 μL of labeled protein to 2 μL of sample buffer and add water to 10 μL. Heat the sample at 95 °C for 10 min and centrifuge the samples at ~3000 × *g* for 1 min.
5. Load the sample on a gel in a pocket next to one loaded with 5 μL prestained molecular weight standard (*see* **Note 43**). Fill the chamber with SDS running buffer (1×) and perform electrophoresis at a constant voltage of 250 V until the bromophenol blue dye reaches the end of the gel. Separate the glass plates and remove the stacking gel.
6. Analyze successful incorporation of the fluorophore by a fluorescence imager (320 nm).
7. Incubate the SDS gel in staining solution (completely covered). Heat the staining solution gel mixture in a microwave until it starts boiling (*see* **Note 44**). Further incubate the gel for approx. 10 min.
8. Remove the staining solution and incubate the SDS gel in destaining solution (completely covered). Heat the destaining

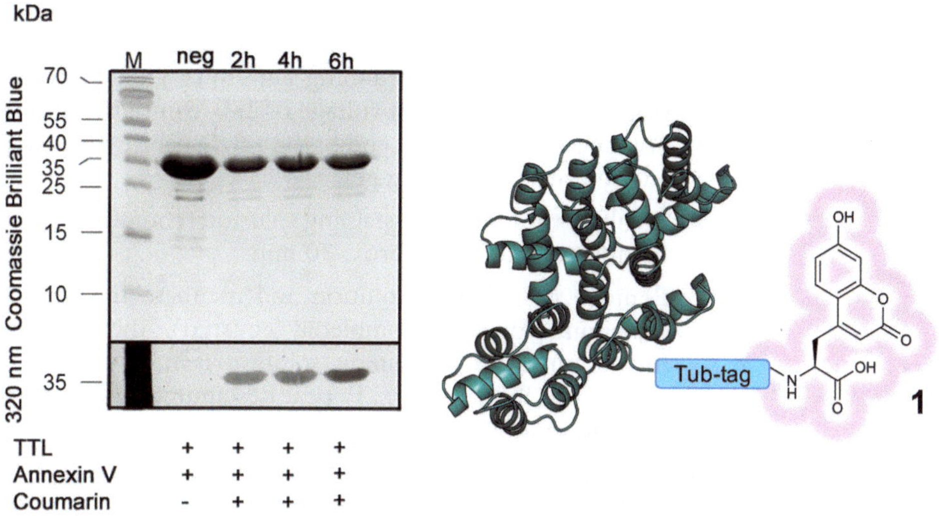

Fig. 4 SDS-PAGE and fluorescent in-gel analysis of the TTL catalyzed one-step coumarin **1** incorporation to Annexin V (PDB Code: 1SAV) [15]. This Figure has been reproduced with permission from [14]—Published by The Royal Society of Chemistry and is licensed under a Creative Commons Attribution 3.0 Unported Licence (https:/creativecommons.org/licenses/by/3.0/)

solution gel mixture in a microwave until it starts boiling (*see* **Note 44**). Incubate the gel further until the destaining solution is fully saturated with Coomassie Brilliant Blue. Repeat destaining until background staining vanished and protein bands are visible.

9. Analyze SDS gel using an imager (Fig. 4) (*see* **Note 45**).

3.7 Tryptic Digest and MSMS Analysis

1. Mix 1.7 mL of resolving gel buffer, 5 mL of acrylamide mixture and 3.1 mL of water in a 15 mL falcon tube. Add 100 μL of SDS solution, 100 μL of ammonium persulfate solution and 10 μL of TEMED and cast gel within a ~7.25 cm × 10 cm × 1.5 mm gel cassette. Allow space for stacking gel and gently overlay with isopropanol (*see* **Note 41**). Polymerize the gel at ambient temperature for 20 min.
2. Carefully remove isopropanol after polymerization.
3. Prepare the stacking gel by mixing 1.25 mL of stacking gel buffer, 0.65 mL of acrylamide mixture and 3 mL of water in a 15 mL falcon tube. Add 50 μL of SDS solution, 50 μL of ammonium persulfate solution and 5 μL of TEMED. Transfer to the resolving gel and immediately insert a 10-well gel comb (*see* **Note 42**).
4. Add 1.5 μL of labeled protein to 2 μL sample buffer and add water to 10 μL. Boil the sample at 95 °C for 10 min and centrifuge the samples at ~3000 × *g* for 1 min.

5. Load the sample on a gel in a pocket next to one loaded with 5 μL prestained molecular weight standard (*see* **Note 43**). Fill the chamber with SDS running buffer (1×) and perform electrophoresis at a constant voltage of 250 V until the bromophenol blue dye reaches the end of the gel. Separate the glass plates and remove the stacking gel.
6. Incubate the SDS gel in staining solution (completely covered). Incubate the gel for approx. 30 min.
7. Remove the staining solution and incubate the SDS gel in destaining solution (completely covered). Incubate the gel further until the destaining solution is fully saturated with Coomassie Brilliant Blue. Repeat destaining until background staining vanished and protein bands are visible.
8. Wash the gel with water 3–4 times. Excise the bands of interest with a clean scalpel (*see* **Note 46**).
9. Cut excised bands in small cubes (~1 × 1 cm), transfer them into a microcentrifuge tube and spin them down at 3000 × *g* for 30 s.
10. Add 100 μL of wash and dehydration buffer and incubate the pieces for ~30 min at 37 °C with occasional vortexing to destain gel pieces (*see* **Note 47**).
11. Add 500 μL acetonitrile and incubate at ambient temperature. Gel pieces will become white and shrink as acetonitrile withdraws remaining water from the pieces.
12. Add 50 μL of reducing buffer (pieces completely covered) and incubate at 56 °C for 30 min in a dry block incubator to reduce all disulfides in the protein. Chill tube to ambient temperature and add 500 μL of acetonitrile. Incubate for 10 min and remove the liquid.
13. Add 50 μL of alkylation buffer (pieces completely covered) and incubate at ambient temperature for 30 min to alkylate all free thiols. Add 500 μL of acetonitrile. Incubate for 10 min and remove the liquid.
14. Add 500 μL acetonitrile and incubate at ambient temperature. Gel pieces will become white and shrink as acetonitrile withdraws remaining water from the pieces. Remove the acetonitrile using a vacuum centrifuge.
15. Equilibrate gel pieces with 100 μL of equilibration and digestion buffer for 10 min. Remove all liquid and cover gel pieces with 40 μL of equilibration and digestion buffer. Add 5 μL of trypsin solution followed by additional 15 μL of equilibration and digestion buffer. Digest the protein at 37 °C in a dry block incubator overnight.

16. Spin down the mixtures at 3000 × *g* for 30 s. Add 20 μL of stop solution.
17. Leave the samples for 10 min in an ultrasonic bath and transfer the solution in a new microcentrifuge tube.
18. Add 50 μL of acetonitrile to the gel pieces and wait until they shrink completely. Transfer the acetonitrile to the other solution in the new microcentrifuge tube.
19. Remove all solvent using a vacuum centrifuge.
20. Dissolve the digested protein in 6 mL MSMS solution and perform MS/MS analysis.

3.8 Detection of Apoptotic Cells by Annexin V-Coumarin 1 and Fluorescence Microscopy

1. Seed ~2×10^4 of adherent cells (e.g., HEK or HeLa cells) in each well of a 96-well clear plate.
2. Induce apoptosis by adding 5 μM staurosporine and incubate cells for 3 h at 37 °C under 5% CO_2.
3. Discard cell culture medium.
4. Wash cells with 200 μL PBS. Repeat this step 3 times.
5. Stain cells by adding 5 μg/1×10^5 cells of Annexin V-coumarin in PBS (100 μL) for 30 min at ambient temperature.
6. Fix cells in 3.7% formaldehyde in PBS for 10 min at ambient temperature.
7. Add 100 μL PBS-T, then discard the fixative.
8. Wash cells with 200 μL PBS-T. Repeat this step 3 times.
9. Permeabilize cells with 0.5% Triton X-100 solution for 10 min at ambient temperature.
10. Add 100 μL PBS-T, then discard Triton X-100.
11. Add 100 μL PBS-T.
12. Counterstain the cells with propidium iodide at a concentration of 100 μg/mL (100 μL) for 10 min in the dark.
13. Wash cells with 100 μL PBS-T. Repeat this step at least 3 times.
14. Analyze the cells using a confocal or widefield microscope (Fig. 5) (*see* **Note 48**).

4 Notes

1. We usually use petri dishes with dimensions of 92 × 16 mm.
2. 50× Rotiphorese buffer is commercially available (Carl Roth GmbH & Co. KG, Germany).
3. Prepare aliquots of T4 DNA Ligase Buffer and store at −20 °C. Do not reuse an aliquot after thawing once.

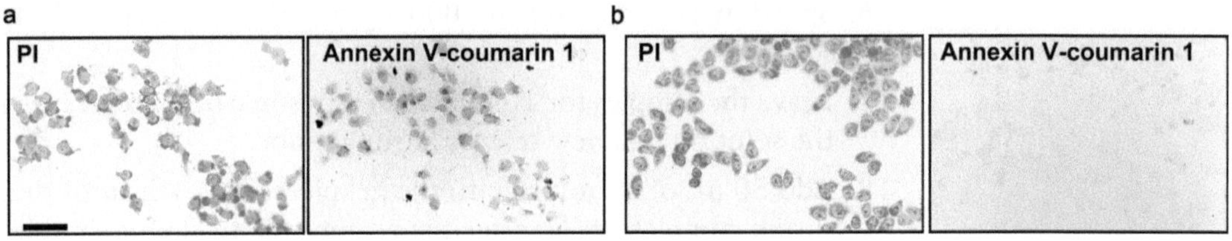

Fig. 5 Detection of apoptotic cells by Tub-tag functionalized Annexin V-coumarin. (**a**) Staurosporine-treated (5 μM) and b) untreated cells were stained with Annexin V-coumarin **1**. Counterstaining was done with propidium iodide. Scale bar is 50 μm. This Figure has been reproduced with permission from [14]—Published by The Royal Society of Chemistry and is licensed under a Creative Commons Attribution 3.0 Unported Licence (https:/creativecommons.org/licenses/by/3.0/)

4. Make sure to not completely close the cap of the Schott bottle to allow for pressure equalization. After termination of the sterilization process, immediately close the cap using heat-resistant gloves.
5. Chemically competent cells can be obtained commercially or prepared using published protocols [16].
6. We use a His-Trap column (GE Healthcare, USA).
7. We use a PD10 column (GE Healthcare).
8. It is advisable to use an *E. coli* strain with the T7 polymerase promoter system to achieve high regulation of expression. Chemically competent cells can be obtained commercially or prepared using published protocols [16].
9. Start by using concentrated HCl (12 N) until getting close to the required pH. From then on use diluted HCl solutions for final adjustment of the pH.
10. Cbz-(L)-Asp-OBn is commercially available (e.g., Sigma-Aldrich Merck KGaA, Germany).
11. Monomethyl malonic acid potassium salt is commercially available (e.g., Alfa Aesar, Thermo Fisher Scientific).
12. 3-Hydroxyphenol is commercially available (e.g., Sigma-Aldrich Merck KGaA, Germany).
13. Methanesulfonic acid is commercially available (e.g., Sigma-Aldrich Merck KGaA, Germany).
14. Prepare 20 μL aliquots of ATP stock solution and store at −20 °C. Do not reuse an aliquot after thawing once.
15. Prepare 10 μL aliquots of DTT stock solution and store at −20 °C. Do not reuse an aliquot after thawing once.
16. This is optional equipment. Heating during staining and destaining of SDS gels decreases time needed for staining of proteins.

17. Make sure to store 10% SDS solution at ambient temperature only.
18. Commercially available.
19. Prepare $(NH_4)HCO_3$ containing buffers freshly and discard after use.
20. Do not re-use an aliquot after thawing once.
21. Staurosporine is commercially available (e.g., Sigma-Aldrich Merck KGaA, Germany).
22. 96-well clear plates are commercially available (e.g., Greiner Bio-One GmbH, Austria).
23. Phosphate-Buffered Saline (PBS) is commercially available (e.g., Thermo Fisher Scientific, Germany).
24. Tween 20 is commercially available (e.g., Sigma-Aldrich Merck KGaA, Germany).
25. You can add Tween 20 to PBS using a standard pipette when removing the top of the pipette tip with a scissors.
26. Propidium iodide (1.0 mg/mL in water) is commercially available (e.g., Sigma-Aldrich Merck KGaA, Germany).
27. You can find exemplarily PCR protocols in the manufacturer's manual of the Phusion DNA polymerase.
28. Try to keep the exposure of your DNA to UV-light short. UV light is damaging the DNA.
29. Do not exceed 4 h of digestion time to prevent star activity of the enzymes.
30. Make sure to streak out the cell suspension evenly to allow for single colony formation.
31. Depending on the protein you are expressing and the vector you are using, you might need to change the antibiotic.
32. The cell suspension will become greyish as soon as the cells have been disrupted successfully.
33. It is not mandatory to use an automated FPLC system and the purification of the protein can be performed manually as described in literature [17]. However, if you perform the Ni-NTA purification of the H_6 tagged TTL manually, make sure that you elute the immobilized protein by slowly increasing the TTL IMAC elution buffer and carefully analyze the purity of the eluting fractions by SDS-PAGE.
34. Depending on the POI you may need to change IPTG concentration, temperature and expression time. If your protein seems to be toxic for *E. coli* cells, you may decrease the IPTG concentration to 50 μM.
35. The THF phase is on top of the aqueous solution.

36. Slowly decrease the pressure until you reach 330 mbar. This way you prevent boiling retardation. Keep the pressure at 330 mbar until no additional solvent is evaporating. Further decrease the pressure to 10 mbar (very slowly). Boiling retardation is very likely if you decrease the pressure too fast. Keep the pressure at 10 mbar for at least 20 min.
37. Slowly decrease the pressure until you reach 200 mbar. This way you prevent boiling retardation. Keep the pressure at 200 mbar until no additional EtOAc is evaporating. Slowly decrease the pressure until you reach 150 mbar and keep the pressure at 150 mbar until no additional EtOH is evaporating. Further decrease the pressure to 10 mbar (very slowly). Boiling retardation is very likely if you decrease the pressure too fast. Keep the pressure at 10 mbar for at least 20 min.
38. Analytical MS and NMR data can be found in the literature [13, 14].
39. Start by using 1 N KOH until getting close to pH 7.0. From then on use 0.1 N KOH for final adjustment of the pH.
40. Depending on the POI you can vary incubation time to achieve full conversion. Do not shake while incubating to prevent aggregation of proteins.
41. We use a Mini-PROTEAN Tetra Cell Casting stand (Bio-Rad, USA). Isobutanol prevents contact with polymerization inhibiting oxygen and levels the resolving gel.
42. Make sure to not introduce air bubbles when adding the gel comb. In our hands, storing the gel in a wet tissue at 4 °C for one night results in better protein separation.
43. Adding 5 μL sample buffer to any empty pocket of the gel ensures a uniform running front and prevents the "smiling effect."
44. If you want to perform tryptic digest and MSMS analysis, do not boil the gel during staining and destaining to increase sensitivity of the MS analysis.
45. The incorporation of single amino acids cannot be resolved in an SDS gel.
46. It is important to wear gloves to reduce keratin background.
47. Although most of the Coomassie should be removed, it is not necessary to destain the pieces completely.
48. We use the Operetta high-content imaging platform (Perkin Elmer, USA), the preset DsRed and a DAPI filter. Optionally, this high-content platform allows for quantitative readouts via automated image analysis.

Acknowledgments

This work was supported by grants from the Bundesministerium für Wirtschaft und Energie (Exist Forschungstransfer I) to D.S. and J.H., the Deutsche Forschungsgemeinschaft (SPP1623) to C.P.R.H. (HA 4468/9-1) and H.L. (LE 721/13-1), the Nanosystems Initiative Munich (NIM) to H.L., the Einstein Foundation Berlin (Leibniz-Humboldt Professorship) and the Boehringer-Ingelheim Foundation (Plus 3 award) to C.P.R.H., and the Fonds der Chemischen Industrie (FCI) to C.P.R.H. and to D.S. (Kekulé-scholarship).

References

1. Hackenberger CPR, Schwarzer D (2008) Chemoselective ligation and modification strategies for peptides and proteins. Angew Chem Int Ed 47(52):10030–10074
2. Lotze J, Reinhardt U, Seitz O, Beck-Sickinger AG (2016) Peptide-tags for site-specific protein labelling in vitro and in vivo. Mol BioSyst:1731–1745
3. Schumacher D, Hackenberger CPR (2014) More than add-on: chemoselective reactions for the synthesis of functional peptides and proteins. Curr Opin Chem Biol 22:62–69
4. Lang K, Chin JW (2014) Cellular incorporation of unnatural amino acids and bioorthogonal labeling of proteins. Chem Rev:4764–4806
5. Drake PM, Rabuka D (2017) Recent developments in ADC technology: preclinical studies signal future clinical trends. BioDrugs:521–531
6. Beck A, Goetsch L, Dumontet C, Corvaia N (2017) Strategies and challenges for the next generation of antibody-drug conjugates. Nat Rev Drug Discov 16(5):315–337
7. Massa S, Xavier C, Muyldermans S, Devoogdt N (2016) Emerging site-specific bioconjugation strategies for radioimmunotracer development. Expert Opin Drug Deliv 13 (8):1149–1163
8. Schumacher D, Helma J, Mann FA, Pichler G, Natale F, Krause E, Cardoso MC, Hackenberger CPR, Leonhardt H (2015) Versatile and efficient site-specific protein functionalization by tubulin tyrosine ligase. Angew Chem Int Ed 54(46):13787–13791
9. Chen X, Wu YW (2016) Selective chemical labeling of proteins. Org Biomol Chem 14 (24):5417–5439
10. Sletten EM, Bertozzi CR (2009) Bioorthogonal chemistry: fishing for selectivity in a sea of functionality. Angew Chem Int Ed 48 (38):6974–6998
11. Patterson DM, Nazarova LA, Prescher JA (2014) Finding the right (bioorthogonal) chemistry. ACS Chem Biol 9(3):592–605
12. Helma J, Leonhardt H, Hackenberger CPR, Schumacher D (2018) Tub-tag labeling; chemoenzymatic incorporation of unnatural amino acids, vol 1728 noncanonical amino acids: methods and protocols. Humana, New York, NY
13. Brun MP, Bischoff L, Garbay C (2004) A very short route to enantiomerically pure coumarin-bearing fluorescent amino acids. Angew Chem Int Ed 43(26):3432–3436
14. Schumacher D, Lemke O, Helma J, Greszonowicz L, Waller V, Stoschek T, Durkin PM, Budisa N, Leonhardt H, Keller B, Hackenberger CP (2017) Broad substrate tolerance of tubulin tyrosine ligase enables one-step site-specific enzymatic protein labeling. Chem Sci 8:3471–3478
15. Huber R, Berendes R, Burger A, Schneider M, Karshikov A, Luecke H, Romisch J, Paques E (1992) Crystal and molecular structure of human annexin V after refinement. Implications for structure, membrane binding and ion channel formation of the annexin family of proteins. J Mol Biol 223(3):683–704
16. Inoue H, Nojima H, Okayama H (1990) High efficiency transformation of *Escherichia coli* with plasmids. Gene 96(1):23–28
17. Bornhorst JA, Falke JJ (2000) Purification of proteins using polyhistidine affinity tags. Methods Enzymol 326:245–254

Acknowledgments

[illegible]

References

[illegible]

Chapter 13

Enzymatically Catalyzed Radiofluorination of Biomolecules

Christopher R. Drake, Michael J. Evans, and Henry F. VanBrocklin

Abstract

There has been significant and rapid growth in the development of amino acid-based molecular imaging agents (e.g., peptides, proteins, and antibody constructs) largely due to facile library preparation and high throughput screening. Positron-emitting fluorine-18 (half-life = 109.7 min) has a unique set of properties that match well with the pharmacokinetics of smaller sized constructs. Several indirect fluorine-18 labeling approaches have been developed yet only a few have advanced to human trials. Enzymatically catalyzed radiofluorination utilizing lipoic acid ligase shows promise as a mild site-specific method for coupling fluorine-18-labeled carboxylate substrates with biomolecules. Methods for preparation of two [^{18}F]fluorocarboxylates and their ligation to a specific peptide sequence (LAP peptide) are presented herein.

Key words PET, Enzymatic radiolabeling, Fluorine-18, Peptide, Biomolecule, Radiofluorination, Lipoic acid ligase, LAP peptide

1 Introduction

Amino acid-based imaging agents represent a broad class of molecules from small peptides to proteins to monoclonal antibodies with varying pharmacokinetic properties. A small number of these agents, mostly labeled with SPECT isotopes such as indium-111, have been approved for clinical application [1, 2]. Most recently the FDA approved the first peptide labelled with a PET radionuclide, gallium-68 (68 min half-life) DOTATATE (NETSPOT®), for clinical use; this cyclic eight amino acid peptide targets somatostatin receptors on neuroendocrine tumors to detect and stage the disease [3]. Biomolecule-based imaging agent discovery and development is gaining prominence, in large part due to the variety of molecules available through contemporary molecular biology methodologies that enable the generation of peptide, protein, and antibody libraries, as well as the ability to screen these libraries against important imaging targets to rapidly identify and subsequently optimize lead imaging agent candidates.

Sam Massa and Nick Devoogdt (eds.), *Bioconjugation: Methods and Protocols*, Methods in Molecular Biology, vol. 2033, https://doi.org/10.1007/978-1-4939-9654-4_13, © Springer Science+Business Media, LLC, part of Springer Nature 2019

To date, while numerous PET-radionuclide labeled biomolecules have been evaluated in preclinical studies, only a relatively small number have advanced to human trials [1, 2]. Thus, despite their potentially advantageous features, the impact of biomolecule-based PET agents remains limited compared to their small-molecule counterparts. For larger molecules, including some antibodies, antibody fragments and proteins, this may be due to their suboptimal pharmacokinetics; they are slow to localize to target tissue and slow to clear from nontarget tissue. Therefore, they must be labeled with longer lived PET isotopes and frequently require a multiple-day imaging protocol that is inconvenient for the patient. Lower molecular weight peptides and antibody fragments, with more rapid clearance times, are pharmacokinetically more attractive, permitting same-day imaging. These molecules can be labeled with shorter-lived PET radionuclides, such as fluorine-18; however, as will be outlined below, the radiochemistry required for labeling is nontrivial and currently impedes the clinical translation of fluorine-18 labeled biomolecules.

Fluorine-18 possesses a unique set of properties that are ideal for nuclear imaging with amino acid-based molecules. Fluorine-18 decays almost exclusively (>97%) by positron emission, with no other radioactive emissions, to a stable daughter product, oxygen-18. It has a half-life of 109.7 min, allowing imaging out to 4–6 h post tracer injection. The positron energy is 635 KeV, meaning that the coincident annihilation photons will be produced closer to the emission nucleus. Taken together the emission properties result in high sensitivity and high-resolution PET images. Additionally, fluorine-18 radiotracers are readily available through global radiopharmacy networks, facilitating their clinical translation.

While fluorine-18 is an attractive isotope, the conditions required for direct formation of a carbon-[^{18}F]fluorine bond, including high temperature and nonaqueous solvents, are incompatible with amino acid-based molecules. Therefore, small molecule fluorine-18-labeled prosthetics have been developed that conjugate to the biomolecules under mild conditions which preserve their bioactivity. The most widely used prosthetic is *N*-succinimidyl-[^{18}F]fluorobenzoate ([^{18}F]SFB), an activated ester that reacts in aqueous solution with free amines on the biomolecule structure [4, 5]. While this prosthetic has been used to label a variety of biomolecules, there are limitations to its successful application in the preparation of clinical tracers. The [^{18}F]SFB preparation is a multistep synthesis with a low (25–40% decay corrected) yield. Subsequent conjugation to the biomolecule is not site specific, may interact with free amines throughout the structure and potentially lower bioactivity, and is typically inefficient with only 40–50% of the [^{18}F]SFB conjugated.

Several alternative prosthetics have been and are being developed to overcome these limitations. Solvent-exposed cysteines incorporated into biomolecules offer a potential for site-specific labeling with thiol conjugating maleimides including [^{18}F]BAM [6] and [^{18}F]FPEGMA [7] or thiol condensation with [^{18}F] fluoro-2-cyanobenzthiazole [8]. Solid phase bound peptides have been terminally labeled with [^{18}F]fluorobenzoic acid and [^{18}F] fluoropropionic acid [9]. Alteration of the peptide structure with nonnatural amino acids enable orthogonal labeling approaches such as click chemistry between tetrazine on the protein and [^{18}F]-*trans*-cyclooctene [10] or addition of [^{18}F]fluoroalkynes to azido-peptides [11]. Recently, direct [^{18}F]fluorination approaches have been reported on proteins that have been tailored with fluorine-18 acceptor molecules. Preconjugation of a NOTA group to the biomolecule provides a receptacle for chelation with Al[^{18}F] and appending a silicon fluoride acceptor enables a fluorine-18-for-fluorine-19 exchange [12, 13]. This is by no means an exhaustive list of all prosthetic approaches that have been explored, however, it is representative of the emerging approaches that still have fundamental limitations that may constrain their clinical translation viability.

The ideal biomolecule prosthetic radiofluorination would be characterized by a high yielding prosthetic synthesis and then a rapid, high yielding, site specific conjugation under mild, aqueous conditions with suitable pH, ambient temperature and minimal organic cosolvents. In an effort to develop a novel approach closer to these ideals, we investigated an enzymatic conjugation. There is precedence for this approach including examples of transferases which append small molecules to proteins. Additionally, enzymes generally work with small amounts of substrate that would favor higher molar activity preparations, and ligations are site-specific and proceed under mild aqueous (i.e., physiological) conditions.

As such, enzymatically catalyzed radiolabeling (ECR) represents an approach that may address some of the current prosthetic radiolabeling limitations. Lipoic acid ligase (LplA) was chosen as the enzyme to ligate a [^{18}F]-labeled prosthetic to a specific lysine residue within a 13-amino acid sequence (GFEID**K**VWYDLDA; termed "LAP"), based on the demonstrated attributes of LplA to ligate fluorescent molecules to LAP-tagged proteins in cells [14–18]. LplA forms a stable amide bond between the substrate and the terminal amine of the LAP lysine, a range of carboxylate substrates may be applied due to the promiscuity of the enzyme and the various engineered variants described, and the low K_M (13.3 μM) may provide high yields of the radiolabeled biomolecule at low protein concentrations. The latter would provide high molar activity, a ratio of the radioactively labeled molecules to the total mass, labeled and unlabeled molecules. This is essential for low concentration in vivo targets and is often difficult to achieve in reactions that require a substantive mass of the biomolecule for labeling, as it is generally not feasible to separate labeled from unlabeled molecules.

We have evaluated two [^{18}F]fluorooctanoate/LplA substrate/enzyme pairs and found that the conjugation of [^{18}F]fluorooctanoic acids to LAP and LAP-tagged biomolecules is rapid (<10 min) and high yielding (greater than 90% in ~10 min) under mild conditions (aqueous reaction, pH 7.4, 30 °C) using ≤10 nmol of peptide [19, 20]. The first generation [^{18}F]fluorooctanoate prosthetic, [^{18}F]fluorooctanoic acid ([^{18}F]FA), is a close structural analog of the substrate for wild-type lipoic acid ligase and was used for proof-of-concept studies. However, it was limited by (1) metabolic defluorination in vivo, resulting in high bone signal from free [^{18}F]fluoride ion; (2) the lack of a chromophore, which prevented facile measurement of chemical purity by high-performance liquid chromatography (HPLC); and (3) a low boiling point, which complicated its synthesis and handling. Hence, a second-generation [^{18}F]fluoroaryl prosthetic, 7-(4-[^{18}F]fluorophenyl)-7-oxyheptanoic acid ([^{18}F]FPOA), was developed. Encouragingly, preliminary studies suggest an improved stability against metabolic defluorination and better physical characteristics that enhanced characterization and handling. Herein, we review the optimal radiosynthetic protocols obtained to date for both [^{18}F]FA and [^{18}F]FPOA.

2 Materials

2.1 General

1. Water used throughout the synthesis for reaction, solid-phase extraction, or HPLC is commercially available, doubly distilled or deionized, and with a conductivity of 18 Ω (e.g., from a laboratory purifier or VWR).
2. Temperatures stated refer to the external medium. Oil baths, in combination with IKA heating mantles and thermocouples, are used to achieve elevated temperatures.

2.2 Azeotropic Drying of [^{18}F] Fluoride Ion

1. Drying station as depicted in Fig. 1. The low volume vacuum pump used for radiosynthetic evaporation steps was a BTC Diaphragm pump 24VDC (Hargraves Technology Corp., Hollis, NH).
2. Heating bath (up to 110 °C).
3. Glass vials (1 dram, with phenolic screw-caps) used for radiochemical reactions can be purchased from VWR (Radnor, PA); approximately 4 are required for a typical protocol.
4. Norm-Ject syringes from VWR (Radnor, PA).
5. [^{18}F]-fluoride ion was purchased from either the University of California Radiopharmaceutical Facility, where it was generated by proton irradiation of enriched [^{18}O]-water by a PETtrace 18 MeV cyclotron (GE Healthcare, Buckinghamshire, UK), or from PETNET solutions (Culver City, CA).

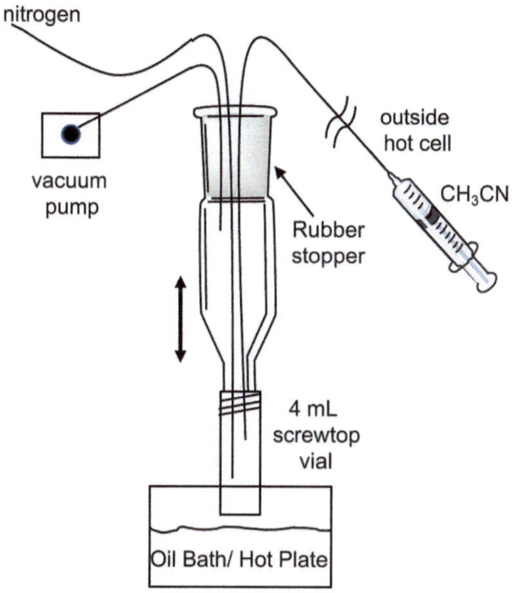

Fig. 1 Drying apparatus/station for azeotropic drying of the [^{18}F]fluoride ion solution

6. Preconditioned QMA cartridges (ABX, Radeberg, Germany), used as received.
7. Potassium carbonate (K_2CO_3; anhydrous, ACS reagent, Sigma-Aldrich, St. Louis, MO).
8. 4,7,13.16,21,21-hexaoxa-1,10-diazabicyclo[8.8.8]hexacosane ($K_{2.2.2}$; 98%, Sigma-Aldrich, St. Louis, MO).
9. Acetonitrile (CH_3CN; anhydrous, 99.8%, Sigma-Aldrich, St. Louis, MO).
10. Fluoride ion release K_2CO_3–$K_{2.2.2}$ solution: Weigh 20 mg K_2CO_3 and 252 mg of $K_{2.2.2}$ into a screw-cap vial. Add 1 mL of water to dissolve the K_2CO_3 followed by 9 mL acetonitrile to produce a clear colorless solution. Cap and store solution at 4 °C (*see* **Note 1**).

2.3 High-Performance Liquid Chromatography and Radio-Thin-Layer Chromatography (TLC)

1. RP-HPLC purification and analysis of radiolabeled compounds was performed using a Waters 600 system (Waters, Milford, MA) with a Shimadzu-10A UV-Vis detector (Shimadzu, Kyoto, Japan) and an in-line CsI(Tl) radiation detector (Carroll & Ramsey, Berkeley, CA).
2. Semipreparative reversed-phase HPLC column for [^{18}F]FA purification: Phenomenex Luna C18(2), 100 Å, 10 μm, 10 × 250 mm.
3. Analytical reversed-phase HPLC column for [^{18}F]FA quality evaluation: Phenomenex Luna C18(2), 100 Å, 5 μm, 4.6 × 250 mm.

4. Semipreparative reversed-phase HPLC column for [^{18}F]FPOA purification: Hamilton PRP-1, 100 Å, 10 μm, 10 × 250 mm.
5. Analytical reversed-phase HPLC column for [^{18}F]FPOA quality evaluation: Hamilton PRP-1, 100 Å, 10 μM, 4.6 × 250 mm.
6. Trifluoroacetic acid (TFA; HPLC-grade, ≥99.0%, Sigma-Aldrich, St. Louis, MO).
7. Acetonitrile (CH_3CN; HPLC-grade, VWR, Radnor, PA).
8. Aqueous HPLC solvent: Mix 10 mL TFA into 990 mL water to make a 0.1% TFA–water solution.
9. Organic HPLC solvent: Mix 10 mL TFA into 990 mL acetonitrile to make a 0.1% TFA–CH_3CN solution.
10. Radio-thin-layer chromatography (TLC) was performed either on a Dual Scan-RAM equipped with a PS plastic/PMT radiation detector (LabLogic; Sheffield, UK) or a Bioscan AR-2000 radio-TLC reader (Eckert & Ziegler; Hopkington, MA) using silica gel 60 matrix aluminum-backed TLC plates (Z740226, Sigma-Aldrich; St. Louis, MO).

2.4 Synthesis Reagents and Products

1. Eppendorf tubes (1.5 mL, polypropylene) for protein labeling with [^{18}F]FA or [^{18}F]FPOA can be purchased from Fisher Scientific (Hampton, NH).
2. Acetonitrile (CH_3CN; anhydrous, 99.8%, Sigma-Aldrich, St. Louis, MO).
3. Potassium hydroxide (KOH; pellets, ≥85% KOH, ACS reagent, Sigma-Aldrich, St. Louis, MO).
4. Glacial acetic acid (ACS reagent, ≥99.7%, Sigma-Aldrich, St. Louis, MO).
5. Anhydrous dimethyl sulfoxide (DMSO; ≥99.9%, Sigma-Aldrich, St. Louis, MO).
6. Ethyl acetate (EtOAc; ACS reagent, ≥99.5%, Sigma-Aldrich, St. Louis, MO).
7. Methanol (anhydrous, 99.8%, Sigma-Aldrich, St. Louis, MO).
8. Concentrated HCl (37%, ACS reagent, Sigma-Aldrich, St. Louis, MO).
9. Phosphate-buffered saline (PBS) was purchased from UCSF's Cell Culture Facility.
10. Adenosine 5′-triphosphate disodium salt hydrate (ATP; 99%, Sigma-Aldrich, St. Louis, MO).
11. Magnesium acetate tetrahydrate ($Mg(OAc)_2$; ACS reagent, Sigma-Aldrich, St. Louis, MO).
12. Ethylenediaminetetraacetic acid (EDTA; anhydrous, BioReagent, Sigma-Aldrich, St. Louis, MO).

13. The 8-tosyl-octanoate precursor for [^{18}F]FA is prepared as previously described in Nagatsugi et al. [21]. Full synthetic protocols for the 8-tosyl-octanoate and nonradioactive 8-fluorooctanoate are available in Drake et al. [19].
14. The precursor for [^{18}F]FPOA is prepared as described in Drake et al. [20].
15. 5 N KOH: weigh 2.8 g of KOH into a graduated cylinder. Add methanol up to the 10 mL mark and completely dissolve the KOH. Store solution in a sealed plastic container.
16. 6 M HCl: dilute 5 mL 37% HCl solution with 5 mL of water. Store in a capped glass vial.
17. 600 μM LAP peptide solution (amino acid sequence Ac-G-FEIDKVWYDLDA-OH): dissolve 0.942 mg of LAP per 1 mL PBS. LAP peptide was custom-synthesized by Wuxi AppTech Co. Ltd. (Hong Kong).
18. 100 μM WT or W37I mutant LplA solution in PBS: dilute as required from a prepared stock solution (*see* **Note 2**). LplA (wild-type) can be produced as described in Fernández-Suárez et al. [14]. ^{W37I}LplA can be produced as described in Cohen et al. [17].
19. 30 mM ATP and 50 mM $Mg(OAc)_2$ reagent mixture in PBS: dissolve 16.53 mg ATP disodium salt hydrate and 10.72 mg magnesium acetate tetrahydrate per 1 mL of PBS (*see* **Note 3**).
20. 360 mM EDTA solution: dissolve 1.05 mg per 10 mL water.
21. CH_3CN–H_2O (1:1, v:v): Mix equal volumes of CH_3CN and water (e.g., 5 mL of each). Store in a capped glass vial.
22. CH_3CN–6 M HCl (1:1, v:v): Mix equal volumes of CH_3CN and 6 M HCl (from Subheading 2.3, **item 3**) (e.g., 5 mL of each). Store in a capped glass vial.
23. C_{18} light and Oasis HLB solid-phase extraction cartridges (Sep-Pak, Waters, Milford, MA). C_{18} light and Oasis HLB cartridges are preconditioned with 5 mL of acetonitrile and rinsed with 10 mL of water prior to use.

3 Methods

3.1 Azeotropic Drying of [^{18}F] Fluoride Ion

1. Pass the [^{18}F]fluoride ion/$H_2{}^{18}O$ solution obtained from the cyclotron through the quaternary methyl ammonium QMA column (*see* **Note 4**).
2. Remove excess $H_2{}^{18}O$ from the QMA column with 0.5 mL of air.
3. Elute QMA column with either 0.5 mL (for [^{18}F]FA) or 1.0 mL (for [^{18}F]FPOA) of the $K_2CO_3/K_{2.2.2}$ solution into a glass vial (1 dram, with phenolic screw-cap).

4. Azeotropically dry the [^{18}F]fluoride ion solution in the drying station depicted in Fig. 1. Attach the vial containing [^{18}F]fluoride ion/K_2CO_3/$K_{2.2.2}$ solution to the drying station. Apply a gently stream of nitrogen and turn on low volume vacuum pump.
5. Load 5 mL syringe with acetonitrile and attach 1/16″ tubing to go into the drying station.
6. Lower vial into heating bath (temperature 100–110 °C).
7. Evaporate [^{18}F]fluoride ion/K_2CO_3/$K_{2.2.2}$ solution to near dry (*see* **Note 5**).
8. Add 0.3–0.5 mL of acetonitrile.
9. Evaporate solution to near dry.
10. Repeat acetonitrile addition and evaporation steps up to 3 times (*see* **Note 6**).
11. Remove from heat for final drying.
12. Turn off vacuum and nitrogen stream.
13. Remove [^{18}F]fluoride ion vial from drying station.

3.2 [^{18}F]Fluorooctanoic Acid ([^{18}F]FA) Preparation

A schematic overview of the [^{18}F]FA synthesis scheme is depicted in Fig. 2.

1. Weigh out 2 mg 8-tosyloxy-octanoate precursor into a glass vial (1 dram, with phenolic screw-cap).
2. Add 0.3 mL of anhydrous acetonitrile to the precursor vial and dissolve the precursor.
3. Pipet the dissolved precursor into the dried [^{18}F]fluoride ion vial from Subheading 3.1, **step 13**.
4. Cap vial and heat at 90 °C for 10 min (*see* **Note 7**).
5. Cool vial over ice for 1–2 min and the dilute contents with ~5 mL water.
6. Load solution into the pHPLC loop.
7. Purify intermediate ethyl-8-[^{18}F]fluorooctanoate via reversed-phase semipreparative HPLC using radiometric detector. A Luna C18(2) column (10 mm × 250 mm, Phenomenex) should be used with gradient eluent from 60:40:0.1 CH_3CN–H_2O–TFA to 90:10:0.1 CH_3CN–H_2O–TFA over 20 min at a flow rate of 4 mL/min (*see* **Note 8**).

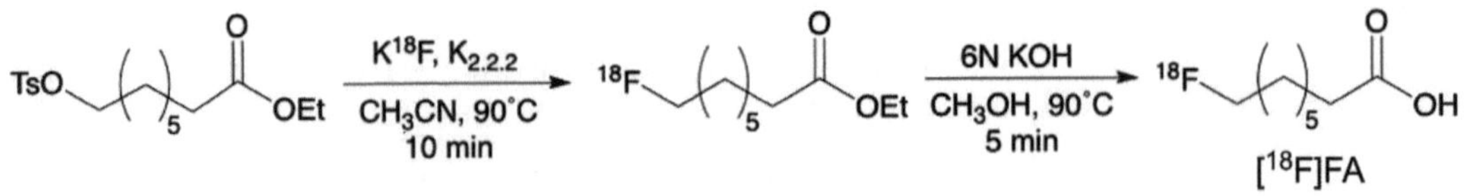

Fig. 2 Synthesis of 8-[^{18}F]fluorooctanoic Acid

8. Collect the purified ethyl-8-[^{18}F]fluorooctanoate peak and dilute with ~30 mL of water (*see* **Note 9**).
9. Load onto a preconditioned C18-light Sep-Pak.
10. Wash Sep-Pak with 10 mL water.
11. Elute the ethyl-8-[^{18}F]fluorooctanoate from the Sep Pak with 1 mL of methanolic 5 N KOH into a glass vial (1 dram, with phenolic screw-cap).
12. Cap vial and heat at 90 °C for 10 min (*see* **Note 10**).
13. Cool vial over ice for 1–2 min.
14. Neutralize solution with 750 μL of glacial acetic acid and dilute to ~30 mL with water.
15. Load the radioactive solution onto a preconditioned Oasis HLB Plus Light Sep Pak.
16. Wash Sep-Pak with 10 mL of water.
17. Elute 8-[^{18}F]fluorooctanoic acid with 2 mL of CH_3CN.
18. Concentrate CH_3CN solution to dryness using the drying station depicted in Fig. 1 at 50 °C under reduced pressure (*see* **Note 11**).
19. Reconstitute the dried 8-[^{18}F]fluorooctanoic acid in 130 μL PBS.
20. Analyze a portion of [^{18}F]-FA (~50–100 μCi) for Quality Control on analytical HPLC equipped with Luna C18(2) column (4.6 mm × 250 mm, Phenomenex), using an eluent of 45:65:0.1 CH_3CN–H_2O–TFA at a flow rate of 1 mL/min. Radiochemical purity is typically ~100%.

3.3 [^{18}F]FPOA Preparation

A schematic overview of the [^{18}F]FPOA synthesis procedure is depicted in Fig. 3.

1. Weigh out 5 mg [^{18}F]FPOA precursor into a glass vial (1 dram, with phenolic screw-cap).
2. Add 0.5 mL of anhydrous DMSO to the glass vial and dissolve the precursor.
3. Pipet the dissolved precursor into the dried [^{18}F]fluoride ion vial from Subheading 3.1, **step 13** (prepared using 1.0 mL of $K_2CO_3/K_{2.2.2}$ solution, *see* Subheading 3.1, **step 3**).

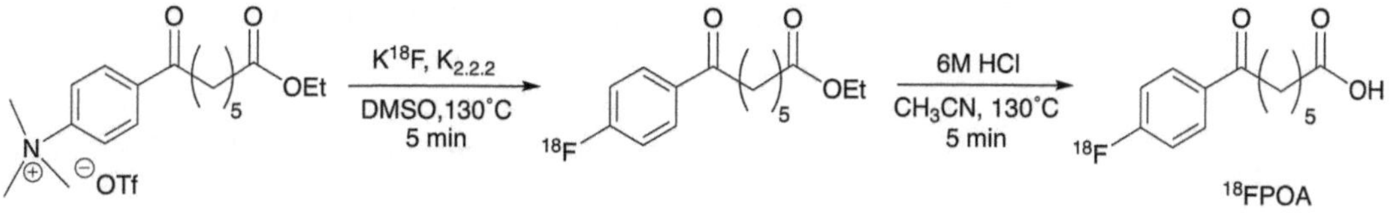

Fig. 3 Synthesis of [^{18}F]FPOA

4. Cap vial and heat at 130 °C for 10 min.
5. Cool vial over ice for 1–2 min and add 9 mL water (*see* **Note 12**).
6. Load onto a preconditioned Oasis HLB plus Sep-Pak.
7. Wash HLB Sep-Pak with 10 mL water.
8. Wash the HLB Sep-Pak with 3 mL CH_3CN–H_2O (1:1, v:v).
9. Elute ethyl-7-(4-[^{18}F]fluorophenyl)-7-oxoheptanoate intermediate from the HLB Sep-Pak with 3 mL of EtOAc into a glass vial (1 dram, with phenolic screw-cap).
10. Concentrate the EtOAc solution to near dryness using the drying station depicted in Fig. 1 at 80 °C under reduced pressure.
11. Rinse the HLB column with an additional 3 mL of EtOAc and add it to the glass vial.
12. Concentrate the EtOAc solution to dryness using the drying station depicted in Fig. 1 at 80 °C under reduced pressure.
13. Add 2 mL of CH_3CN–6 M HCl (1:1, v:v) to the glass vial.
14. Cap and heat the vial at 130 °C for 5 min (*see* **Note 10**).
15. Cool vial over ice for 1–2 min and add 3 mL of water to the vial.
16. Load solution onto the pHPLC loop.
17. Purify [^{18}F]FPOA via reversed-phase semipreparative HPLC with UV (254 nm) and radiometric detectors. A PRP-1 column (10 mm × 250 mm, Hamilton) should be used with gradient eluent from 20:80:0.1 CH_3CN–H_2O–TFA to 50:50:0.1 CH_3CN–H_2O–TFA over 30 min at a flow rate of 4 mL/min.
18. Collect the second half of the product [^{18}F]FPOA peak, visualized on radiometric chromatogram, and dilute with ~30 mL of water (*see* **Note 13**).
19. Load onto a preconditioned C18-light Sep-Pak.
20. Wash Sep-Pak with 20 mL water.
21. Elute the [^{18}F]FPOA from the Sep-Pak with 2 mL of CH_3CN.
22. Concentrate CH_3CN solution to dryness using the drying station depicted in Fig. 1 at 80 °C under reduced pressure.
23. Reconstitute [^{18}F]FPOA in 130 μL of PBS (*see* **Note 14**).
24. Analyze a portion of [^{18}F]FPOA for Quality Control on analytical HPLC equipped with a PRP-1 column (4.6 mm × 250 mm, Phenomenex) and UV detector (254 nm), using an eluent of 50:50:0.1 CH_3CN–H_2O–TFA at a flow rate of 1 mL/min (*see* **Note 15**).

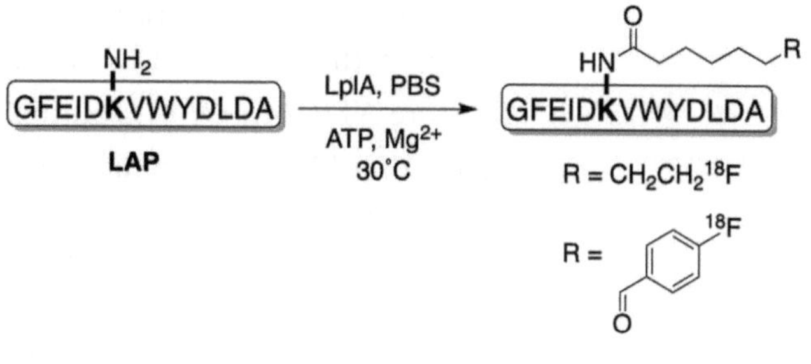

Fig. 4 Conjugation of 8-[^{18}F]fluorooctanoic acids to the LAP peptide catalyzed by LplA

3.4 Enzymatic Radiofluorination of LAP Peptide

In this section the experimental details are provided for radiolabeling of a model LAP peptide, as schematically illustrated in Fig. 4. Radiolabeling of other LAP-tagged peptides and proteins follows analogous conditions. Protocols to generate LAP-tagged species are dependent on the peptide/protein to be labeled. Suggested techniques include solid-phase synthesis (for peptides) and molecular cloning (for proteins; *see* Drake et al. [19] for example).

1. Add 130 μL of either [^{18}F]FA (from Subheading 3.2, **step 19**) or [^{18}F]FPOA (from Subheading 3.3, **step 23**) in PBS solution to a 1.5 mL eppendorf tube (*see* **Note 16**).
2. Add 20 μL of the LAP peptide solution (600 μM in PBS) to the eppendorf tube (final concentration = 60 μM) (*see* **Note 17**).
3. Add LplA to the eppendorf tube to a final concentration of 5 μM (volume added dependent on concentration of stock LplA solution) (*see* **Note 18**).
4. Add 20 μL of 30 mM ATP and 50 mM $Mg(OAc)_2$ reagent mixture in PBS to the tube (final concentration = 3 mM ATP, 5 mM $Mg(OAc)_2$).
5. Warm at 30 °C for 10–15 min.
6. Quench reaction by addition of 200 μL of 360 mM EDTA solution.
7. Analyze the radiolabeled peptide product (*see* **Note 19**).

4 Notes

1. Potassium carbonate/Kryptofix solution may be stored at 4 °C for up to 1 week.
2. LplA stock solution is currently available from University of California San Francisco (UCSF) upon request. The final volume of the labeling solution will be constant so the LplA solution volume will vary depending on the concentration of

the stock LplA solution. Aliquot stock solution to minimize freeze–thaw cycles and store at −80 °C.

3. Store ATP–$Mg(OAc)_2$ solution at −20 °C.
4. *Caution!* Minimize radiation exposure by using proper shielding, including a lead-lined hot cell, automated processes to maintain distance, and reduced direct radioisotope handling time.
5. It is best to not let the solution completely evaporate while the vial is in the heating bath to avoid excess residual fluorine-18 remaining "stuck" to a glass container. Leave a little solution volume and add acetonitrile.
6. After the last evaporation, remove the vial from the heat bath before all solution has evaporated and allow to completely dry under nitrogen stream and vacuum.
7. After the vial is cooled a radio-TLC (eluent 9:1, v:v, EtOAc–hexane) may be used to measure radiofluorination yield. Values of >70% are expected.
8. The UV detector can be used (set at 220 nm) but is not critical as ethyl-8-fluorooctanoate does not have a chromophore so it is typically not observed on the UV chromatograph. The radiolabeled product is easily detected by the radiometric detector. Under the described conditions, ethyl-8-[^{18}F]fluorooctanoate has a retention time of ~17 min.
9. It is expected that ~25–30% of the starting activity is collected off from the pHPLC as ethyl-8-[^{18}F]fluorooctanoate.
10. Conversion of ethyl ester to analogous carboxylic acid is quantitative.
11. Carefully evaporate the solvent as the 8-[^{18}F]fluorooctanoic acid is volatile. Vigorous evaporation may cause significant losses of the radioactive product.
12. After the vial is cooled a radio-TLC (eluent 9:1, v:v, EtOAc–hexane) may be used to measure radiofluorination yield. Values of 75–85% are expected.
13. In order to ensure high chemical purity of [^{18}F]FPOA, only the second half of the radiometric peak should be collected and carried forward. The first half of the radiometric peak should be collected separately and not used.
14. Up to 10 μL of DMSO can be added to aid [^{18}F]FPOA solubilization. The total volume of the solution (PBS + DMSO) should always be 130 μL.
15. Radiochemical purity is typically >97% and chemical purity should be >90%. A known by-product of [^{18}F]FPOA is 7-[4-(dimethylamino)phenyl]-7-oxyheptanoic acid, presumably formed via demethylation of the [^{18}F]FPOA precursor

and subsequent ester hydrolysis. It elutes immediately prior to [^{18}F]FPOA on HPLC. Due to the substrate promiscuity of LplA enzymes, this impurity is likely to compete with [^{18}F] FPOA for LAP-tag labeling, potentially suppressing conjugation yields. If chemical purities of <80% are observed for [^{18}F] FPOA due to the presence of this impurity, delay collection of the [^{18}F]FPOA peak from semipreparative HPLC.

16. If one wants to aliquot the [^{18}F]-labelled prosthetic to perform multiple labeling studies or use less activity, then add desired volume of activity solution and dilute to 130 μL with PBS.
17. Optimal peptide/protein concentration for radiolabeling must be determined individually for each species. In general, to maximize molar activities, the lowest possible concentration of peptide/protein which gives high conjugation yields (>80% is a reasonable benchmark) should be used.
18. [^{18}F]FA should be used in conjunction with wild-type LplA; [^{18}F]FPOA should be used in conjunction with engineered LplA mutant ^{W37I}LplA. Use of incorrect prosthetic/enzyme pairing may result in suppressed conjugation yields.
19. Conjugation yields may be measured using radio-TLC, silica TLC plates, and an eluent of 7:3:0.1, v:v, EtOAc–hexane–acetic acid. Under these conditions, unconjugated prosthetic has an R_f value of 0.6–0.7, radiofluorinated bioconjugates have an R_f value of <0.1. To ensure accurate results, TLC samples (~3 μL is sufficient) must be treated with SDS-PAGE reducing buffer (~1 μL) and heated at 95 °C for 5 min prior to analysis. It is presumed that both [^{18}F]FA and [^{18}F]-FPOA can bind noncovalently to higher molecular weight proteins (including LplA), potentially resulting in false positives for the radio-TLC assay. Denaturing all proteins prior to TLC analysis ensures accurate measurements.

    Purification schemes must be designed and tailored to the specific peptide/protein being radiolabeled. Semipreparative HPLC is suggested for low molecular weight peptides. Size-exclusion chromatography may be suitable for higher molecular weight proteins, however care must be taken to ensure LplA is adequately separated. An alternative approach for higher molecular weight species, utilizing nickel-affinity chromatography, is described in Drake et al. [19] and requires the use of LplA enzymes engineered with no His_6-tags.

    A straightforward SDS-PAGE assay to measure residual LplA in radiotracer formulations in described in Drake et al. [19].

Acknowledgments

This work was supported by a seed grant from the UCSF Department of Radiology and Biomedical Imaging, the UCSF Resource Allocation Program, the US Department of Energy (DOE DE-SC002061), and NIH National Institute of Biomedical Imaging and Bioengineering SBIR (R43EB023782). Additional support for C.R.D. was received from the National Cancer Institute (HHSN261201400041C) and National Institute of Mental Health (2R44MH097271). M.J.E. was supported by the National Institute of Mental Health (R01MH115043) and the National Institute of Biomedical Imaging and Bioengineering (R01EB025207).

References

1. Fani M, Maecke HR, Okarvi SM (2012) Radiolabeled peptides: valuable tools for the detection and treatment of cancer. Theranostics 2:481–501
2. Jackson IM, Scott PJH, Thompson S (2017) Clinical applications of radiolabeled peptides for PET. Semin Nucl Med 47:493–523
3. Mullard A (2016) FDA approvals for the first 6 months of 2016. Nat Rev Drug Disc 15:523
4. Vaidyanathan G, Zalutsky MR (2006) Synthesis of N-succinimidyl 4-[^{18}F]fluorobenzoate, an agent for labeling proteins and peptides with ^{18}F. Nat Protoc 1:1655–1661
5. Vaidyanathan G, Zalutsky MR (1992) Labeling proteins with fluorine-18 using N-succinimidyl 4-[^{18}F]fluorobenzoate. Nucl Med Biol 19:275–281
6. Berndt M, Pietzsch J, Wuest F (2007) Labeling of low-density lipoproteins using the ^{18}F-labeled thiol-reactive reagent N-[6-(4[^{18}F] fluorobenzylidene)aminooxy-hexyl]maleimide. Nuc Med Biol 34:5–15
7. Gill HS, Tinianow JN, Ogasawara A et al (2009) A modular platform for the rapid site-specific radiolabeling of proteins with ^{18}F exemplified by quantitative positron emission tomography of human epidermal growth factor receptor 2. J Med Chem 52:5816–5825
8. Jeon J, Shen B, Xiong L et al (2012) Efficient method for site-specific ^{18}F-labeling of biomolecules using the rapid condensation reaction between 2-cyanobenzothiazole and cysteine. Bioconjug Chem 23:1902–1908
9. Hausner SH, Marik J, Gagnon MKJ, Sutcliffe JL (2008) In vivo positron emission tomography (PET) imaging with an $\alpha_v\beta_6$ specific peptide radiolabeled using ^{18}F-"click" chemistry: evaluation and comparison with the corresponding 4-[^{18}F]Fluorobenzoyl- and 2-[^{18}F]Fluoropropionyl- peptides. J Med Chem 51:5901–5904
10. Liu S, Hassink M, Selvaraj R et al (2013) Efficient ^{18}F labeling of cysteine-containing peptides and proteins using tetrazine-trans-cyclooctene ligation. Mol Imag 12:121–128
11. Marik J, Sutcliffe JL (2006) Click for PET: rapid preparation of [F-18]fluoropeptides using Cu-I catalyzed 1,3-dipolar cycloaddition. Tet Lett 47:6681–6684
12. Wängler C, Niedermoser S, Chin J et al (2012) One-step ^{18}F-labeling of peptides for positron emission tomography imaging using the SiFA methodology. Nat Protoc 7:1946–1955
13. Wängler C, Waser B, Alke A et al (2010) One-step ^{8}F-labeling of carbohydrate-conjugated octreotate-derivatives containing a silicon-fluoride-acceptor (SiFA): in vitro and in vivo evaluation as tumor imaging agents for positron emission tomography (PET). Bioconjug Chem 21:2289–2296
14. Fernández-Suárez M, Baruah H, Martínez-Hernández L et al (2007) Redirecting lipoic acid ligase for cell surface protein labeling with small-molecule probes. Nat Biotech 25:1483–1487
15. Uttamapinant C, White KA, Baruah H et al (2010) A fluorophore ligase for site-specific protein labeling inside living cells. PNAS 107:10914–10919
16. Puthenveetil S, Liu DS, White KA et al (2009) Yeast display evolution of a kinetically efficient 13-amino acid substrate for lipoic acid ligase. J Am Chem Soc 131:16430–16438

17. Cohen JD, Zou P, Ting AY (2012) Site-specific protein modification using lipoic acid ligase and bis-aryl hydrazone formation. Chembiochem 13:888–894
18. Cohen JD, Thompson S, Ting AY (2011) Structure-guided engineering of a Pacific blue fluorophore ligase for specific protein imaging in living cells. Biochemist 50:8221–8225
19. Drake CR, Sevillano N, Truillet C et al (2016) Site-specific radiofluorination of biomolecules with 8-[^{18}F]-Fluorooctanoic acid catalyzed by Lipoic acid ligase. ACS Chem Biol 11:1587–1594
20. Drake CR, Sevillano N, Blecha J, et al (Submitted) Second-generation ^{18}F-labeled prosthetic for automated enzymatically catalyzed radiolabeling of proteins on the ELIXYS platform.
21. Nagatsugi F, Sasaki S, Maeda M (1994) 8-[F-18] Fluorooctanoic acid and its Beta-substituted derivatives as potential agents for cerebral fatty-acid studies—synthesis and biodistribution. Nucl Med Biol 21:809–817

Chapter 14

Site-Selective Enzymatic Labeling of Designed Ankyrin Repeat Proteins Using Protein Farnesyltransferase

Yi Zhang, Shelby Auger, Jonas V. Schaefer, Andreas Plückthun, and Mark D. Distefano

Abstract

Affinity agents coupled to a functional moiety play an ever-increasing role in modern medicine, ranging from radiolabeled selective binders in diagnosis to antibody–drug conjugates in targeted therapies. In biomedical research, protein coupling to fluorophores, surfaces and nanoparticles has become an integral part of many procedures. In addition to antibodies, small scaffold proteins with similar target binding properties are being widely explored as alternative targeting moieties. To label these binders of interest with different functional moieties, conventional chemical coupling methods can be employed, but often result in heterogeneously modified protein products. In contrast, enzymatic labeling methods are highly site-specific and efficient. Protein farnesyltransferase (PFTase) catalyzes the transfer of an isoprenoid moiety from farnesyl diphosphate (FPP) to a cysteine residue in a C-terminal CaaX motif at the C-terminus of a protein substrate. The addition of only four amino acid residues minimizes the influence on the native protein structure. In addition, a variety of isoprenoid analogs containing different bioorthogonal functional groups, including azides, alkynes, and aldehydes, have been developed to enable conjugation to various cargos after being incorporated onto the target protein by PFTase. In this protocol, we present a detailed procedure for labeling Designed Ankyrin Repeat Proteins (DARPins) engineered with a C-terminal CVIA sequence using an azide-containing FPP analog by yeast PFTase (yPFTase). In addition, procedures to subsequently conjugate the labeled DARPins to a TAMRA fluorophore using strained-promoted alkyne–azide cycloaddition (SPAAC) reactions as well as the sample preparation to evaluate the target binding ability of the conjugates by flow cytometry are described.

Key words Enzymatic protein labeling, Protein farnesyltransferase, Isoprenoid analogs, Site-specific conjugation, DARPins, SPACC, Flow cytometry

1 Introduction

With the rapid development of targeted therapeutics and diagnostic agents in modern medicine, and the large number of procedures in biomedical research requiring specific binding reagents, there is an increasing demand for protein-based conjugates. To label a protein of interest, conventional chemical modification methods exploit the side-chain functional groups of certain amino acids. Two of

Sam Massa and Nick Devoogdt (eds.), *Bioconjugation: Methods and Protocols*, Methods in Molecular Biology, vol. 2033, https://doi.org/10.1007/978-1-4939-9654-4_14,

the commonly targeted residues include cysteines that can undergo Michael-type addition with maleimides [1] and lysines reacting with *N*-hydroxylsuccinimide esters [2]. However, due to the lack of specificity, when multiple copies of the same amino acid are present on the target protein, which often is the case for lysine residues, a heterogeneously labeled product is obtained. As to reactions with maleimides, although they are highly specific toward cysteine residues under acidic and neutral conditions, alkylation of primary amines, such as the side chain amines in the lysine residues, can also occur at a higher pH (pH > 8) [3] or excess reaction time, resulting in a mixture of differently modified proteins. While such complex compositions obviously complicate the characterization, an uncontrolled conjugation can also result in reduced efficacy of therapeutic proteins compared to counterparts modified using site-specific labeling methods [4].

To reliably achieve site-specific conjugation, enzymatic labeling methods have been developed and are widely utilized [5]. Since most of the enzymes employed for protein labeling have a preferred recognition sequence [5], the modification site as well as the labeling stoichiometry can be precisely controlled. In addition, reactions catalyzed by enzymes occur under mild conditions with fast rates. To date, several enzymes have been employed for protein labeling purpose, including sortase A [6, 7], transglutaminase [8, 9], protein farnesyltransferase (PFTase) [10, 11], and tubulin tyrosine ligase [12] as well as related methods such as expressed protein ligation [13].

PFTase catalyzes the transfer of the isoprenoid unit from farnesyl diphosphate (FPP) to a cysteine residue near the C-terminus of the protein substrate, resulting in thioether bond formation. The enzyme recognition sequence is composed of a C-terminal CaaX motif, where “C” is a cysteine residue, “a” represents aliphatic amino acids, and “X” determines whether the protein is a substrate for PFTase (X = Ala, Ser, or Met) or geranylgeranyltransferase type I (X = Leu, Ile, or Phe) [14]. Since the enzyme is highly selective toward its recognition sequence, only the cysteine residue within the CaaX motif will be modified even in the presence of other cysteines in the protein of interest. It has been shown that the incorporation of the CaaX motif, such as the CVIA sequence which is found in a naturally prenylated peptide [15], onto the C-terminus of a given protein of interest makes it a substrate for PFTase [16, 17]. The addition of only four amino acids minimizes potential perturbation to the structure and function of the genetically engineered target proteins. While the PFTase active site cannot accommodate large moieties such as biotin, drugs or fluorophores (although it can be modified to do so via protein engineering [18, 19]), the enzyme does accept smaller modifications of the isoprenoid substrate. To apply the PFTase reaction for protein modification, our laboratory and others have developed a variety of isoprenoid analogs containing different bioorthogonal functional groups, including azides [20, 21], alkynes [22, 23], and aldehydes [10, 24]. These analogs can be efficiently incorporated to targeted proteins bearing a C-terminal CaaX motif, enabling subsequent conjugation for different applications.

Using the PFTase labeling strategy, protein conjugation to various functionalities has been achieved, including fluorophores [10], DNA oligonucleotides [25], PEG polymers [26], and solid surfaces [27]. More recently, protein–drug conjugates created via the PFTase labeling method have also been reported as potential targeted therapeutics [28]: a repebody was engineered with a C-terminal CVIM sequence, enzymatically labeled with a ketone functional group by PFTase and subsequently reacted with aminooxy-functionalized monomethyl auristatin F (MMAF) to yield a stable and homogeneous protein–drug conjugate with anti-tumor activities in vivo.

As a robust and general labeling method, we applied the PFTase enzyme to modify Designed Ankyrin Repeat Proteins (DARPins) (Fig. 1) which can be selected in vitro for different target binding with high affinity and specificity and that have been used for a variety of applications [29]. As a class of small binding protein scaffolds, DARPins are extremely stable, can easily be engineered as well as expressed in *E. coli* in high yield [30].

In this protocol, we describe in detail the procedure to label DARPins bearing an engineered C-terminal CVIA sequence with an azide-containing isoprenoid analog using PFTase. The subsequent conjugation of the azide-modified DARPins to a dibenzocyclooctyne (DBCO)-functionalized fluorophore is also explained. Finally, we present the sample preparation procedure to examine the target binding ability of the DARPin-fluorophore conjugates by flow cytometry using an epithelial cell adhesion molecule (EpCAM)-binding DARPin [31] as an example.

2 Materials

2.1 Reagents and Buffers for Protein Labeling

1. DARPin engineered with a C-terminal CVIA sequence, stored in PBS_{400} supplemented with 10% glycerol.
2. C10-Azi analog **1** (Fig. 1). The synthesis procedure has been previously published [32]. Analog **1** is stored in 25 mM NH_4HCO_3 solution with concentrations typically ranging from 1 mM to 10 mM.
3. Yeast farnesyltransferase (yPFTase). The expression and purification methods have been previously published [18]. The yPFTase is stored in a storage buffer (25 mM Tris–HCl, pH 7.5, 100 mM·NaCl, 2.5 μM·$ZnCl_2$, 2.5 mM·$MgCl_2$, 1 mM β-mercaptoethanol, 50% glycerol (v/v)) with concentrations typically ranging from 200 μM to 500 μM.

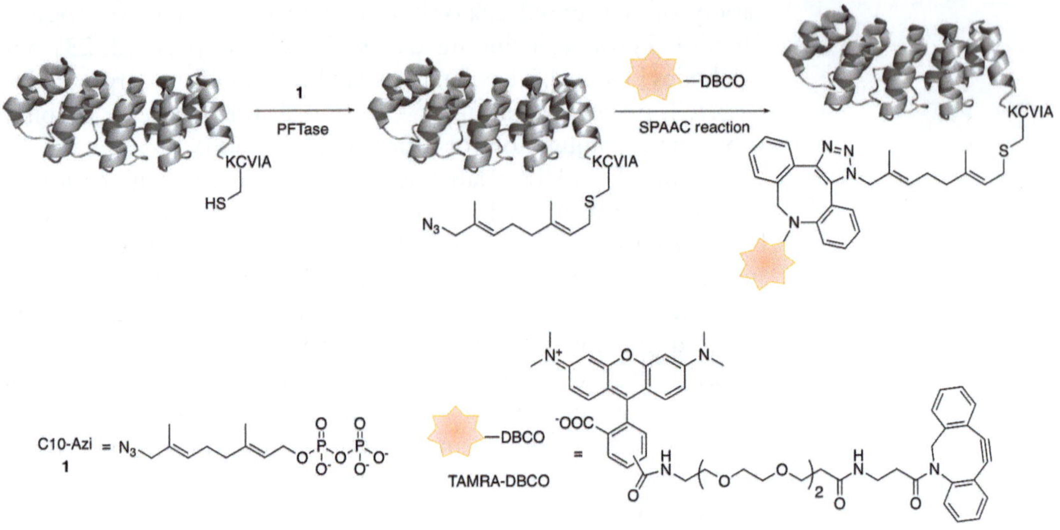

Fig. 1 Schematic summary of the enzymatic labeling of DARPins engineered with a C-terminal CVIA sequence to C10-Azi analog **1** by using PFTase and subsequent conjugation with DBCO-TAMRA using the SPAAC reaction

4. Bradford assay dye (Bio-Rad).
5. TAMRA-DBCO (Click Chemistry Tools LLC.).
6. Prenylation buffer (6×): 300 mM Tris–HCl, pH 7.5, 120 mM·KCl, 60 mM $MgCl_2$, 30 mM dithiothreitol (DTT), 60 μM·$ZnCl_2$ (*see* **Note 1**).
7. Phosphate-buffered saline (1× PBS): 8.1 mM·Na_2HPO_4, 1.5 mM·KH_2PO_4, pH 7.4, 137 mM·NaCl, 2.7 mM·KCl.
8. PBS_{400}: 1× PBS, 400 mM NaCl (total NaCl concentration).
9. Buffers for LC-MS analysis. Buffer A: H_2O with 0.1% formic acid (FA); Buffer B: acetonitrile with 0.1% FA.
10. 4× Laemmli loading buffer: 8% (w/v) sodium dodecyl sulfate, 40% (v/v) glycerol, 0.04% (w/v) bromophenol blue, 200 mM Tris–HCl, pH 6.8 (no reducing agent present).
11. 0.5 M DTT (*see* **Note 1**).
12. Molecular weight ladder (e.g. precision plus protein dual color standards (Bio-Rad)).
13. Coomassie Blue stain solution: 0.2% (w/v) Coomassie Brilliant Blue R250, 45% (v/v) H_2O, 45% (v/v) methanol, 10% (v/v) acetic acid.
14. Destaining buffer: 60% H_2O, 30% methanol and 10% acetic acid (v/v).
15. PBSA buffer: 1× PBS containing 1 mg/mL bovine serum albumin (BSA).

2.2 Reagents and Buffers for Mammalian Cell Cultures

1. DMEM: Dulbecco's Modified Eagle's Medium.
2. FBS: fetal bovine serum.
3. Pen/Strep: 10,000 U penicillin, 10 mg/mL streptomycin.
4. 10× trypsin (2.5%), without EDTA.
5. Versene buffer: 1× PBS, 0.6 mM EDTA.

2.3 Instruments

1. Amicon Ultra centrifugal filters (0.5 mL, 3 kDa molecular weight cut off, Millipore Sigma).
2. NAP-5 columns (GE Healthcare).
3. Isotemp water bath.
4. Microcentrifuge.
5. Microplate reader.
6. UV-Vis spectrometer.
7. Fluorescence scanner: Typhoon FLA 9500 with 532 laser and LPG (575 LP) filter.
8. LC-MS instrument: Agilent 1100 Series LC/MSD trap SL (equipped with a UV-Vis detector); column: Zorbax 300 SB-C8 capillary 0.3 × 100 mm, 3.5 μm (Agilent); and sample vials: interlocked vials with 300 μL fused glass insert (Chrome Tech).
9. CO_2 incubator, microscope and biosafety hood.
10. Flow cytometer (equipped with 561 nm laser): BD LSRFortessa H0081.

3 Methods

3.1 Enzymatic Labeling of DARPin with C10-Azi Analog 1

The following protocol describes a 2 mL-scale prenylation reaction.

1. Reduction: Prepare the reduction reaction in a 5 mL conical tube. Dilute the DARPin engineered with a C-terminal CVIA sequence to 15 μM in prenylation buffer (6×), which contains DTT. Incubate the tube on ice for 30 min. After the reduction reaction, dilute the reaction mixture 6× by adding 1.65 mL of H_2O. This will result in a 2 mL substrate mixture containing 2.5 μM of reduced DARPin (*see* **Note 2**).
2. Prenylation: Add analog **1** (15 μM final concentration, *see* **Note 3**) and yPFTase (300 nM final concentration, *see* **Notes 4–6**) to initiate the reaction. Incubate the tube in a 32 °C water bath for 6 h (*see* **Note 7**).
3. Analyze the product by injecting the reaction mixture directly into LC-MS without prior purification to confirm the formation of the desired product (*see* an example in Fig. 2b) *see* Subheading 3.4 for detailed LC-MS method.

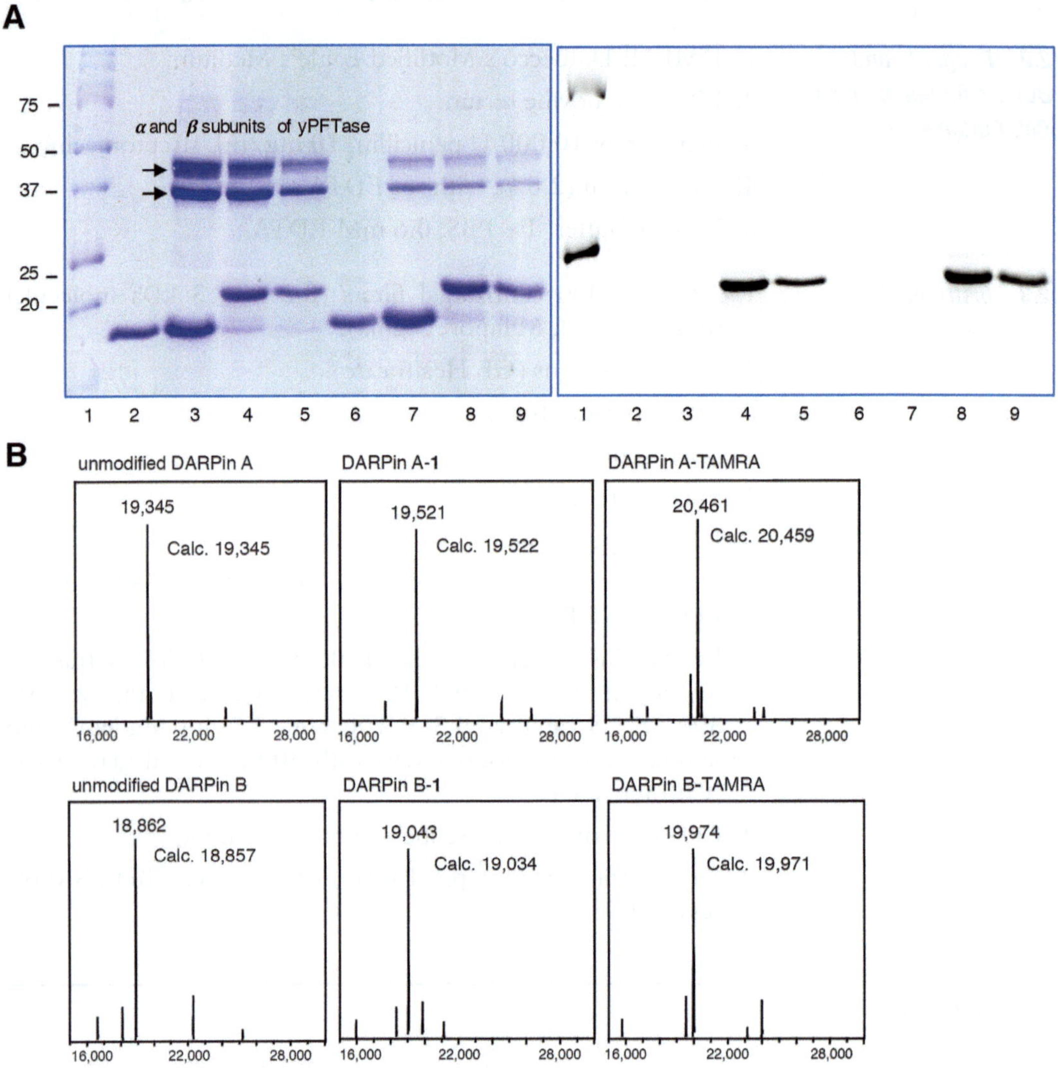

Fig. 2 (**a**) Coomassie-stained gel from SDS-PAGE analysis (left) and in-gel fluorescence imaging (right) of two azide-modified DARPin variants A-**1** and B-**1** and DARPin-TAMRA conjugates A-TAMRA and B-TAMRA. Lane 1: Precision plus protein dual color standards. Lane 2: unmodified DARPin B. Lane 3: DARPin B-**1** prenylation mixture. Lane 4: DARPin B-TAMRA reaction mixture. Lane 5: DARPin B-TAMRA subjected to partial purification (buffer exchange and removal of TAMRA reagent excess) using a NAP-5 column. Lane 6: unmodified DARPin A. Lane 7: DARPin A-**1** prenylation mixture. Lane 8: DARPin A-TAMRA reaction mixture. Lane 9: DARPin A-TAMRA partially purified with NAP-5 column (buffer exchange and removal of TAMRA reagent excess). The fluorescence band from lane 4/5 and lane 8/9 confirmed successful conjugation. It should be noted that the yPFTase concentration in DARPin B labeling was increased from 300 nM to 800 nM to ensure efficient modification. (**b**) ESI-MS of the unmodified DARPins, azide-modified DARPins, and the DARPin-TAMRA conjugates. The observed masses closely match to the calculated values, suggesting successful modifications. The enzymatic prenylation efficiency was quantified based on the absorbance chromatogram at 280 nm obtained in the LC-MS analysis by comparing the peak areas for the unmodified DARPin and the DARPin-**1**. For DARPin A protein, labeling efficiency ranges from 85% to 90% using 300 nM to 500 nM yPFTase. The prenylation for DARPin B protein is less efficient with conversion around 70% using 500 nM yPFTase. Concerning the SPAAC reactions, when both DARPin A-TAMRA and DARPin B-TAMRA conjugates were subjected to LC-MS analysis, no mass corresponding to the unreacted DARPin A-**1** or DARPin B-**1** conjugates could be detected, suggesting high reaction yield for the SPAAC reactions

4. To concentrate the solution, load 500 μL of the reaction mixture into a 0.5 mL centrifugal filter (3 kDa cutoff) and centrifuge at 13,000 × *g* for 5 min. Refill the centrifugal filter with another 300 μL of the reaction mixture and centrifuge for 5 min. Repeat this procedure until all the reaction mixture solution is concentrated. Concentrate the final solution to approximately 100 μL.
5. To remove excess isoprenoid analogs present in the concentrated solution, wash the centrifugal filter with 400 μL PBS_{400} buffer and centrifuge for 5 min at 13,000 × *g*. Repeat this procedure two more times. Concentrate the final solution to approximately 100 μL.
6. To determine the concentration of the azide-modified DARPin (DARPin-**1**), several steps are employed (*see* **Note 8**).
   (a) Measure the total protein concentration (total protein in mg/mL) in the solution using a Bradford assay with a microplate reader.
   (b) Calculate the theoretical mass ratio of the DARPin protein to the total amount of protein present in the prenylation reaction (**step 2**) based on the amounts of proteins added: R = [mass of DARPin]/([mass of DARPin] + [mass of yPFTase]). Note that the molecular masses of DARPin and yPFTase are approximately 20 and 80 kDa, respectively.
   (c) Determine the prenylation efficiency (E) by comparing the peak area of UV absorption at 280 nm for the azide-modified DARPin and the unreacted DARPin from LC-MS analyses (*see* an example in Fig. 3).
   (d) Calculate the concentration of DARPin-**1** conjugate:

$$[\text{DARPin-}\mathbf{1}] = [\text{total protein}] \times R \times E.$$

3.2 Construction of DARPin-Fluorophore Conjugates Using SPAAC Reaction

1. Prepare a 10 mM stock solution of DBCO-TAMRA in DMSO.
2. To the resulting DARPin-**1** solution (100 μL) in **step 5** of Subheading 3.1 (*see* **Note 9**), add a fivefold (molar ratio) excess of DBCO-TAMRA based on the calculated concentration of DARPin-**1** conjugate.
3. Cover the reaction tube with aluminum foil and incubate it at room temperature for 3 h.
4. To remove excess fluorophore, load the reaction mixture (100 μL) onto a NAP-5 column (*see* **Note 10**), pre-equilibrated with 1× PBS. Add 400 μL of 1× PBS and then elute the proteins with 500 μL of 1× PBS.
5. Analyze the mass of the DARPin-TAMRA conjugates by LC-MS to verify the formation of the desired product.

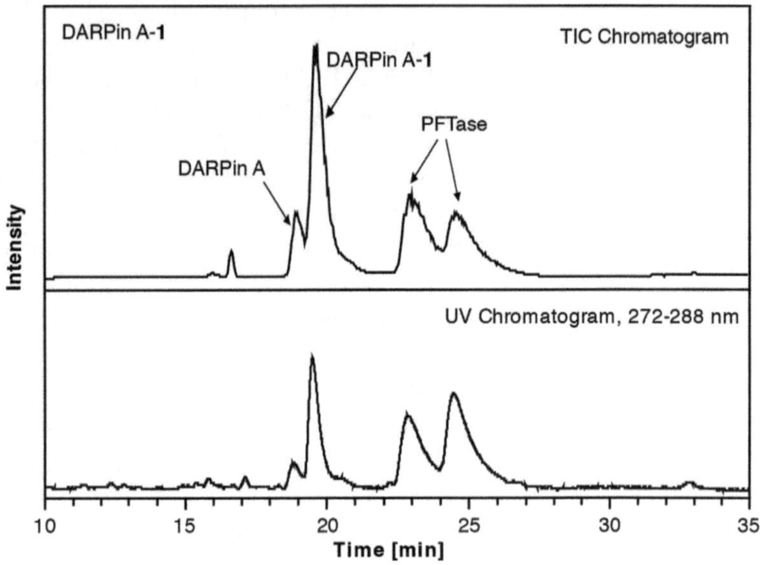

Fig. 3 Total ion chromatogram (TIC) and UV chromatogram of DARPin A-**1** conjugates. The prenylation efficiency estimated from UV chromatogram is approximately 85%

6. Determine the concentration of the DARPin-TAMRA conjugates by measuring the absorbance of the solution in 1× PBS at 555 nm using a UV-Vis spectrophotometer. (ε_{555} = 65,000 $M^{-1} \cdot cm^{-1}$).

3.3 Procedure for Characterization of DARPin-TAMRA Conjugates by SDS-PAGE and In-Gel Fluorescence Imaging

1. Based on the concentration of DARPin-TAMRA conjugates, calculate the volume for 3 μg of conjugates. If the volume is smaller than 10 μL, add H_2O to reach 10 μL.
2. Aliquot the desired amount of DARPin-TAMRA into an Eppendorf tube. Add 5 μL of 4× Laemmli loading buffer and 5 μL of 0.5 M DTT to reach a final volume of 20 μL.
3. Heat the samples at 95 °C for 10 min and load onto a 15% acrylamide (3″ × 4″) SDS-PAGE gel for electrophoresis at a constant voltage of 120 V.
4. Before staining the gel with Coomassie Blue, scan the gel for TAMRA fluorescence (553/575 nm excitation/emission).
5. Stain the gel with Coomassie Blue stain solution for 20 min, followed by destaining in destaining buffer for 2 to 4 h (*see* an example in Fig. 2a).

3.4 Procedure for LC-MS Analysis of Protein Samples

1. Dilute the protein samples to be analyzed to 3 μM (*see* **Note 11**) and centrifuge at 10,000 × *g* for 2 min to remove any precipitate.
2. Aliquot 20 μL of solution into sample vials. Inject 8 μL for LC-MS analysis.

3. LC gradient for LC-MS analysis: 0 to 2 min, 10% buffer B; 2 to 23 min, 10 to 70% buffer B; 23 to 25 min, 70 to 90% buffer B; 25 to 35 min, 90% buffer B. Flow rate: 9 μL/min. UV-Vis detection: 280 and 555 nm.

3.5 Procedure for Sample Preparation for Flow Cytometry Analysis to Examine DARPin Binding to Cell Surface EpCAM

1. Prepare TAMRA conjugates of the EpCAM-binding DARPin (DARPin A-TAMRA) as well as of a control DARPin (not binding to any target, DARPin B-TAMRA) using the methods described above.
2. Culture MCF-7 (EpCAM overexpressing cells) and U87-MG cells (EpCAM negative) in DMEM media supplied with 10% FBS and 1% Pen/Strep. Incubate the cells at 37 °C in a moisture-controlled environment with a supply of 4% CO_2.
3. When the cells reach 80% confluence, detach the cells using 1× trypsin solution (diluted in Versene buffer from 10× stock). Count and aliquot 400,000 cells into separate vials (a total of four vials for MCF-7 cells and three vials for U87-MG cells, *see* **Note 12**). These samples should be kept on ice at all times.
4. Centrifuge the cells at 300 × *g* for 3 min. Discard the media. Wash the cells by resuspending in 1 mL of ice-cold PBSA, followed by centrifugation to remove the supernatant.
5. Add 400 μL of the sample solution to each vial so that the concentration of cells is 1×10^6 cells/mL. For the control sample with no conjugate treatment, add PBSA. This sample is used to measure the background fluorescence for each cell line. For the positive binding sample, add 400 μL of 100 nM DARPin A-TAMRA (diluted in PBSA). For the control binding sample, add 400 μL of 100 nM DARPin B-TAMRA (diluted in PBSA).
6. Incubate the cells at 4 °C on a rotary shaker in the dark for 45 min.
7. For blocking experiments with MCF-7 cells (*see* **Note 12**), first incubate the cells in 200 μL of 2 μM unlabeled DARPin A at 4 °C on a rotary shaker in dark. After 15 min, add 200 μL of 200 nM DARPin A-TAMRA for an additional 30 min incubation.
8. Wash the cells with 1 mL ice-cold PBSA and resuspend them in 400 μL PBSA for flow cytometry analysis (*see* an example in Fig. 4).

4 Notes

1. To maintain the reducing ability of DTT, a stock solution of 0.5 M in H_2O is prepared, aliquoted into small volumes (30 μL) and stored at −20 °C. Thaw individual aliquots when needed and discard the tubes after one-time usage.

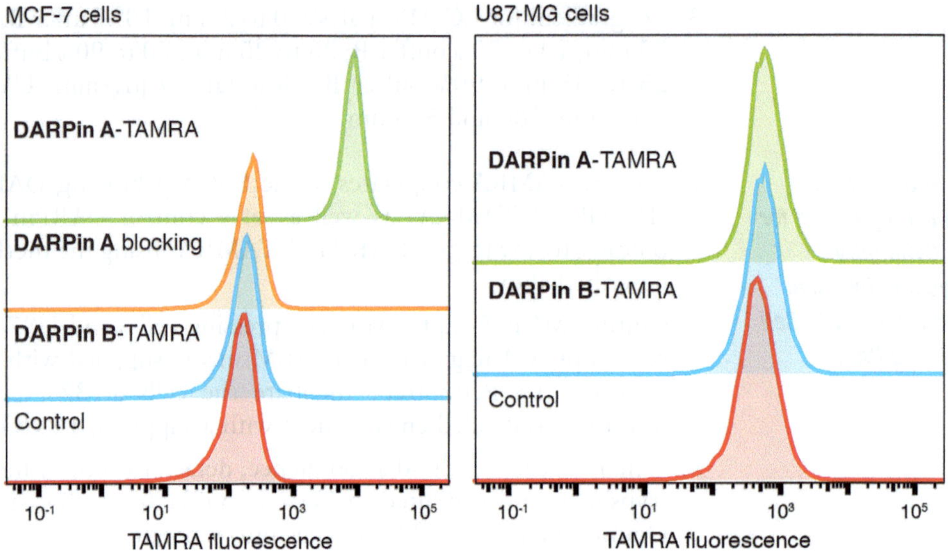

Fig. 4 Evaluation of DARPin A-TAMRA binding to cell surface EpCAM by flow cytometry. Two cell lines were utilized, MCF-7 cells with EpCAM overexpression (left) and U87-MG cells as negative control (right). DARPin A binds to cell surface EpCAM while DARPin B is a control protein that does not bind to any target. The control sample for each cell line contains cells treated with only PBSA buffer. The results confirmed the ability and specificity of DARPin A-TAMRA binding to cell surface EpCAM without non-specific interactions

2. It is critical to incubate the protein substrate with DTT on ice before the addition of the yPFTase enzyme to fully reduce any disulfide bonds involving the CaaX box and make sure that the cysteine residue is accessible for the prenylation reaction. Unlike in reactions with maleimides, the excess amount of DTT does not need to be removed prior to the prenylation reaction. To maximize the concentrations of DTT and the DARPin protein during reduction (prior to prenylation), the reaction is performed in a concentrated prenylation buffer (6×). Therefore, the tube contains Tris–HCl, pH 7.5 (200 μL of 0.5 M stock), KCl (80 μL of 0.5 M stock), $MgCl_2$ (20 μL of 1 M stock), DTT (20 μL of 0.5 M stock), $ZnCl_2$ (10 μL of 2 mM stock), and DARPin (22 μL of 224 μL stock).

   Alternatively, TCEP can also be used as the reducing agent [33].

3. In addition to the C10-Azi analog **1** employed here, a C15-Azi analog with similar structure but a farnesyl backbone has also been previously reported to be a good substrate for yPFTase and can be used for protein labeling [21]. The precursor to make the C15-Azi analog (one-step esterification reaction of the alcohol precursor to generate the diphosphate compound) is commercially available from Cayman Chemical (farnesyl alcohol azide).

4. In addition to the yPFTase enzyme used in this protocol, rat PFTase (rPFTase) can also be employed to catalyze a prenylation reaction, which is commercially available from Jena Bioscience. However, yPFTase has a higher activity [34] and is more promiscuous toward different isoprenoid analogs than rPFTase.
5. Both the analog **1** and the yPFTase are added directed to the reaction mixture from the stock solution. Since both reagents are diluted approximately 1000-fold, the effect of their corresponding stock solutions to the composition of the prenylation buffer is negligible.
6. If the reaction efficiency results from LC-MS analysis (**step 3**) are not satisfying, increase the amount of yPFTase added to the prenylation reaction. A labeling yield of 90% or higher can be achieved.
7. Since the isoprenoid analog **1** is relatively hydrophobic, protein precipitation might occur after the conjugation depending on the target protein. To determine if there is precipitation, centrifuge the tube after prenylation reaction to check for any visible precipitate.
8. This method can only be used to estimate the concentration of the DARPin-**1** conjugate since it assumes that the DARPin and yPFTase give similar signals in the Bradford assay. However, this is not critical since the purpose of this calculation is to obtain an approximate value to determine the correct amount of DBCO-TAMRA to add for subsequent conjugation. As an alternative, the concentration of the DARPin-**1** conjugate can be estimated from the initial amount used in the prenylation reaction assuming complete prenylation and recovery.
9. The SPAAC reaction is performed in PBS_{400} buffer with high salt concentration that is used to store DARPin proteins (to prevent any potential aggregation of binders at very high concentrations). Alternatively, the reaction can also be carried out in 1× PBS buffer.
10. The NAP-5 column is a simple gel filtration column used to remove excess TAMRA fluorophore from the SPAAC reaction. To remove the yPFTase from DARPin-TAMRA conjugates more quantitatively, preparative size-exclusion columns, such as the Superdex 75 Increase column (GE Healthcare), can be employed on a fast protein liquid chromatography (FPLC) system.
11. Depending on the sensitivity of the LC-MS instrument employed, the concentration and volume of the protein samples being analyzed might have to be adjusted accordingly.

12. In order to check for specific binding, we recommend not just to test the binding of the labeled DARPins on both cell types but also to include a sample where the receptor is blocked in advance with a tenfold excess of unlabeled DARPins in a competition setup.

Acknowledgments

This work was supported by National Institutes of Health (GM084152) and the National Science Foundation (CHE-1359181).

References

1. Gunnoo SB, Annemieke M (2016) Chemical protein modification through cysteine. Chembiochem 17(7):529–553. https://doi.org/10.1002/cbic.201500667
2. Baslé E, Joubert N, Pucheault M (2010) Protein chemical modification on endogenous amino acids. Chem Biol 17(3):213–227. https://doi.org/10.1016/j.chembiol.2010.02.008
3. Brewer CF, Riehm JP (1967) Evidence for possible nonspecific reactions between N-ethylmaleimide and proteins. Anal Biochem 18(2):248–255. https://doi.org/10.1016/0003-2697(67)90007-3
4. Dozier JK, Distefano MD (2015) Site-specific PEGylation of therapeutic proteins. Int J Mol Sci 16(10):25831–25864. https://doi.org/10.3390/ijms161025831
5. Rashidian M, Dozier JK, Distefano MD (2013) Enzymatic labeling of proteins: techniques and approaches. Bioconjug Chem 24 (8):1277–1294. https://doi.org/10.1021/bc400102w
6. Zhang Y, Blanden MJ, Sudheer C et al (2015) Simultaneous site-specific dual protein labeling using protein prenyltransferases. Bioconjug Chem 26(12):2542–2553. https://doi.org/10.1021/acs.bioconjchem.5b00553
7. Glasgow JE, Salit ML, Cochran JR (2016) In vivo site-specific protein tagging with diverse amines using an engineered sortase variant. J Am Chem Soc 138(24):7496–7499. https://doi.org/10.1021/jacs.6b03836
8. Strop P, Liu SH, Dorywalska M et al (2013) Location matters: site of conjugation modulates stability and pharmacokinetics of antibody drug conjugates. Chem Biol 20(2):161–167. https://doi.org/10.1016/j.chembiol.2013.01.010
9. Rachel NM, Toulouse JL, Pelletier JN (2017) Transglutaminase-catalyzed bioconjugation using one-pot metal-free bioorthogonal chemistry. Bioconjug Chem 28(10):2518–2523. https://doi.org/10.1021/acs.bioconjchem.7b00509
10. Rashidian M, Song JM, Pricer RE et al (2012) Chemoenzymatic reversible immobilization and labeling of proteins without prior purification. J Am Chem Soc 134(20):8455–8467. https://doi.org/10.1021/ja211308s
11. Seo JS, Lee S, Poulter CD (2013) Regioselective covalent immobilization of recombinant antibody-binding proteins A, G, and L for construction of antibody arrays. J Am Chem Soc 135(24):8973–8980. https://doi.org/10.1021/ja402447g
12. Schumacher D, Helma J, Mann FA et al (2015) Versatile and efficient site-specific protein functionalization by tubulin tyrosine ligase. Angew Chem Int Ed 54(46):13787–13791. https://doi.org/10.1002/anie.201505456
13. Muir TW (2003) Semisynthesis of proteins by expressed protein ligation. Annu Rev Biochem 72(1):249–289. https://doi.org/10.1146/annurev.biochem.72.121801.161900
14. Palsuledesai CC, Distefano MD (2015) Protein prenylation: enzymes, therapeutics, and biotechnology applications. ACS Chem Biol 10(1):51–62. https://doi.org/10.1021/cb500791f
15. Diaz-Rodriguez V, Hsu ET, Ganusova E et al (2018) A-factor analogues containing alkyne- and azide-functionalized isoprenoids are efficiently enzymatically processed and retain wild-type bioactivity. Bioconjug Chem 29 (2):316–323. https://doi.org/10.1021/acs.bioconjchem.7b00648

16. Duckworth BP, Chen Y, Wollack JW et al (2007) A universal method for the preparation of covalent protein-DNA conjugates for use in creating protein nanostructures. Angew Chem Int Ed 46(46):8819–8822. https://doi.org/10.1002/anie.200701942
17. Gauchet C, Labadie GR, Poulter CD (2006) Regio- and chemoselective covalent immobilization of proteins through unnatural amino acids. J Am Chem Soc 128(29):9274–9275. https://doi.org/10.1021/ja061131o
18. Dozier JK, Khatwani SL, Wollack JW et al (2014) Engineering protein farnesyltransferase for enzymatic protein labeling applications. Bioconjug Chem 25(7):1203–1212. https://doi.org/10.1021/bc500240p
19. Nguyen UT, Guo Z, Delon C et al (2009) Analysis of the eukaryotic prenylome by isoprenoid affinity tagging. Nat Chem Biol 5 (4):227–235. https://doi.org/10.1038/nchembio.149
20. Labadie GR, Viswanathan R, Poulter CD (2007) Farnesyl diphosphate analogues with ω-bioorthogonal azide and alkyne functional groups for protein farnesyl transferase-catalyzed ligation reactions. J Org Chem 72 (24):9291–9297. https://doi.org/10.1021/jo7017747
21. Rose MW, Rose ND, Boggs J et al (2005) Evaluation of geranylazide and farnesylazide diphosphate for incorporation of prenylazides into a CAAX box-containing peptide using protein farnesyltransferase. J Pept Res 65 (6):529–537. https://doi.org/10.1111/j.1399-3011.2005.00261.x
22. Hosokawa A, Wollack JW, Zhang Z et al (2007) Evaluation of an alkyne-containing analogue of farnesyl diphosphate as a dual substrate for protein-prenyltransferases. Int J Pept Res Ther 13(1-2):345–354. https://doi.org/10.1007/s10989-007-9090-3
23. Duckworth BP, Zhang Z, Hosokawa A et al (2007) Selective labeling of proteins by using protein farnesyltransferase. Chembiochem 8 (1):98–105. https://doi.org/10.1002/cbic.200600340
24. Rashidian M, Dozier JK, Lenevich S et al (2010) Selective labeling of polypeptides using protein farnesyltransferase via rapid oxime ligation. Chem Commun 46(47):8998–9000. https://doi.org/10.1039/c0cc03305g
25. Yeo JE, Wickramaratne S, Khatwani S et al (2014) Synthesis of site-specific DNA-protein conjugates and their effects on DNA replication. ACS Chem Biol 9(8):1860–1868. https://doi.org/10.1021/cb5001795
26. Rashidian M, Kumarapperuma SC, Gabrielse K et al (2013) Simultaneous dual protein labeling using a triorthogonal reagent. J Am Chem Soc 135(44):16388–16396. https://doi.org/10.1021/ja403813b
27. Seo JS, Poulter CD (2014) Sandwich antibody arrays using recombinant antibody-binding protein L. Langmuir 30(22):6629–6635. https://doi.org/10.1021/la500822w
28. Lee JJ, Choi HJ, Yun M et al (2015) Enzymatic prenylation and oxime ligation for the synthesis of stable and homogeneous protein-drug conjugates for targeted therapy. Angew Chem Int Ed 54(41):12020–12024. https://doi.org/10.1002/anie.201505964
29. Plückthun A (2015) Designed ankyrin repeat proteins (DARPins): binding proteins for research, diagnostics, and therapy. Annu Rev Pharmacol Toxicol 55(1):489–511. https://doi.org/10.1146/annurev-pharmtox-010611-134654
30. Tamaskovic R, Simon M, Stefan N et al (2012) Designed ankyrin repeat proteins (DARPins) from research to therapy. Methods Enzymol 503:101–134. https://doi.org/10.1016/B978-0-12-396962-0.00005-7
31. Stefan N, Martin-Killias P, Wyss-Stoeckle S et al (2011) DARPins recognizing the tumor-associated antigen EpCAM selected by phage and ribosome display and engineered for multivalency. J Mol Biol 413(4):826–843. https://doi.org/10.1016/j.jmb.2011.09.016
32. Rose MW, Xu J, Kale TA et al (2005) Enzymatic incorporation of orthogonally reactive prenylazide groups into peptides using geranylazide diphosphate via protein farnesyltransferase: implications for selective protein labeling. Biopolymers 80(2-3):164–171. https://doi.org/10.1002/bip.20239
33. Blanden MJ, Suazo KF, Hildebrandt ER et al (2018) Efficient farnesylation of an extended C-terminal C(x)3X sequence motif expands the scope of the prenylated proteome. J Biol Chem 293(8):2770–2785. https://doi.org/10.1074/jbc.M117.805770
34. Mathis JR, Poulter CD (1997) Yeast protein farnesyltransferase: a pre-steady-state kinetic analysis. Biochemistry 36(21):6367–6376. https://doi.org/10.1021/bi9629182

Chapter 15

Homogeneous Antibody–Drug Conjugates via Glycoengineering

Feng Tang, Wei Shi, and Wei Huang

Abstract

Conventional antibody–drug conjugates (ADCs) randomly assemble small-molecule drugs onto Lys or Cys residues of a tumor-targeting antibody, featured with heterogeneity in payload numbers and conjugation positions. Glycosite-specific ADCs (gsADCs) link payload drugs onto IgG Fc *N*-glycans with high homogeneity that facilitates structural optimization and quality control for ADC drug development. In this protocol, we report two strategies for preparation of homogeneous ADCs via chemoenzymatic glycoengineering. First, an azido-tagged unnatural *N*-glycan substrate is transferred onto Fc glycosites of a therapeutic antibody through Endo-*S*-catalyzed glycoremodeling, followed by click reaction with an alkyne-tagged payload drug to give a well-defined gsADC. In an alternative way, glycoengineering of antibody with a natural sialylated *N*-glycan and successive selective oxidation of sialic acid moieties using sodium periodate provided an aldehyde handle on the glycans for conjugation with an aminooxy-assembled payload. These two strategies both enable gsADCs with high homogeneity in their conjugation sites, payload numbers, and glycoforms, which are characterized of a single mass under mass-spectral detection.

Key words Glycosite-specific antibody–drug conjugates, Chemoenzymatic glycoengineering, Homogeneous ADCs, Biotherapeutics

1 Introduction

Antibody–drug conjugates (ADCs), a complex composed of drugs, antibody, and variable linkers for coupling, have emerged as a significant therapeutic antibody subtype in the past decade [1, 2]. Harnessing antibody as the tumor-targeting warhead, ADCs greatly improve drug delivery with decreased system cytotoxicity of small-molecule payloads, benefited from the postrelease of drugs intracellularly [3]. The successful launch of four ADCs in clinic inspired the scientists to develop new ADCs for treatment of various malignancies such as solid tumors, lymphoma, and severe bacterial infections. However, conventional ADCs usually conjugate the cytotoxic payloads onto the Lys or Cys residues of an antibody in a random manner, resulting in heterogeneity in drug

Sam Massa and Nick Devoogdt (eds.), *Bioconjugation: Methods and Protocols*, Methods in Molecular Biology, vol. 2033, https://doi.org/10.1007/978-1-4939-9654-4_15, © Springer Science+Business Media, LLC, part of Springer Nature 2019

loadings and conjugation positions. Therefore, random conjugation strategy of an ADC apparently led to difficult quality control in preparation and challenging reproducibility in pharmacodynamics and pharmacokinetics [4, 5].

Site-specific conjugation strategy for ADC development has become a hot topic in this field. New methods such as THIOMAB technology, nonnatural amino acid incorporation, and Sortase-A ligation have been reported [1, 6–19]. These site-specific approaches greatly promoted the homogeneity and PK and PD characteristics of ADCs [4, 5, 20]. Besides these techniques, glycosite-specific ADCs (gsADCs) utilize the conserved *N*-glycosylation site Asn-297 on Fc domain as a specific conjugation location for assembly of small molecules [8, 9, 14, 18, 21–23]. To attach a bioorthogonal and reactive handle on the glycans, antibody glycoengineering is indispensable for gsADCs. Zhou et al. employed galactosyl and sialyl transferases to engineer the IgG glycan and achieved full-length sialylated *N*-glycans, followed by successive $NaIO_4$ oxidation of the sialic acid motif and oxime ligation to obtain a gsADC [9]. Alternatively, Boons group chose the azido-tagged cytidine monophosphate sialic acid as an unnatural substrate for sialyltransferase-catalyzed glycoengineering and synthesized a homogeneous gsADC with cyclooctyne derivatives via strain-promoted azide–alkyne cycloadditions (SPAAC) [8]. We have recently reported a different strategy for gsADC development using Endoglycosidase-catalyzed antibody glycoengineering with an azido-tagged glycan oxazoline substrate and followed with payload assembly using dibenzocyclooctyne (DBCO) derivatives via click chemistry [22, 23]. Endo-S from *Streptococcus pyogenes* is an endoglycosidase with excellent specificity for IgG deglycosylation by cleaving the GlcNAc-β1,4-GlcNAc moiety, whereas Endo-S D233Q mutant possesses transglycosylation activity to assemble an unnatural glycan substrate onto IgG glycosites without hydrolytic activity, therefore enabling efficient IgG glycoengineering for gsADCs [21, 24–27]. With consecutive use of both enzymes, we have prepared a series of gsADCs in high homogeneity with well-defined conjugation site and payload numbers [22, 23, 28]. Herein, we report the detailed procedures for synthesizing glycosite-specific ADCs via Endo-*S*-catalyzed IgG glycoengineering.

We choose the therapeutic antibody Herceptin (trastuzumab) as a model to describe this procedure (Fig. 1). The heterogeneous *N*-glycans on Herceptin (**1**) are removed by wild-type Endo-S and gave the deglycosylated Herceptin (**2**) carrying the innermost GlcNAc-Fuc disaccharide at the glycosylation site. Next, the deglycosylated Herceptin (**2**) is reglycosylated with a natural sialylated complex type glycan oxazoline (SCT-ox, **7**) or an unnatural azido-tagged SCT-ox (Az-SCT-ox, **10**) in the presence of the glycosynthase Endo-S D233Q mutant. Thereafter, the transglycosylation product azido-tagged Herceptin (Az-S2G2F-Herceptin, **3a**) is incubated with DBCO-linked MMAE (**4a**) to give gsADC (**5a**) via

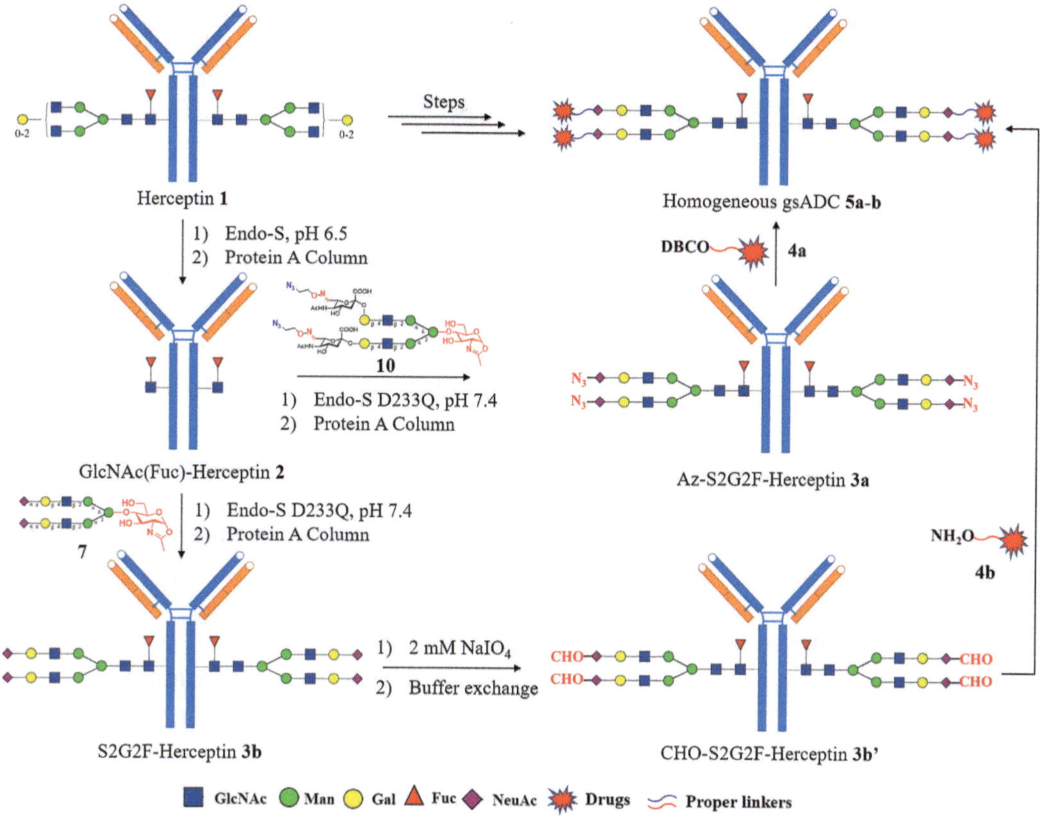

Fig. 1 Site-specific antibody drug conjugates based on glycoengineered Herceptin

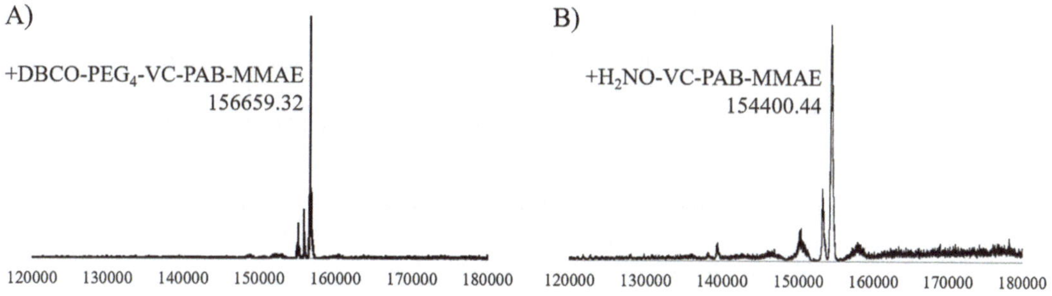

Fig. 2 Mass spectra of gsADC **5a** (panel **a**) and gsADC **5b** (panel **b**)

SPAAC reaction. In an alternative way, the glycoengineered S2G2F-Herceptin (**3b**) is selectively oxidized by 2 mM·$NaIO_4$ to give Herceptin carrying an aldehyde handle on its terminal sialic acid motif (CHO-S2G2F-Herceptin, **3b'**), which reacted with hydroxylamine-containing MMAE (**4b**) to give the oxime gsADC (**5b**).

These gsADCs are purified by one-step affinity chromatography with a protein A column and analyzed by SDS-PAGE and LC-MS (the LC-MS spectra of gsADC **5a** and **5b** are illustrated in Fig. 2), indicating excellent purity and homogeneity. The semisynthesis of sialylated and azido-tagged sialylated complex type glycan oxazoline (**7**, **10**) from egg-yolk sialylglycopeptide (SGP), and the synthesis of MMAE derivatives (**4a-b**) are also described in Subheadings 3.1–3.4 respectively.

2 Materials

2.1 Enzymes, Antibodies, and Substrates

1. Endo-M WT, Endo-S WT, and Endo-S D233Q are expressed in *Escherichia coli* and purified by Ni chromatography respectively (*see* **Note 1**, [22]).
2. Commercial monoclonal antibody (mAb) Herceptin (*see* **Note 2**).
3. Sialylglycopeptide (SGP, extracted and purified from egg yolks, *see* **Note 3**, [22, 29]).
4. *O*-(2-azidoethyl)-hydroxylamine hydrochloride (**17**) ([22], *see* **Note 4**).

2.2 Buffers, Columns and Reagents

1. Protein A beads (Yeli Bio, Shanghai, China).
2. Sephadex G-25 (Sigma-Aldrich, Shanghai, China).
3. Analytical C18 column 5 μm 4.6 × 250 mm (Thermo Scientific).
4. C18 Prep column 5 μm 19 × 250 mm (Waters Corporation).
5. Protein A wash buffer (A): 0.2 M sodium phosphate buffer, pH 7.4. The phosphate buffer is prepared by mixing 0.2 M NaH_2PO_4 and 0.2 M·Na_2HPO_4 until reaching the target pH value.
6. Protein A wash buffer (B): 0.2 M sodium phosphate buffer, pH 5.0. The phosphate buffer is prepared by mixing 0.2 M·NaH_2PO_4 and 0.2 M·Na_2HPO_4 until reaching the target pH value.
7. Protein A elution buffer (C): 0.1 M glycine–HCl, 0.2 M·NaH_2PO_4, 0.01% NaN_3, pH 2.7.
8. BCA kit (Yeason Bio, Shanghai, China).
9. HPLC-grade trifluoroacetic acid (TFA; J&K Chemical).
10. HPLC-grade acetonitrile (J&K Chemical).
11. HPLC Solvent A: 0.1% TFA (v/v) in ddH_2O.
12. HPLC Solvent B: 0.1% TFA (v/v) in acetonitrile.
13. HPEAC-PAD Solvent A: 100 mM NaOH.
14. HPEAC-PAD Solvent B: 100 mM NaOH, 1 M·NaOAc.
15. 50 mM sodium phosphate buffer, pH 7.5: the phosphate buffer is prepared by mixing 0.2 M·NaH_2PO_4 and 0.2 M·Na_2HPO_4 until reaching the target pH value. Then

the 0.2 M phosphate buffer is diluted with H_2O to the target concentration (50 mM).

16. 50 mM sodium phosphate buffer, pH 6.5: the phosphate buffer is prepared by mixing 0.2 M·NaH_2PO_4 and 0.2 M·Na_2HPO_4 until reaching the target pH value. Then the 0.2 M phosphate buffer is diluted with H_2O to the target concentration (50 mM).
17. 0.2 M sodium phosphate buffer, pH 7.1: the phosphate buffer is prepared by mixing 0.2 M·NaH_2PO_4 and 0.2 M·Na_2HPO_4 until reaching the target pH value.
18. 50 mM Tris–HCl, pH 7.4.
19. 100 mM NaOAc–HOAc, pH 5.6.
20. 50 mM sodium citrate, 50 mM·NaCl, pH 5.5.
21. 2-Chloro-1,3-dimethylimidazolinium chloride (DMC; Sigma-Aldrich).
22. 2-Chloro-1,3-dimethyl-^{1}H-benzimidazol-3-ium chloride (CDMBI, [30]).
23. Triethylamine (TEA).
24. Sodium periodate.
25. *N*,*N*-dimethylformamide (DMF).
26. *N*,*N*-dimethylacetamide (DMA).
27. 1-[Bis(dimethylamino)methylene]-^{1}H-1,2,3-triazolo[4,5-b] pyridinium3-oxid hexafluorophosphate (HATU).
28. 1-Hydroxybenzotriazole (HOBt).
29. Piperidine.
30. *N*,*N*-diisopropylethylamine (DIPEA).
31. DBCO-PEG_4-CO_2Su (ChemPartner, Shanghai).
32. (Boc-aminooxy) acetic acid.
33. 10% sulfuric acid containing methanol: 10 mL of concentrated sulfuric acid is slowly added to 90 mL of methanol in a wide-mouth bottle on the ice.
34. Sodium dodecyl sulfate–polyacrylamide gel electrophoresis (SDS-PAGE): 20 mL of running gel (10%) is composed of 7.9 mL of pure water, 6.7 mL of 30% Acryl-Bis (29:1), 5.0 mL of 1.5 M Tris–HCl, pH 8.8, 0.2 mL of 10% SDS, 0.2 mL of 10% ammonium persulfate (AP), and 0.008 mL of *N*,*N*,*N*′,*N*′-Tetramethylethylenediamine (TEMED). Every 4.5 mL of above solution is transferred to one Bio-Rad gel casting frame (totally four frames), which is followed by slowly adding 1 mL of pure water to stack the top edge of running gel to a smooth line. 8 mL of stacking gel is composed of 5.5 mL of pure water, 1.3 mL of 30% Acryl-Bis (29:1), 1.0 mL of

1.0 M Tris–HCl, pH 6.8, 0.08 mL of 10% SDS, 0.08 mL of 10% AP, 0.008 mL of TEMED. The water in the frame is discarded when the running gel is solidified. Afterward, 1.5 mL of stacking gel solution is added to the top of each running gel and a 15-tooth Bio-Rad gel comb is inserted for making a 10% SDS-PAGE gel with 15 lanes.

35. 1× Protein loading buffer for SDS-PAGE (protein denature buffer).
36. Coomassie blue buffer: 1 g Coomassie brilliant blue R250 is dissolved in 300 mL of ethanol, 100 mL of acetic acid, 600 mL of H_2O.
37. SDS-PAGE destaining solution: 500 mL of ethanol, 100 mL of acetic acid, 400 mL of H_2O.
38. 10× SDS-PAGE running buffer: 30 g Tris base, 144 g glycine, and 10 g SDS are dissolved in 1 L of pure water. It should be stored at room temperature and diluted to 1× with pure water before use.
39. Thin-Layer-Chromatography plate (TLC plate, silica gel plate).

2.3 Equipment and Setup

1. LC3000 analytical HPLC (ChuangXinTongHeng, Beijing).
2. LC3000 preparative HPLC (ChuangXinTongHeng, Beijing).
3. HPLC Ultimate 3000 (Thermo Scientific).
4. 6230 TOF mass spectrometer (Agilent).
5. ICS 5000 (Thermo Scientific).
6. Lyophilizer.
7. Fraction collector.
8. Temperature controllable shaker (0–100 °C).
9. Bio-Rad ChemiDoc™ MP imaging system.
10. Centrifuge (Sorvall ST 8R; Thermo Fisher).
11. Bio-Rad PowerPac™ Basic.
12. Analytical HPLC method A: a linear elution gradient of 2–11% HPLC solvent B for 10 min followed by a linear elution gradient of 11–90% solvent B for 20 min at a flow rate of 1 mL/min. Set column temperature at 40 °C.
13. Analytical HPLC method B: a linear elution gradient of 2–90% HPLC solvent B for 30 min at a flow rate of 1 mL/min. Set column temperature at 40 °C.
14. Preparative HPLC method: a linear elution gradient of 2–90% HPLC solvent B for 40 min at a flow rate of 8 mL/min. Set column temperature at room temperature.
15. LC-MS analysis of antibodies and gsADCs: setup the method according to published literature (*see* **Note 5**, [22, 28]).

16. SDS-PAGE analysis of antibodies and gsADCs: prepare or remove 1 μg antibody from each reaction mixture, denature it in 10 μL of 1× protein loading buffer, and boil the solution at 100 °C for 5 min to fully denature the antibody. Add all the samples into separate 10% SDS-PAGE lanes and run it in 1× SDS running buffer at 100 V for 10 min followed by 180 V (Bio-Rad PowerPac Basic) until the blue loading buffer is run out from the gel. Visualize the heavy chain and light chain by staining the gel with Coomassie blue buffer and destaining the background with SDS-PAGE destaining buffer. Finally, analyze the gel with Bio-Rad imaging system.
17. HPAEC-PAD: elute the column with 100% HPEAC-PAD Solvent A for 2 min followed by a gradient elution of 0–15% HPEAC-PAD Solvent B for 20 min. Then wash the column with 25% HPAEC-PAD Solvent B for 5 min and equilibrate the column with HPAEC-PAD Solvent A for another 5 min. Set the flow rate at 1 mL/min and the column temperature at 40 °C.
18. Sephadex G-25 column: swell 100 g solid Sephadex G-25 in 600 mL of pure water overnight and pour the mixture into a 2.6 cm × 100 cm glass column until the column is 80% full (at the 80-cm mark). The G-25 column is eluted with 3 column volumes of water, 0.2% triethylamine aqueous solution, or other suitable solvents by gravity before use.

3 Methods

3.1 "One-Pot" Synthesis of Sialylated Complex Type Oxazoline (SCT-Ox) (Fig. 3)

1. Weigh 100 mg of SGP (34.9 μmol, *see* **Note 6**), dissolve it with 900 μL of 50 mM sodium phosphate buffer, pH 6.5 and adjust the pH to 6.5 if pH is altered.
2. Add 100 μL of 2.5 mg/mL Endo-M (final concentration in reaction is 0.5 mg/mL) to the mixture and incubate the mixture at 30 °C for 4 h. Monitor the reaction with HPLC (Method A) until the hydrolysis is complete (*see* **Note 7**).
3. Add 219 μL of 99% triethylamine (TEA) (1.57 mmol, 45 equiv.) and 88.5 mg of DMC (523.5 μmol, 15 equiv.) to the reaction tube (*see* **Note 8**). Place the reaction mixture on the shaker (precooled to 4 °C) immediately and shake vigorously (900 rpm, 4 °C) for 60 min (*see* **Note 9**). Monitor the reaction with HPAEC-PAD (Fig. 4, *see* **Note 10**).
4. Load the reaction mixture onto a Sephadex G-25 column (preequilibrated with 0.2% TEA aqueous solution). Elute the column with 0.2% TEA in ddH_2O and collect all the fractions at a rate of 3.5 min/tube (about 6–8 mL) by a fraction collector. Load 0.5 μL of solutions from each fraction onto a slice of TLC plate and stain the plate with 10% sulfuric acid in

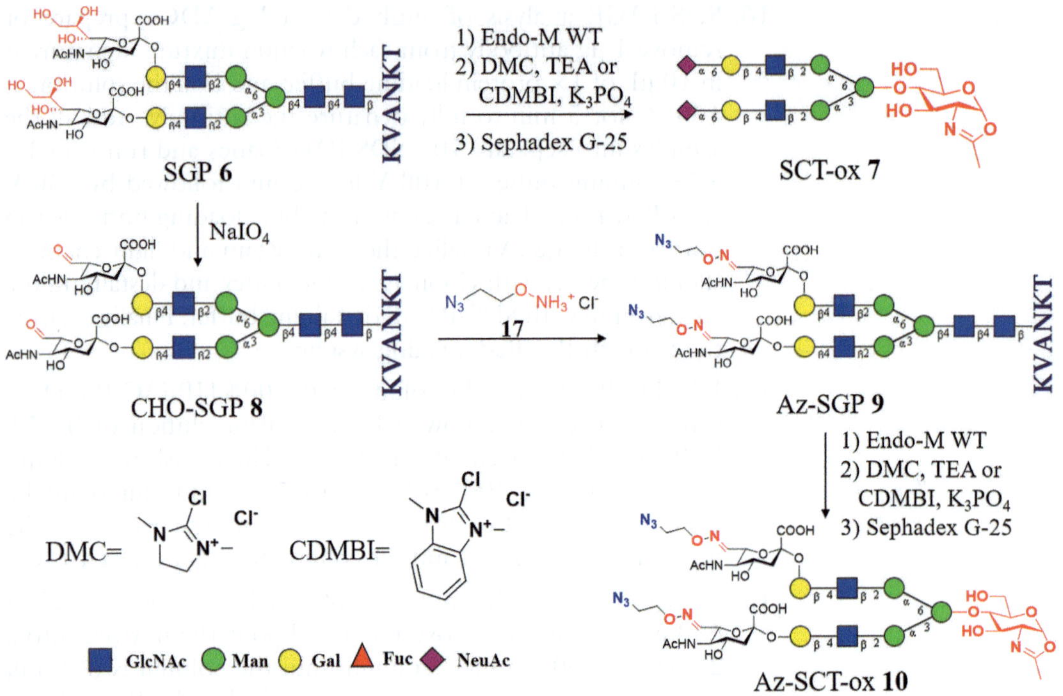

Fig. 3 Semisynthesis of sialylated complex type glycan oxazoline (SCT-ox **7**) and the azido tagged sialylated complex glycan oxazoline (Az-SCT-ox **10**)

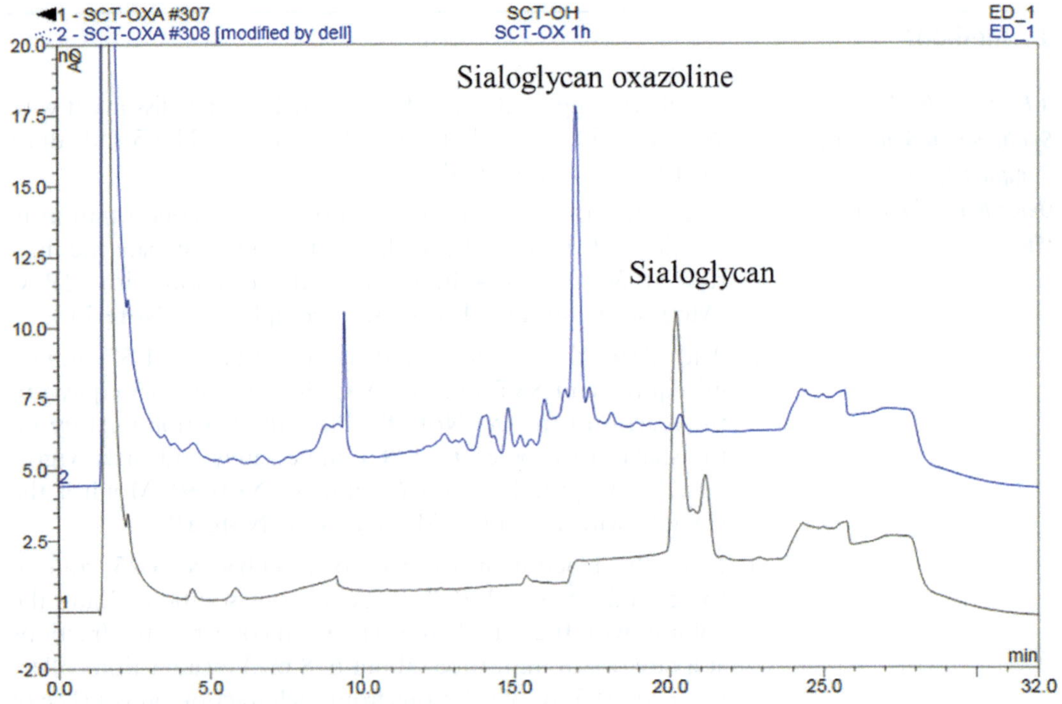

Fig. 4 HPAEC-PAD profile of SCT-ox synthesis

methanol after heated with a heat gun. The fractions containing glycans indicate black spots on the TLC plate after staining.

5. Collect and combine the fractions containing SCT-ox. Add 6 mg NaOH, freeze the solution via liquid nitrogen and lyophilize it in a lyophilizer (−45 °C, 60 Pa) to give compound **7** as a white solid (expected yield: 65 mg, 92.2%) (*see* **Note 11**).

3.2 Synthesis of Azido-Oxime-SCT Oxazoline (Az-SCT-Ox)

1. Dissolve 100 mg of SGP in 5 mL of 0.2 M sodium phosphate buffer, pH 7.1 and cool it to 0 °C in an ice bath.
2. Add 5 mL of 30 mM·$NaIO_4$ to above mixture and stir it at 0 °C for 15 min. Monitor the reaction with HPLC (analytical HPLC method A) and LCMS (*see* **Note 12**), as described in Subheading 2.3.
3. Load the reaction mixture onto a Sephadex G-25 column (preequilibrated with pure water) and elute the column with pure water to remove the salts. Analyze fractions and collect the product CHO-SGP as described in Subheading 3.1, **step 4** (CHO-SGP, **8**, expected yield 90 mg, 94%).
4. Weigh 50 mg of CHO-SGP (**8**) in a 2-mL tube and add 960 μL of 50 mM sodium phosphate buffer, pH 7.5.
5. Add 7.59 mg of *O*-(2-azidoethyl)-hydroxylamine hydrochloride **17** (54.77 μmol, 3 eq), adjust the pH to 7.5 with 1 N·NaOH, and add extra 50 mM sodium phosphate buffer, pH 7.5 until a total volume of 1 mL is reached.
6. Shake the reaction mixture at room temperature for 4 h and monitor the reaction with HPLC (Method A).
7. Load the mixture solution onto a Sephadex G-25 column (preequilibrated with pure water) and elute the column with pure water. Collect and lyophilize the fractions containing product azido-SGP to obtain a white powder (expected yield 50 mg, 94%) (*see* **Note 13**).
8. Dissolve the lyophilized solid Az-oxime-SGP (50 mg) in 450 μL of 50 mM sodium phosphate buffer, pH 6.5 and adjust the pH to 6.5 if the pH is altered.
9. Add 50 μL of 2.5 mg/mL Endo-M (final concentration in reaction is 0.5 mg/mL) to the mixture and incubate the mixture at 30 °C for 4 h. Monitor the reaction with HPLC (Method A) until the hydrolysis is complete.
10. Add 108 μL of 99% TEA (774 μmol, 45 equiv.) and 43.6 mg of DMC (258 μmol, 15 equiv.) to the reaction mixture. Place the reaction mixture on a shaker (precooled to 4 °C) immediately and shake it vigorously (900 rpm, 4 °C) for 60 min.
11. Purify, collect, and lyophilize the product Az-SCT-ox as described in **steps 4** and **5** in Subheading 3.1 to give a white solid (expected yield 32 mg, 91.4%).

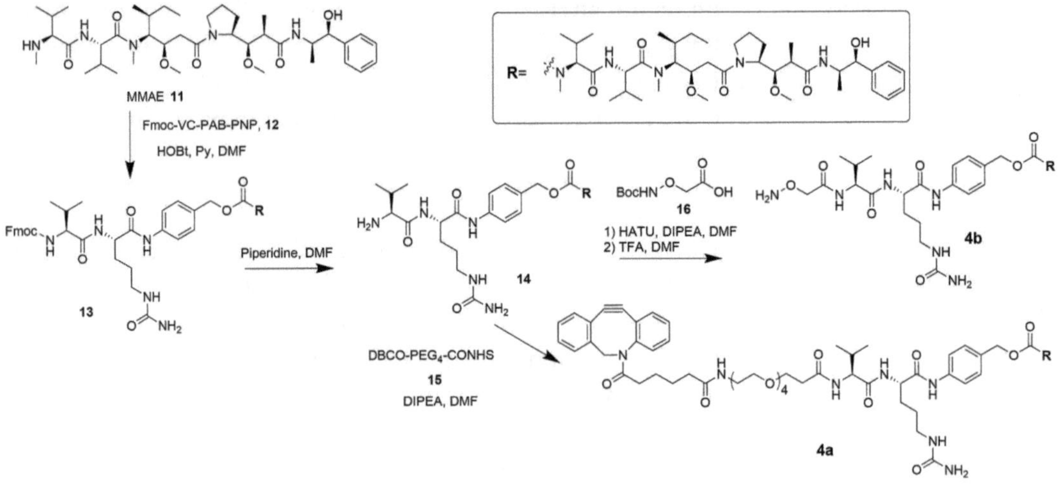

Fig. 5 Synthesis of MMAE derivatives with DBCO (**4a**) and hydroxylamine (**4b**)

3.3 Synthesis of DBCO-PEG$_4$-VC-PAB-MMAE (Compound 4a in Fig. 5)

1. Dissolve 40 mg of Fmoc-VC-PAB-PNP (**12**, 52.2 μmol, 1 equiv.) and 48.8 mg of MMAE (**11**, 68 μmol, 1.3 equiv.), 1.6 mg of HOBt (10.4 μmol, 0.2 equiv.) and 120 μL of pyridine in 8.0 mL of DMF. Stir the reaction mixture at room temperature for 24 h, monitor the reaction by HPLC (Method B) and purify the product with preparative HPLC equipped with a C18 column (Method C). Collect and lyophilize the product **13** to obtain a white powder (expected yield 58 mg, 93%).
2. Dissolve 58 mg of Fmoc-VC-PAB-MMAE (**13**) in 800 μL of DMF and 200 μL of piperidine. Stir the reaction mixture at room temperature for 30 min and monitor the reaction mixture with HPLC (Method B). Purify the product with preparative HPLC (Method C), collect and lyophilize the fractions to obtain **14** as a white powder (46 mg, 96%).
3. Weigh 10 mg of NH_2-VC-PAB-MMAE (**14**, 8.9 μmol, 1 equiv.) and dissolve it in 1.0 mL of DMF.
4. Dissolve 50 mg of DBCO-PEG$_4$-CO_2Su (**15**) in 1 mL of DMSO giving a 50 mg/mL stock solution and store it at −20 °C. Transfer 132.6 μL of the stock solution (**15**, 9.8 μmol, 1.1 equiv.) and 7.66 μL of DIPEA (44.5 μmol, 5 equiv.) to above mixture from **step 3**.
5. Stir the reaction mixture at room temperature overnight and monitor the reaction with HPLC (Method B). Purify the product with preparative HPLC (Method C), collect and lyophilize the fractions to obtain a stick oil product **4a** (Fig. 6a, expected yield 14 mg, 95%).

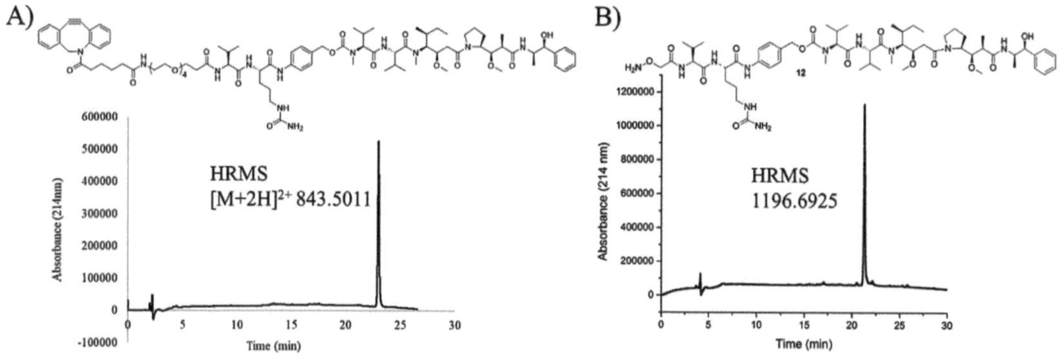

Fig. 6 HPLC profiles of MMAE derivatives. (**a**) HPLC profile of DBCO-PEG$_4$-VC-PAB-MMAE. (**b**) HPLC profile of aminooxy-VC-PAB-MMAE

3.4 Synthesis of Aminooxy-VC-PAB-MMAE (Compound 4b in Fig. 5)

1. Weigh 10 mg of NH_2-VC-PAB-MMAE (**14**, 8.9 μmol, 1 equiv.) and dissolve it in 1.0 mL of DMF. Add 6.77 mg of HATU (17.8 μmol, 2 equiv), 1.7 mg of Boc-aminooxy acetic acid (**16**, 8.9 μmol, 1 equiv), and 4.42 μL of DIPEA (26.7 μmol, 3 equiv) to the above mixture.
2. Stir the reaction mixture at room temperature overnight and monitor the reaction with HPLC (Method B). Purify the product with preparative HPLC (Method C), collect and lyophilize the fractions to obtain a white powder.
3. Dissolve the white powder from **step 2** in 1 mL of 50% TFA in DCM and stir the reaction mixture for 30 min at room temperature.
4. Monitor the reaction with HPLC (Method B) and purify the product with preparative HPLC (Method C). Collect and lyophilize the fractions to obtain **4b** as a white powder (Fig. 6b, expected yield 3.9 mg, 32%).

3.5 Deglycosylation of Herceptin by Endo-S

1. Prepare a solution of 33 mg of Herceptin in 1.0 mL of 50 mM sodium phosphate buffer, pH 6.5.
2. Add 1 μL of 0.5 mg/mL WT Endo-S (final concentration: 0.5 μg/mL, *see* **Note 14**) and incubate the reaction mixture at 37 °C.
3. Monitor the reaction with LC-MS until the complete deglycosylation of Herceptin (*see* **Note 15**). The mass spectra of Herceptin before (A) and after (B) Endo-S digestion are recorded in Fig. 7.
4. Assemble a protein A column with 1 mL solid beads (*see* **Note 16**). Wash the column with 10 mL of Protein A elution buffer (C) and equilibrate it with successive 10 mL of Protein A wash buffer (A) by gravity.

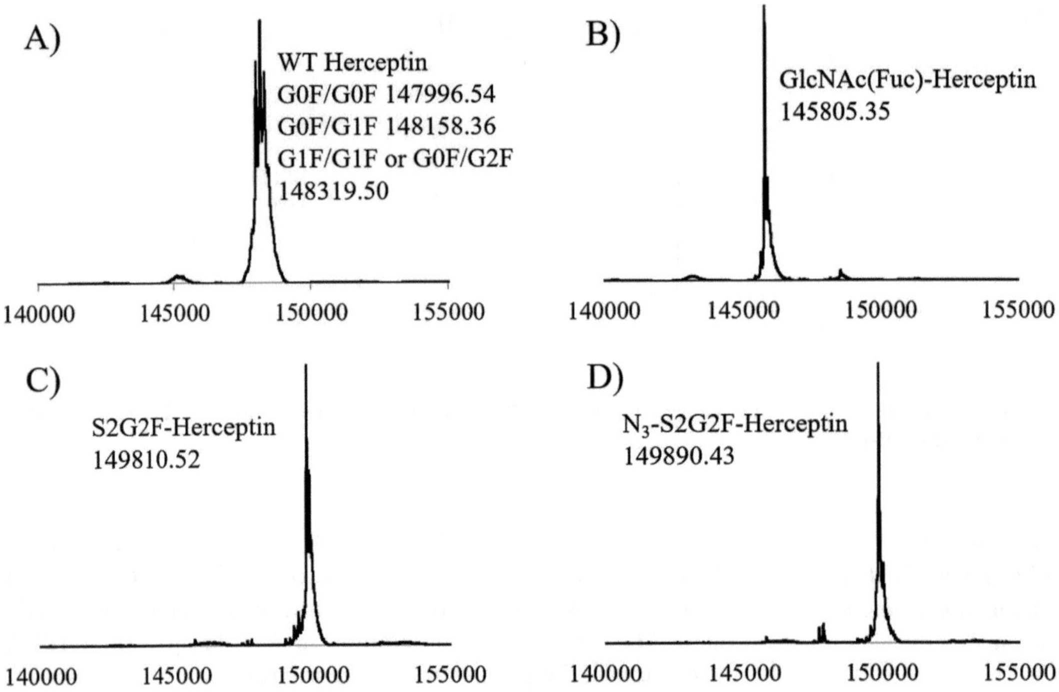

Fig. 7 Mass spectra of Herceptin samples. (**a**) commercial Herceptin, (**b**) deglycosylated Herceptin; (**c**) S2G2F-Herceptin, (**d**) Azido-tagged Herceptin

5. Load the deglycosylated Herceptin onto above column and reload flow-through solution for 5 times until the complete adhesion of the antibody onto Protein A.
6. Wash the column with 30 mL of Protein A wash buffer (A) and 30 mL of Protein A wash buffer (B).
7. Elute the deglycosylated Herceptin with 20 mL Protein A elution buffer (C) by gravity. Collect the elution fractions containing 1 mL volume in each fraction, neutralize each fraction to pH ~8 with 20 μL of 10 N·NaOH immediately.
8. Take 1–5 μL solution from each fraction and detect the aliquots with a Bradford protein chromogenic solution.
9. Combine and concentrate the fractions containing GlcNAc (Fuc)-Herceptin. Concentrate the mixture and exchange its buffer with a 50 mM Tris–HCl, pH 7.4 through an Amicon ultracentrifugal filter (30 kDa cutoff). Measure the concentration using a BCA kit to give GlcNAc(Fuc)-Herceptin (concentrate to more than 20 mg/mL).

3.6 Synthesis of Glycoengineered Herceptin by Endo-S D233Q

1. Weigh 23 mg of Az-SCT-ox (prepared from Subheading 3.2) and dissolve it in 200 μL of ddH_2O to give a 50 mM oxazoline stock solution. Prepare a 50 mM SCT-ox stock solution (prepared from Subheading 3.1) by the same procedure (*see* **Note 17**).

2. Add 500 μL of 50 mM Tris–HCl, pH 7.4 and 30 μL of 50 mM Az-SCT-ox or SCT-ox to a 1.5-mL tube. Adjust the pH to 7.4 with 0.1 N HCl (*see* **Note 18**).
3. Add 250 μL of 20 mg/mL GlcNAc(Fuc)-Herceptin (prepared from Subheading 3.5), 6.3 μL of 16 mg/mL Endo-S D233Q and additional Tris–HCl until the final volume is 1 mL (final concentration: 5 mg/mL deglycosylated Herceptin, 1.5 mM oxazoline, 0.15 mg/mL Endo-S D233Q).
4. Mix the solution gently and incubate the reaction mixture at 30 °C for 2 h. Monitor the reaction at 30 min, 60 min and 120 min with SDS-PAGE (1 μg) and LC-MS (5 μg) until the transglycosylation process is complete.
5. Purify the products by Protein A column immediately following above procedures described in **steps 4–8** from Subheading 3.5 (*see* **Note 19**).
6. Combine the product fractions and concentrate the product as described in **step 9** from Subheading 3.5. Exchange its buffer to a 50 mM sodium phosphate buffer, pH 7.5) or 100 mM NaOAc–HOAc, pH 5.6 to give Az-S2G2F-Hercpetin (10 mg/mL, Fig. 7d) and S2G2F-Herceptin (6 mg/mL, Fig. 7c) respectively. The mass spectra can be found in Fig. 7 (c refers to S2 G2F-Herceptin and d refers to Az-S2G2F-Hercpetin).

3.7 Preparation of gsADC 5a from Az-S2G2F-Hercpetin via SPAAC

1. Add 1 mL of 10 mg/mL Az-S2G2F-Hercpetin into 8 mL of 50 mM sodium phosphate buffer, pH 7.5 containing 0.87 mL of DMSO in a 15-mL tube (*see* **Note 20**).
2. Prepare a 10 mM stock solution of DBCO-PEG$_4$-VC-PAB-MMAE in DMSO and transfer 0.13 mL of the solution into above tube from **step 1**. Mix the solution gently.
3. Shake and incubate the reaction mixture at 30 °C for 24 h. Monitor the reaction with SDS-PAGE and LC-MS until the conjugation is complete.
4. Purify the product as described in **steps 4–8** from Subheading 3.5.
5. Combine and concentrate the fractions containing gsADC **5a**.

3.8 Preparation of gsADC 5b from S2G2F-Hercpetin via Oxime Ligation

1. Add 450 μL of 6 mg/mL S2G2F-Her into 2115 μL of 100 mM NaOAc–HOAc, pH 5.6.
2. Add 135 μL of 30 mM·$NaIO_4$ into the above solution and place the reaction mixture at 0 °C for 30 min (final concentrations: 1 mg/mL S2G2F-Hercpetin, 1.5 mM·$NaIO_4$).
3. Add 10 mL of a dilution buffer (50 mM sodium citrate, 50 mM·NaCl, pH 5.5) into the reaction, concentrate the solution using an Amicon ultracentrifugal filter (30 kDa cutoff). Measure the concentration using BCA kit (CHO-S2G2F-Herceptin, 25 mg/mL).

4. Prepare a stock of 100 mM aminooxy-VC-PAB-MMAE in DMA.
5. Mix 40 μL of 25 mg/mL CHO-S2G2F-Herceptin, 3.7 μL of DMA and 1.3 μL of 100 mM aminooxy-VC-PAB-MMAE stock with 55 μL of reaction buffer (50 mM sodium citrate, 50 mM·NaCl, pH 5.5) in a 1.5-mL tube (final concentration: 10 mg/mL CHO-S2G2F-Herceptin, 1.3 mM MMAE-VC-ONH_2, 5% DMA v/v).
6. Shake the reaction mixture at 37 °C for 8 h. Monitor the reaction with SDS-PAGE and LC-MS until the reaction is complete.
7. Purify the target gsADC product using a Protein A column following the procedures described in **steps 4–8** from Subheading 3.5.

4 Notes

1. For wild-type Endo-M, *E. coli BL21* (DE3) containing Endo-M vector is grown in 1 L of LB medium supplemented with ampicillin at 37 °C, 220 rpm until the OD600 has reached 0.6–0.8. Then the culture temperature is set to 20 °C and the bacteria are incubated for 3 days. Cells are harvested by centrifugation at 3200 × *g* for 10 min at, 4 °C. The collected cells are suspended in 40 mL of 50 mM sodium phosphate buffer, pH 7.5 and are lysed by ultrasonication followed by obtaining supernatant by centrifugation at 24,000 × *g* for 10 min at 4 °C. The supernatant is mixed with 5 mL of Ni-NTA agarose overnight at 4 °C and the agarose is packed into a 10-mL size column which is washed by successive 50 mM sodium phosphate buffer, pH 7.5, 5 mM imidazole in 50 mM sodium phosphate buffer, pH 7.5, 10 mM imidazole in 50 mM sodium phosphate buffer, pH 7.5, 20 mM imidazole in 50 mM sodium phosphate buffer, pH 7.5 and 100 mM imidazole in 50 mM sodium phosphate buffer, pH 7.5. Target protein can be eluted from Ni column by 100 mM imidazole and the solution is concentrated by an Amicon ultracentrifugal filter (30 kDa cutoff) with following buffer-exchange into 50 mM sodium phosphate buffer, pH 7.5.

   For wild-type Endo-S and its mutant Endo-S D233Q, cells are incubated in 1 L of LB medium supplemented with kanamycin at 37 °C, 220 rpm until the OD600 has reached 0.6–0.8. Then 0.1 mM IPTG (final concentration) is added and the cells are further incubated at 16 °C for 12 h. Cells are harvested and lysed as the steps in Endo-M purification. The Ni-NTA column with target enzyme is washed with 50 mM sodium phosphate buffer, pH 7.5, 50 mM imidazole in 50 mM sodium phosphate buffer, pH 7.5 and is eluted with 200 mM imidazole in 50 mM sodium phosphate buffer, pH 7.5. The 200 mM eluate is concentrated by an Amicon ultracentrifugal filter (30 kDa cutoff) with following buffer-exchange into 50 mM sodium phosphate buffer, pH 7.5.

2. In this chapter, we use Herceptin as a model for preparation of glycosite-specific antibody drug conjugates. The procedure is also suitable for gsADC synthesis with other IgG antibodies.
3. Briefly, the lyophilized egg yolk powder (1.2 kg) is washed with methyl tertiary butyl ether (4 L, twice) and 70% acetone (4 L, twice) successively. The remaining cake is extracted with 40% acetone (4 L, twice) and the solution is concentrated to a small volume (200 mL) before subjected to the active carbon–celite (1:1, w/w, total 200 g) column which had been prewashed with 50% acetonitrile (ACN) and preequilibrated with pure water. The column is washed with 1 L of pure water, 1 L of 5% ACN, 1 L of 10% ACN, 1 L of 20% ACN, and 1 L of 30% ACN successively. All these solutions are analyzed by HPLC (HPLC method A) and the fractions (250 mL/bottle) containing SGP are collected, concentrated, and lyophilized to give a crude SGP, which is further purified by semipreparative HPLC and Sephadex G-25 to give high-purity SGP.
4. Other hydroxylamine-containing compounds are also applicable for derivation of CHO-SGP.
5. LC-MS method for antibodies and gsADCs: ESI-MS spectra are measured on an Agilent 6230 LC–TOF MS spectrometer with a THERMO MAbPac RP column (4 μm, 3.0 × 100 mm) at 80 °C. The column is eluted with an isocratic mobile phase of 20% acetonitrile and 80% water containing 0.1% formic acid for the first 3 min at a flow rate of 0.4 mL/min, then it is successively eluted at the same flow rate with a linear gradient of 20–50% acetonitrile for additional 2.5 min, an isocratic 50% acetonitrile for 2 min, another linear gradient of 50–90% acetonitrile 0.5 min, and an isocratic 90% acetonitrile for 2 min followed by 20% acetonitrile for last 2 min for equilibration. The mass spectra of antibodies are collected under the extended mass range mode (high 20,000 m/z, 1 GHz) in the mass range of 800–5000. The multiple charged peaks of the antibody are deconvoluted using the Agilent MassHunter BioConfirm software (deconvolution for protein, Agilent Technologies) with the deconvolution range from 100 kDa to 200 kDa.
6. Sialylglycopeptide extracted from egg yolks is a mixture of monosialyl and disialyl SGP. To obtain pure disialyl SGP, repeated (≥3 times) Sephadex G-25 purification is required combined with fraction analysis using HPAEC-PAD.

7. We suggest to monitor the reaction at 1 h, 2 h, 3 h, and 4 h if it is possible, while the 4 h time point monitoring is preferred when the technician is experienced at ENGase-catalyzed hydrolysis.
8. Here we use DMC and TEA as reagents for oxazoline formation. Alternatively, CDMBI and K_3PO_4 could also be used for the synthesis of glycan oxazoline, especially when using the "in situ oxazoline" in "one-pot" IgG glycoengineering for the reason that CDMBI will turn into DMBI and precipitate from the reaction mixture while synthesizing oxazoline [22, 30].
9. Since the oxazoline is formed and stable under basic conditions, the pH of reaction mixture should be controlled strictly above 10.
10. For HPAEC-PAD analysis, an aliquot of about 1 μg of glycan is injected.
11. The additional NaOH is added to the oxazoline for stabilizing the product.
12. The Rt (retention time) of CHO-SGP is about 8.0 min while Rt of SGP is 7.5 min. Meanwhile, the peak of CHO-SGP is blunter than SGP.
13. G-25 but not C18 preparative column purification is recommended because of the oxime hydrolysis under acidic condition.
14. Endo-S is a highly efficient endoglycosidase to remove the *N*-glycan on Fc domain. Trace amounts of Endo-S in the IgG solution after Protein A purification can lead to the deglycosylation of glycoengineered antibodies in next step. So, we use only small amount of Endo-S for deglycosylation of Herceptin.
15. Theoretically, the deglycosylation of 33 mg/mL Herceptin by 0.5 μg/mL Endo-S can be achieved within 4 h. We suggest to monitor the deglycosylation process with a 1 h interval.
16. To confirm the complete removal of wild-type Endo-S from purified GlcNAc(Fuc)-Herceptin, we mix the WT Herceptin and purified GlcNAc(Fuc)-Herceptin (1:1, together 20 mg/mL, pH 7.0) in a 20 μL solution and incubate the mixture at 37 °C for 48 h. The mixture is subjected to SDS-PAGE to confirm the WT Herceptin is not hydrolyzed which implicated no Endo-S remained in purified GlcNAc(Fuc)-Herceptin. On the contrary, if WT Herceptin is hydrolyzed, another round of purification is necessary to completely remove WT Endo-S.
17. The oxazoline stock solution can be stored at 4 °C for 1–2 days. Storage at −20 °C or −80 °C is recommended for longer time.

18. Glycan oxazolines are stored under basic condition, so the addition of oxazoline into chemoenzymatic reactions will result in pH increase. Meanwhile, the acidic condition can lead to the hydrolysis of oxazoline ring. In our experience, pH 7.2–7.4 is an optimal condition for the Endo-*S*-catalyzed antibody glycoengineering.
19. The glycoengineered Herceptin should be purified immediately. A nonenzymatic glycation on Herceptin is observed after long-term storage and during the thawing process (data not shown).
20. The SPAAC reaction occurs rapidly at pH around 7.5 and the reaction rate decreases dramatically when the pH value is more than 8.

Acknowledgments

This work is supported by the National Natural Science Foundation of China (NNSFC, No. 21572244), the SIMM Institute Fund (CASIMM0120153004), and the "Personalized Medicines—Molecular Signature-based Drug Discovery and Development," Strategic Priority Research Program of the Chinese Academy of Sciences, Grant No. XDA12020311.

References

1. Beck A, Goetsch L, Dumontet C et al (2017) Strategies and challenges for the next generation of antibody-drug conjugates. Nat Rev Drug Discov 16(5):315–337. https://doi.org/10.1038/nrd.2016.268
2. Lopus M (2011) Antibody-DM1 conjugates as cancer therapeutics. Cancer Lett 307 (2):113–118. https://doi.org/10.1016/j.canlet.2011.03.017
3. Chari RV, Miller ML, Widdison WC (2014) Antibody-drug conjugates: an emerging concept in cancer therapy. Angew Chem Int Ed Engl 53(15):3796–3827. https://doi.org/10.1002/anie.201307628
4. Malik P, Phipps C, Edginton A et al (2017) Pharmacokinetic considerations for antibody-drug conjugates against cancer. Pharm Res 34 (12):2579–2595. https://doi.org/10.1007/s11095-017-2259-3
5. Shen BQ, Xu K, Liu L et al (2012) Conjugation site modulates the in vivo stability and therapeutic activity of antibody-drug conjugates. Nat Biotechnol 30(2):184–189. https://doi.org/10.1038/nbt.2108
6. Beck A (2013) Review of antibody-drug conjugates, methods in molecular biology series. MAbs 6(1):30–33. https://doi.org/10.4161/mabs.27005
7. Okeley NM, Toki BE, Zhang X et al (2013) Metabolic engineering of monoclonal antibody carbohydrates for antibody-drug conjugation. Bioconjug Chem 24(10):1650–1655. https://doi.org/10.1021/bc4002695
8. Li X, Fang T, Boons GJ (2014) Preparation of well-defined antibody-drug conjugates through glycan remodeling and strain-promoted azide-alkyne cycloadditions. Angew Chem Int Ed Engl 53(28):7179–7182. https://doi.org/10.1002/anie.201402606
9. Zhou Q, Stefano JE, Manning C et al (2014) Site-specific antibody-drug conjugation through glycoengineering. Bioconjug Chem 25(3):510–520. https://doi.org/10.1021/bc400505q
10. Behrens CR, Liu B (2014) Methods for site-specific drug conjugation to antibodies. MAbs 6(1):46–53. https://doi.org/10.4161/mabs.26632

11. Zimmerman ES, Heibeck TH, Gill A et al (2014) Production of site-specific antibody-drug conjugates using optimized non-natural amino acids in a cell-free expression system. Bioconjug Chem 25(2):351–361. https://doi.org/10.1021/bc400490z
12. Tian F, Lu Y, Manibusan A et al (2014) A general approach to site-specific antibody drug conjugates. Proc Natl Acad Sci U S A 111(5):1766–1771. https://doi.org/10.1073/pnas.1321237111
13. Dennler P, Chiotellis A, Fischer E et al (2014) Transglutaminase-based chemo-enzymatic conjugation approach yields homogeneous antibody–drug conjugates. Bioconjug Chem 25(3):569–578. https://doi.org/10.1021/bc400574z
14. Zhu Z, Ramakrishnan B, Li J et al (2014) Site-specific antibody-drug conjugation through an engineered glycotransferase and a chemically reactive sugar. MAbs 6(5):1190–1200. https://doi.org/10.4161/mabs.29889
15. Qasba PK (2015) Glycans of antibodies as a specific site for drug conjugation using Glycosyltransferases. Bioconjug Chem 26 (11):2170–2175. https://doi.org/10.1021/acs.bioconjchem.5b00173
16. Kudirka R, Barfield RM, McFarland J et al (2015) Generating site-specifically modified proteins via a versatile and stable nucleophilic carbon ligation. Chem Biol 22(2):293–298. https://doi.org/10.1016/j.chembiol.2014.11.019
17. Beerli RR, Hell T, Merkel AS et al (2015) Sortase enzyme-mediated generation of site-specifically conjugated antibody drug conjugates with high in vitro and in vivo potency. PLoS One 10(7):e0131177. https://doi.org/10.1371/journal.pone.0131177
18. Thompson P, Ezeadi E, Hutchinson I et al (2016) Straightforward glycoengineering approach to site-specific antibody-pyrrolobenzodiazepine conjugates. ACS Med Chem Lett 7(11):1005–1008. https://doi.org/10.1021/acsmedchemlett.6b00278
19. Carter PJ, Lazar GA (2017) Next generation antibody drugs: pursuit of the 'high-hanging fruit'. Nat Rev Drug Discov 17(3):197–223. https://doi.org/10.1038/nrd.2017.227
20. Junutula JR, Raab H, Clark S et al (2008) Site-specific conjugation of a cytotoxic drug to an antibody improves the therapeutic index. Nat Biotechnol 26(8):925–932. https://doi.org/10.1038/nbt.1480
21. Parsons TB, Struwe WB, Gault J et al (2016) Optimal synthetic glycosylation of a therapeutic antibody. Angew Chem Int Ed Engl 55(7):2361–2367. https://doi.org/10.1002/anie.201508723
22. Tang F, Yang Y, Tang Y et al (2016) One-pot N-glycosylation remodeling of IgG with non-natural sialylglycopeptides enables glycosite-specific and dual-payload antibody-drug conjugates. Org Biomol Chem 14 (40):9501–9518. https://doi.org/10.1039/c6ob01751g
23. Tang F, Wang LX, Huang W (2017) Chemoenzymatic synthesis of glycoengineered IgG antibodies and glycosite-specific antibody-drug conjugates. Nat Protoc 12 (8):1702–1721. https://doi.org/10.1038/nprot.2017.058
24. Huang W, Giddens J, Fan SQ et al (2012) Chemoenzymatic glycoengineering of intact IgG antibodies for gain of functions. J Am Chem Soc 134(29):12308–12318. https://doi.org/10.1021/ja3051266
25. Lin CW, Tsai MH, Li ST et al (2015) A common glycan structure on immunoglobulin G for enhancement of effector functions. Proc Natl Acad Sci U S A 112(34):10611–10616. https://doi.org/10.1073/pnas.1513456112
26. Goodfellow JJ, Baruah K, Yamamoto K et al (2012) An endoglycosidase with alternative glycan specificity allows broadened glycoprotein remodelling. J Am Chem Soc 134 (19):8030–8033. https://doi.org/10.1021/ja301334b
27. Guo W, Tang F, Qin K et al (2017) Glycoengineering and glycosite-specific labeling of serum IgGs from various species. Carbohydr Res 446–447:32–39. https://doi.org/10.1016/j.carres.2017.05.001
28. Tang Y, Tang F, Yang Y et al (2017) Real-time analysis on drug-antibody ratio of antibody-drug conjugates for synthesis, process optimization, and quality control. Sci Rep 7(1):7763. https://doi.org/10.1038/s41598-017-08151-2
29. Sun B, Bao W, Tian X et al (2014) A simplified procedure for gram-scale production of sialylglycopeptide (SGP) from egg yolks and subsequent semi-synthesis of Man3GlcNAc oxazoline. Carbohydr Res 396:62–69. https://doi.org/10.1016/j.carres.2014.07.013
30. Noguchi M, Fujieda T, Huang WC et al (2012) A practical one-step synthesis of 1,2-Oxazoline derivatives from unprotected sugars and its application to chemoenzymatic β-N-Acetylglucosaminidation of Disialo-oligosaccharide. Helv Chim Acta 95(10):1928–1936

Chapter 16

Labeling Proteins at Site-Specifically Incorporated 5-Hydroxytryptophan Residues Using a Chemoselective Rapid Azo-Coupling Reaction

Partha Sarathi Addy, Sarah B. Erickson, James S. Italia, and Abhishek Chatterjee

Abstract

Chemoselective protein labeling is a valuable tool in the arsenal of modern chemical biology. The unnatural amino acid mutagenesis technology provides a powerful way to site-specifically introduce nonnatural chemical functionalities into recombinant proteins, which can be subsequently functionalized in a chemoselective manner. Even though several strategies currently exist to selectively label recombinant proteins in this manner, there is considerable interest for the development of additional chemoselective reactions that are fast, catalyst-free, use readily available reagents, and are compatible with existing conjugation chemistries. Here we describe a method to express recombinant proteins in *E. coli* site-specifically incorporating 5-hydroxytryptophan, followed by the chemoselective labeling of this residue using a chemoselective rapid azo-coupling reaction.

Key words Bioconjugation, Unnatural amino acid, 5-Hydroxytryptophan, Genetic code expansion

1 Introduction

The ability to chemoselectively label recombinant proteins has catalyzed numerous applications, from the incorporation of biophysical probes to the creation of well-defined therapeutic antibody–drug conjugates [1–6]. The genetic code expansion technology provides an exciting avenue to achieve this by allowing site-specific incorporation of uniquely reactive chemical functionalities into recombinant proteins, which can be subsequently functionalized in a chemoselective manner [1, 7–11]. This technology uses an engineered aminoacyl-tRNA synthetase–tRNA pair, which does not cross-react with its counterparts from the host cell, to incorporate an unnatural amino acid (Uaa) of interest in response to a repurposed nonsense codon. Several Uaas have been genetically encoded in various domains of life that harbor bioconjugation

Sam Massa and Nick Devoogdt (eds.), *Bioconjugation: Methods and Protocols*, Methods in Molecular Biology, vol. 2033,
https://doi.org/10.1007/978-1-4939-9654-4_16,

handles and enable site-specific protein labeling via established conjugation chemistries [1, 7–11]. However, there is continued interest in further extending the scope of this strategy by genetically encoding additional chemical functionalities that can be labeled using rapid catalyst-free conjugation chemistries.

Recently, our group developed a unique tryptophanyl-tRNA synthetase (TrpRS)–tRNA pair which can be used for Uaa incorporation in both an engineered strain of *Escherichia coli*, as well as in eukaryotes [12]. The endogenous TrpRS–tRNA pair of *E. coli* (EcTrpRS–tRNA) was first functionally replaced by its yeast counterpart, liberating this pair to be used as an orthogonal nonsense suppressor in the resulting strain (ATMW1). Due to its bacterial origin, the EcTrpRS–tRNA pair is also orthogonal in eukaryotic cells. Furthermore, it was possible to engineer this pair to selectively incorporate various tryptophan analogs using a facile *E. coli* based directed evolution system. In addition to tryptophan analogs with established bioconjugation handles—such as an alkyne or azide—we were also able to genetically encode 5-hydroxytryptophan (5HTP; Fig. 1) [12]. We have taken advantage of the electron-rich nature of this Uaa to develop new bioconjugation strategies that target 5HTP in a selective manner [13–15]. For example, we have demonstrated that various aromatic diazonium compounds show dramatically higher azo-coupling rates with

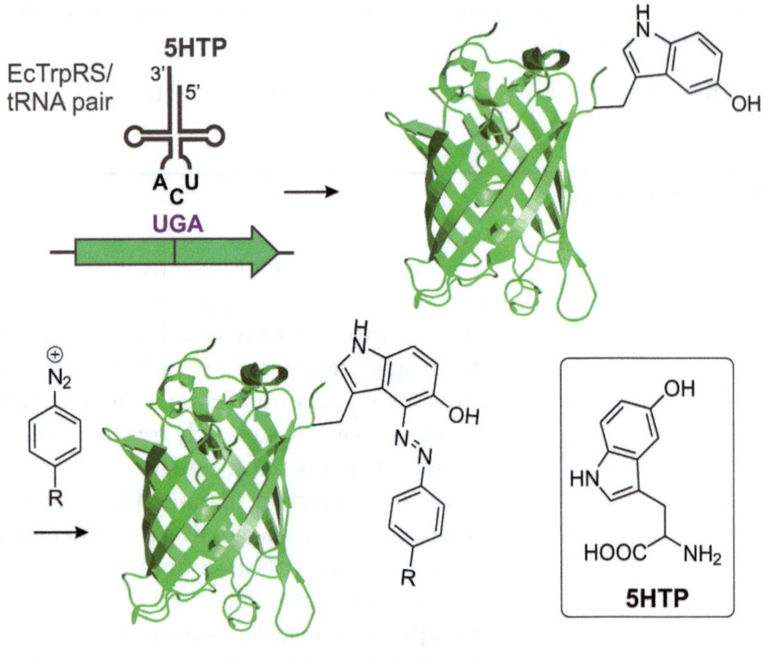

Fig. 1 An engineered opal (UGA) suppressing EcTrpRS–tRNA pair can be used to site-specifically incorporate 5HTP into proteins recombinantly expressed in the ATMW1 strain. The 5HTP residue can be selectively functionalized by a chemoselective rapid azo-coupling reaction using various aryldiazonium ions

5HTP relative to tyrosine [13]. The azo-coupling reaction has been previously used to label tyrosine residues in proteins [16–19]. However, highly electrophilic diazonium compounds (e.g., 4-nitrobenzenediazonium) are needed to perform the labeling reaction at neutral pH. To use less reactive diazonium compounds, significantly higher pH was required, presumably to facilitate the deprotonation of tyrosine residues. Due to its enhanced reactivity, 5HTP was found to undergo rapid azo-coupling reaction with a variety of aromatic diazonium compounds at physiological pH, providing the molecular basis for developing a chemoselective bioconjugation strategy (Fig. 1).

The chemoselective rapid azo-coupling reaction (CRACR) targeted to 5HTP residues offers several advantages. 5HTP is an inexpensive and readily accessible Uaa that can be efficiently and site-specifically incorporated into proteins expressed in both *E. coli* and eukaryotes. Aryldiazonium species attached to various entities can be prepared from the corresponding aromatic amines using the straightforward diazotization reaction. Alternatively, if the diazotization condition is incompatible with the cargo of interest, aryltriazabutadienes can be used as a precursor [13, 20–22]. This group is quite stable and readily generates the corresponding aryldiazonium upon irradiation at 365 nm. The kinetics of CRACR is highly sensitive to the electrophilicity of the aryldiazonium used; the bimolecular rate constant varies from 2 M^{-1} s^{-1} for 4-methoxydiazonium to 6×10^4 M^{-1} s^{-1} for *p*-nitrodiazonium [13]. The ability to tune the rate of coupling over such a large range, simply by introducing rational substitutions on the aryldiazonium, is another attractive feature of this reaction.

In the following, we describe a method to site-specifically incorporate 5HTP into proteins expressed in the ATMW1 strain followed by its site-specific labeling using CRACR. Two different proteins are used as model systems: the superfolder green fluorescent protein (sfGFP) that is expressed in the cytoplasm, and the Fab fragment of the αHer2 antibody that is expressed in the periplasm (to ensure proper folding).

2 Materials

Unless otherwise mentioned, all reagents are used as purchased from the commercial sources.

2.1 Recombinant Expression of Proteins in ATMW1 Cells Incorporating 5HTP

1. Homemade electrocompetent ATMW1 cells (*see* **Note 1**).
2. The pEvoltac-EcW-TGA-h14 plasmid that inducibly expresses the engineered EcTrpRS–tRNA pair (*see* **Note 2**).
3. pET22-T5lac-sfGFP-151TGA and pBK-aHer2-Fab-K169TGA plasmids to express sfGFP-151TGA and αHer2-Fab-K169TGA, inducible by IPTG and arabinose, respectively (*see* **Note 3**).

4. LB medium: 10 g/L casein peptone, 5 g/L yeast extract, 10 g/L NaCl.
5. LB-Agar plates supplemented with appropriate antibiotics.
6. 1000× antibiotics stock solutions: 50 mg/mL spectinomycin in water; 100 mg/mL ampicillin in water; 50 mg/mL kanamycin in water; 10 mg/mL gentamycin in water; 35 mg/mL chloramphenicol in ethanol.
7. 1000× IPTG stock solution: 1 M IPTG in water.
8. 20% L-arabinose in water.
9. 100 mM 5HTP (Chem-Impex International) in water (*see* **Note 4**).
10. Shaking incubator with variable temperature control.
11. Spectrophotometer and cuvettes to measure OD_{600} of cultures.

2.2 Purification of sfGFP by Immobilized Metal Ion Affinity Chromatography (IMAC)

1. HisPur Ni-NTA resin (ThermoFisher Scientific).
2. Cytoplasmic lysis buffer: prepared by adding 1× Halt Protease Inhibitor Cocktail (ThermoFisher Scientific) and 0.01% Pierce Universal Nuclease (ThermoFisher Scientific) into B-PER Bacterial Protein Extraction Reagent (ThermoFisher Scientific).
3. IMAC equilibration buffer: 20 mM·Na_2HPO_4, 300 mM·NaCl, 10 mM imidazole, pH 7.4.
4. IMAC wash buffer: 20 mM·Na_2HPO_4, 300 mM·NaCl, 25 mM imidazole, pH 7.4.
5. IMAC elution buffer: 20 mM·Na_2HPO_4, 300 mM·NaCl, 300 mM imidazole, pH 7.4.
6. Empty gravity-flow disposable plastic columns (Bio-Rad).
7. Phosphate buffer: 100 mM·Na_2HPO_4, 150 mM·NaCl, pH 7.
8. 3 kDa MWCO Slide-A-Lyzer™ MINI Dialysis devices (ThermoFisher Scientific).

2.3 Purification of αHer2-Fab-169-5HTP by Protein G Affinity Chromatography (PGAC)

1. Pierce Protein G Agarose (can be purchased from ThermoFisher Scientific).
2. Periplasmic lysis buffer: 20% sucrose, 30 mM Tris–HCl pH 8, 1 mM EDTA, 0.2 mg/mL lysozyme, 1× Halt Protease Inhibitor Cocktail (ThermoFisher Scientific). Stock solutions of 50% sucrose; 1 M Tris–HCl, pH 8; and 500 mM EDTA were diluted to make this solution (*see* **Note 5**).
3. 1 mL disposable plastic syringe (1 × 100 NORM-JECT).
4. Cotton wool.
5. PGAC binding buffer: 50 mM sodium acetate, pH 5.2.
6. PGAC elution buffer: 100 mM glycine-HCl, pH 2–3.

7. Neutralization buffer: 1 M Tris–HCl, pH 8.
8. Amicon Ultra centrifugal filter units (Millipore, 3 K MWCO).
9. Phosphate buffer: 100 mM·Na_2HPO_4, 150 mM·NaCl, pH 7.
10. 3 kDa MWCO Slide-A-Lyzer™ MINI Dialysis devices (ThermoFisher Scientific).

2.4 Preparation of Fluorescent Diazonium Compound

1. Sodium nitrite.
2. 6-Aminofluorescein (Sigma-Aldrich).
3. 10 mM HCl.
4. 6 mM 5HTP in water.
5. 10 kDa molecular weight cutoff Amicon Ultra-0.5 mL, centrifugal filters.

2.5 Analysis of Protein Samples

1. Bradford assay Reagent (Coomassie Protein Assay Reagent, ThermoFisher Scientific).
2. Polyacrylamide gradient gels, such as 4–15% TGX gels from Bio-Rad.
3. Compatible electrophoresis chamber.
4. 6× SDS gel loading buffer: To make 10 mL, add 1.5 g sodium dodecyl sulfate (SDS), 3 mL β-mercaptoethanol, 3.5 mL of a 1 M Tris–HCl, pH 6.3 solution, 120 μL bromophenol blue, and glycerol to a final volume of 10 mL.
5. Tris-glycine protein gel running buffer: To make 10× buffer, dissolve 30.3 g Tris base, 114.1 g glycine, and 10 g SDS in 1 L ddH_2O. For running gels, dilute the 10× buffer tenfold.
6. Molecular weight markers, such as Bio-Rad Kaleidoscope ladder.
7. Coomassie protein gel stain (such as PageBlue protein stain from ThermoFisher Scientific).
8. Gel imager capable of detecting Coomassie stain and fluorescent label (we use Bio-Rad ChemiDoc MP imaging system).
9. HPLC-coupled mass-spectrometer (we use Agilent Technologies, 1260 Infinity ESI-TOF).

3 Methods

The engineered EcTrpRS–tRNA pair can be used for 5HTP incorporation into proteins expressed in the ATMW1 strain but not in other strains of *E. coli*, since it would cross-react with the endogenous Trp-pair of the host [12]. ATMW1 was made from the EcNR1 strain, which is a close derivative of the wild-type MG1655 strain. This strain does not encode a T7 RNA polymerase and, consequently, would not support protein expression from the T7 promoter. However, other common promoters such as P_{BAD}, T5-lac, and Tac, which use the *E. coli* RNA polymerase, work well in this strain for recombinant expression of target proteins. Additionally, it is important to note that the EcTrpRS–tRNA pair is used as an opal suppressor for Uaa incorporation (*see* **Note 6**).

Cytoplasmic expression of sfGFP-151-TGA from the pET22-T5lac-sfGFP-151-TGA plasmid is driven by the IPTG-inducible T5-lac promoter. The two subunits of αHer2-Fab, the heavy and the light chain, are each expressed from the arabinose-inducible P_{BAD} promoter from the pBK-aHer2-Fab plasmid. An STII leader sequence is included at the N-terminus of each subunit to facilitate their export to the periplasmic space, where the Fab can fold properly.

3.1 Expression and Purification of sfGFP-151-5HTP

1. Cotransform 100 ng each of pET22-T5lac-sfGFP-151-TGA and pEvoltac-EcW-TGA-h14 into 50 μL of electrocompetent ATMW1, recover with 1 mL LB and incubate at 37 °C for 1 h. Plate the cells on LB-Agar supplemented with 50 μg/mL spectinomycin, 100 μg/mL ampicillin, 35 μg/mL chloramphenicol, and 10 μg/mL gentamycin, and incubate overnight at 37 °C. The pET22 and the pEvoltac plasmids encode a beta-lactamase and a chloramphenicol acetyl transferase marker, respectively, while the ATMW1 strain encodes markers conferring resistance to gentamycin and spectinomycin.
2. Use a single colony from the plate from **step 1** to inoculate 5 mL LB media (supplemented with 50 μg/mL spectinomycin, 100 μg/mL ampicillin, 35 μg/mL chloramphenicol, and 10 μg/mL gentamycin) in a sterile plastic tube and grow at 37 °C with shaking (250 rpm) to saturation (*see* **Note 7**).
3. Use 200 μL of the starter culture from **step 2** to inoculate 20 mL LB (supplemented with 50 μg/mL spectinomycin, 100 μg/mL ampicillin, 35 μg/mL chloramphenicol, and 10 μg/mL gentamycin) and grow at 37 °C with shaking (250 rpm) until OD_{600} reaches 0.6 (*see* **Note 8**).
4. At OD_{600} 0.6, add IPTG and 5HTP to a final concentration of 1 mM each (*see* **Note 9**).
5. Incubate the culture at 30 °C with shaking (250 rpm) for 12–16 h (*see* **Note 10**).
6. Harvest the cells by centrifugation (10 min, 5000 × *g*). The cell pellet can be stored at −80 °C for future use.
7. Thaw the cell pellet in 0.5 mL of cytoplasmic lysis buffer, resuspend thoroughly by pipetting, and incubate on ice for 15 min.
8. Dilute the cell lysate with 1.5 mL of equilibration buffer, and clarify by centrifuging (10 min, 14,000 × *g*, 4 °C).

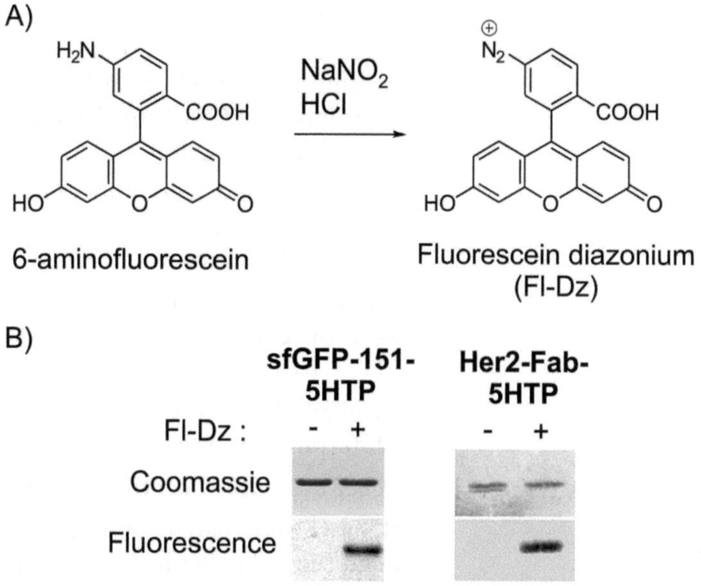

Fig. 2 (**a**) Scheme showing the preparation of fluorescein diazonium by the diazotization of 6-aminofluorescein. (**b**) Labeling of sfGFP-151-5HTP and Her2-Fab-169-5HTP by fluorescein diazonium visualized by fluorescence imaging following reducing SDS-PAGE

9. Add 0.2 mL HisPur Ni-NTA resin into a disposable plastic column and wash once with 0.5 mL water followed by twice with 0.5 mL IMAC equilibration buffer each.
10. Gently add the clarified lysate from **step 8** on to equilibrated HisPur resin and allow it to flow through by gravity.
11. Wash beads three times with 5 mL of IMAC wash buffer.
12. Elute the protein using 0.6 mL of IMAC elution buffer (*see* **Note 11**).
13. Dialyze the protein against phosphate buffer using a Slide-A-Lyzer™ MINI Dialysis device.
14. Assess protein purity by SDS-PAGE (Fig. 2b). Confirm the incorporation of 5HTP by verifying the mass of the protein by mass spectrometry. The expected yields for wild-type and the 5HTP mutant sfGFP are 140 mg/L and 90 mg/L, respectively.

3.2 Expression and Purification of αHer2-Fab-169-5HTP

1. Cotransform pBK-αHer2-Fab-169TGA (*see* **Note 12**) and pEvoltac-EcW-TGA-h14 into electrocompetent ATMW1 strain, recover with 1 mL LB and incubate at 37 °C for 1 h with shaking at 250 rpm. Plate the cells on LB-Agar supplemented with 50 μg/mL spectinomycin, 50 μg/mL kanamycin, and 35 μg/mL chloramphenicol, and incubate overnight at 37 °C.

2. Inoculate a single colony into a 5 mL starter culture of LB supplemented with 50 μg/mL spectinomycin, 50 μg/mL kanamycin, and 35 μg/mL chloramphenicol, and incubate it overnight at 37 °C with shaking at 250 rpm.
3. Inoculate 500 mL LB (supplemented with spectinomycin, kanamycin, and chloramphenicol) in a 2 L Erlenmeyer flask with the 5 mL starter culture and incubate at 37 °C with shaking (250 rpm) until OD_{600} reaches 0.6.
4. Induce protein expression by adding IPTG and L-arabinose to final concentrations of 1 mM and 0.02%, respectively. Also, add 5HTP to a final concentration of 1 mM.
5. Incubate the culture at room temperature (22 °C) with shaking (250 rpm) for 16 h.
6. Pellet the cells by centrifugation at 5000 × *g* for 10 min.
7. Remove supernatant by pouring it off completely and resuspend the cell pellet in 40 mL of periplasmic lysis buffer.
8. Incubate at 37 °C for 30 min.
9. Clarify the lysate by centrifugation at 17,000 × *g* for 30 min at 4 °C.
10. Dilute sample 1:1 in PGAC binding buffer and keep on ice.
11. Pack a 1 mL disposable plastic syringe with cotton wool to the 500 μL mark.
12. Pack column with 50 μL protein G agarose resin ensuring even distribution (*see* **Note 13**).
13. Equilibrate the resin by washing with 500 μL of PGAC binding buffer. Pipette liquid slowly and carefully down the side of the column, making sure that the resin bed is not disturbed.
14. Gently load clarified lysate onto column and let it flow through by gravity (*see* **Note 14**).
15. After sample has completely gone through column, wash with 1 mL of PGAC binding buffer (*see* **Note 15**).
16. Elute antibody with 0.5 mL of PGAC elution buffer in 50 μL fractions. Immediately neutralize each elution using 10 μL neutralization buffer (*see* **Note 16**).
17. To determine which fractions contain the antibody, add 5 μL from each elution fraction to 50 μL of Bradford assay reagent. Visually compare the intensity of the blue color to a control, where 5 μL PGAC elution buffer was added to 50 μL of Bradford assay reagent.
18. Pool the fractions containing the antibody and concentrate down to 100 μL using a 3 kDa MWCO centrifugal filter following the manufacturer's instruction.

19. Dialyze against phosphate buffer using a Slide-A-Lyzer™ MINI Dialysis device.
20. Determine protein concentration by Bradford assay, and purity by SDS-PAGE analysis (Fig. 2b). Incorporation of the 5HTP residue can also be verified by mass-spectrometry analysis. The expected yields for wild-type and the 5HTP mutant Fab are 0.35 mg/L and 0.17 mg/L, respectively.

3.3 Preparation of the Fluorescent Diazonium Compound and Protein Labeling

1. Make a 100 μL 10 mM solution of 6-aminofluorescein in 10 mM HCl.
2. Separately make a 100 μL 60 mM sodium nitrite solution in water (*see* **Note 17**).
3. Cool both solutions on ice. Add 20 μL of 60 mM·$NaNO_2$ into 100 μL of 10 mM fluorescein, mix by brief vortexing, then incubate on ice for 5 min to result in a 8.33 mM fluorescein--diazonium solution (Fig. 2a) (*see* **Note 18**).
4. Use 880 μL of ice-cold water to dilute this solution to reach a final concentration of 1 mM fluorescein–diazonium.
5. Add 1 μL of 1 mM fluorescein–diazonium solution to a 10 μL solution of sfGFP-151-5HTP or αHer2-Fab-169-5HTP (ideally, 4–10 μM protein concentration) in phosphate buffer, mix immediately by pipetting and incubate on ice for 30 min (*see* **Note 19**).
6. Quench the unreacted diazonium by adding 2 μL of 6 mM 5HTP and mix by pipetting.
7. To remove excess reagent and desalt the protein back to phosphate buffer, dilute the reaction by adding 250 μL of phosphate buffer and concentrate back to 25 μL using 10 kDa MWCO Amicon Ultra-0.5 mL centrifugal filters (*see* **Note 20**).
8. Resolve the modified protein as well as an unmodified control by SDS-PAGE and verify fluorescence labeling using an imager (we use Bio-Rad ChemiDoc MP imaging system).
9. Verify the degree of labeling by mass spectrometry analysis of the protein with or without modification.

4 Notes

1. The ATMW1 strain is available from the Chatterjee group upon request. It can be used to prepare homemade competent cells using routine protocols. The competency of this strain was found to be comparable to other common laboratory strains of *E. coli.* This strain contains the yeast TrpRS–tRNA pair encoded in a plasmid with a spectinomycin resistance marker and CloDF13 origin of replication.

2. The pEvoltac-EcW-TGA-h14 plasmid is available from the Chatterjee group upon request. It encodes a mutant EcTrpRS, engineered to selectively charge 5HTP, from an IPTG-inducible TacI promoter and the $tRNA^{EcTrp}{}_{UCA}$ from a constitutive proK promoter.
3. The pET22-T5lac-sfGFP-151-TGA and pBK-aHer2-Fab-K169TGA plasmids are available from the Chatterjee group upon request.
4. Dissolve 5HTP in water; if necessary, add 2 M sodium hydroxide in small volumes followed by vigorous shaking to fully dissolve. Store the amino acid solution frozen at −20 °C.
5. Preparation of 500 mM EDTA requires addition of NaOH to go into solution.
6. Although Uaa incorporation in *E. coli* using all other systems is performed using the suppression of the TAG nonsense codon, the EcTrpRS/tRNA must be used as a TGA suppressor instead. This is because the TAG suppressor variant of the $tRNA^{EcTrp}$ cross-reacts with the *E. coli* glutaminyl-tRNA synthetase. In contrast, the TGA suppressor variant of the same tRNA is orthogonal yet functional in ATMW1.
7. Typically the culture is started in the evening and is allowed to grow overnight (approximately 16 h), which is sufficient to reach saturation.
8. The protein expression can be induced at OD_{600} 0.5–0.7 without significantly affecting the yield.
9. We have previously shown that significantly reduced expression of sfGFP-151-TGA is observed if 5HTP is not added to the culture. To verify this observation, or when extending this 5HTP incorporation protocol to a different target, it is necessary to set up an identical culture which is induced with IPTG, but no 5HTP is added.
10. When extending this protocol to other target proteins, the temperature and the length of expression must be empirically optimized to maximize expression yield.
11. For efficient recovery, collect 50 μL fractions and pool only the fractions containing sfGFP, which is easily detectable by its characteristic fluorescence (using a handheld UV lamp). For other nonfluorescent proteins, Bradford assay can be used to evaluate the protein concentration in eluted fractions.
12. Uaas have been incorporated into several other sites on the surface of the light chain of the αHer2-Fab [23, 24]. In addition to the site described here, these sites could also be used for 5HTP incorporation.

13. The bead capacity is 11–15 mg human IgG per mL of settled resin. Typical yield obtained for the αHer2-Fab-169-5HTP mutant was around 0.17 mg/L bacterial culture.
14. Yield can be improved by applying the flow-through over beads a second time.
15. The wash step can be speed up by applying gentle air pressure to the top of the column. However, make sure not to push air through the resin bed and dry it out.
16. Elute directly into 10 μL neutralization buffer. Check the pH of the solution and add more neutralization buffer if needed.
17. Sodium nitrite solution must be prepared fresh.
18. Aromatic diazonium compounds are not stable and would hydrolyze over time. So, these must be made fresh and used immediately. To confirm the generation of the fluorescein diazonium species, an excess of free 5HTP can be added to rapidly and quantitatively trap this unstable species to generate the corresponding stable azo compound. The reaction mixture can be subsequently analyzed by HPLC-coupled mass-spectrometry analysis to evaluate the amount of azo compound generated.
19. Even though the aryldiazonium compounds react with 5HTP at a significantly higher rate relative to competing canonical amino acids such as tyrosine, it is important to carefully optimize the concentration of the diazonium species and the reaction time to ensure selective and complete labeling of the 5HTP residue, while avoiding nonspecific labeling elsewhere. This is particularly important when extending this labeling strategy to different aryldiazonium compounds, which exhibit dramatically different rates of azo-coupling. Subjecting the corresponding wild-type protein (lacking a 5HTP residue) to identical labeling conditions provides a useful control experiment during such optimization efforts. Additionally, we have recently encountered some cases, where a wild-type protein reacts with aryldiazonium compounds at a much faster rate than what is anticipated from the reactivity of free canonical amino acids. This may be due to the local environment of an amino acid residue which "activates" it for the azo-coupling reaction. This aspect should be carefully investigated for labeling a new target protein using CRACR.
20. Removing the unwanted small molecules is not essential for analyzing the protein by SDS-PAGE or HPLC-coupled mass spectrometry. However, it is necessary for downstream applications of the protein conjugate. For example, we used fluorescence-activated cell sorting to demonstrate that the fluorophore-labeled αHer2-Fab can bind SK-BR-3 cells that overexpress the HER2 receptor.

Acknowledgments

This work was supported by NIH (R01GM124319 to A.C.).

References

1. Lang K, Chin JW (2014) Bioorthogonal reactions for labeling proteins. ACS Chem Biol 9:16–20. https://doi.org/10.1021/cb4009292
2. McKay CS, Finn MG (2014) Click chemistry in complex mixtures: bioorthogonal bioconjugation. Chem Biol 21:1075–1101. https://doi.org/10.1016/j.chembiol.2014.09.002
3. Ramil CP, Lin Q (2013) Bioorthogonal chemistry: strategies and recent developments. Chem Commun (Camb) 49:11007–11022. https://doi.org/10.1039/c3cc44272a
4. Shih HW, Kamber DN, Prescher JA (2014) Building better bioorthogonal reactions. Curr Opin Chem Biol 21:103–111. https://doi.org/10.1016/j.cbpa.2014.07.002
5. Sletten EM, Bertozzi CR (2009) Bioorthogonal chemistry: fishing for selectivity in a sea of functionality. Angew Chem Int Ed Engl 48:6974–6998. https://doi.org/10.1002/anie.200900942
6. Stephanopoulos N, Francis MB (2011) Choosing an effective protein bioconjugation strategy. Nat Chem Biol 7:876–884. https://doi.org/10.1038/nchembio.720
7. Dumas A, Lercher L, Spicer CD, Davis BG (2015) Designing logical codon reassignment–expanding the chemistry in biology. Chem Sci 6:50–69. https://doi.org/10.1039/C4SC01534G
8. Kim CH, Axup JY, Schultz PG (2013) Protein conjugation with genetically encoded unnatural amino acids. Curr Opin Chem Biol 17:412–419. https://doi.org/10.1016/j.cbpa.2013.04.017
9. Lang K, Chin JW (2014) Cellular incorporation of unnatural amino acids and bioorthogonal labeling of proteins. Chem Rev 114:4764–4806. https://doi.org/10.1021/cr400355w
10. Young DD, Schultz PG (2018) Playing with the molecules of life. ACS Chem Biol 13:854–870. https://doi.org/10.1021/acschembio.7b00974
11. Italia JS, Zheng Y, Kelemen RE, Erickson SB, Addy PS, Chatterjee A (2017) Expanding the genetic code of mammalian cells. Biochem Soc Trans 45:555–562. https://doi.org/10.1042/BST20160336
12. Italia JS, Addy PS, Wrobel CJ, Crawford LA, Lajoie MJ, Zheng Y, Chatterjee A (2017) An orthogonalized platform for genetic code expansion in both bacteria and eukaryotes. Nat Chem Biol 13:446–450. https://doi.org/10.1038/nchembio.2312
13. Addy PS, Erickson SB, Italia JS, Chatterjee A (2017) A chemoselective rapid azo-coupling reaction (CRACR) for unclickable bioconjugation. J Am Chem Soc 139:11670–11673. https://doi.org/10.1021/jacs.7b05125
14. Addy PS, Italia JS, Chatterjee A (2018) An oxidative bioconjugation strategy targeted to a genetically encoded 5-hydroxytryptophan. Chembiochem 19:1375–1378. https://doi.org/10.1002/cbic.201800111
15. Italia JS, Addy PS, Erickson SB, Peeler JC, Weerapana E, Chatterjee A (2019) Mutually orthogonal nonsense-suppression systems and conjugation chemistries for precise protein labeling at up to three distinct sites. J Am Chem Soc 141(15):6204–6212
16. Bruckman MA, Kaur G, Lee LA, Xie F, Sepulveda J, Breitenkamp R, Zhang X, Joralemon M, Russell TP, Emrick T, Wang Q (2008) Surface modification of tobacco mosaic virus with “click” chemistry. Chembiochem 9:519–523. https://doi.org/10.1002/cbic.200700559
17. Gavrilyuk J, Ban H, Nagano M, Hakamata W, Barbas CF 3rd (2012) Formylbenzene diazonium hexafluorophosphate reagent for tyrosine-selective modification of proteins and the introduction of a bioorthogonal aldehyde. Bioconjug Chem 23:2321–2328. https://doi.org/10.1021/bc300410p
18. Hooker JM, Kovacs EW, Francis MB (2004) Interior surface modification of bacteriophage MS2. J Am Chem Soc 126:3718–3719. https://doi.org/10.1021/ja031790q
19. Schlick TL, Ding Z, Kovacs EW, Francis MB (2005) Dual-surface modification of the tobacco mosaic virus. J Am Chem Soc 127:3718–3723. https://doi.org/10.1021/ja046239n
20. He J, Kimani FW, Jewett JC (2015) A photobasic functional group. J Am Chem Soc 137:9764–9767. https://doi.org/10.1021/jacs.5b04367

21. Jensen SM, Kimani FW, Jewett JC (2016) Light-activated triazabutadienes for the modification of a viral surface. Chembiochem 17:2216–2219. https://doi.org/10.1002/cbic.201600508
22. Kimani FW, Jewett JC (2015) Water-soluble triazabutadienes that release diazonium species upon protonation under physiologically relevant conditions. Angew Chem Int Ed Engl 54:4051–4054. https://doi.org/10.1002/anie.201411277
23. Hutchins BM, Kazane SA, Staflin K, Forsyth JS, Felding-Habermann B, Schultz PG, Smider VV (2011) Site-specific coupling and sterically controlled formation of multimeric antibody fab fragments with unnatural amino acids. J Mol Biol 406:595–603. https://doi.org/10.1016/j.jmb.2011.01.011
24. Hutchins BM, Kazane SA, Staflin K, Forsyth JS, Felding-Habermann B, Smider VV, Schultz PG (2011) Selective formation of covalent protein heterodimers with an unnatural amino acid. Chem Biol 18:299–303. https://doi.org/10.1016/j.chembiol.2011.01.006

Chapter 17

Facile Site-Specific Multiconjugation Strategies in Recombinant Proteins Produced in Bacteria

Hannes Merten, Jonas V. Schaefer, Fabian Brandl, Uwe Zangemeister-Wittke, and Andreas Plückthun

Abstract

For biomedical applications, proteins may require conjugation to small and large molecules. Typical examples are dyes for imaging, cytotoxic effector molecules for cell killing, or half-life extension modules for optimized pharmacokinetics. Although many conjugation strategies are straightforward to apply, most of them do not enable site-specific and orthogonal conjugation, and do not yield a defined stoichiometry. Moreover, techniques offering these desirable features often rely on complex expression procedures and suffer from low production yields. A more promising manufacturing strategy for flexible, site-specific and stoichiometrically defined payloading of proteins is the combination of click chemistry and thiol–maleimide conjugation, which even enables dual labeling when used consecutively. Here, we describe as an example the production of Designed Ankyrin Repeat Proteins (DARPins), a non-IgG binding scaffold, in a specific *E. coli* strain to obtain high yields of protein carrying both a thiol and an azide group. We provide straightforward protocols for strain-promoted azide–alkyne cycloaddition (SPAAC) and thiol–maleimide conjugation, and furthermore compare these conjugation chemistries with existing alternatives like copper-catalyzed azide–alkyne cycloaddition (CuAAC). Finally, detailed instructions for reactivity analysis and yield estimations of the reactions are provided.

Key words DARPins, Thiol–maleimide conjugation, Click chemistry, CuAAC, SPAAC, Bacterial expression, PEG stain, Polyethylene glycol, Site-specific conjugation, Bioorthogonal

1 Introduction

To investigate pharmacokinetic properties and biodistribution of proteins in in vivo models, or determine their subcellular localization, proteins can be coupled with various chemical moieties such as dyes for visualization and imaging, or radiotracers. For other applications, such moieties must bring specific functional properties to the protein like an increased hydrodynamic radius for an elevated circulation half-life or cytotoxic activity for cancer therapy. The underlying chemistries for the modification of proteins ideally combine both site specificity and a controlled stoichiometry and

Sam Massa and Nick Devoogdt (eds.), *Bioconjugation: Methods and Protocols*, Methods in Molecular Biology, vol. 2033, https://doi.org/10.1007/978-1-4939-9654-4_17,

hence need to be fully orthogonal to the functional groups in the protein (e.g., thiols, alcohols, and amines) to prevent unspecific conjugation that might interfere with target binding or protein function. The components used in such conjugations must be stable and small, while the reaction must be fast and conditions mild enough to be tolerated by the protein [1]. A popular concept employs specific enzymes, either for the conjugation reaction itself or for the incorporation of unique functional groups into proteins. These approaches have several benefits but also drawbacks as described elsewhere [2]. Most importantly, they usually require extensive optimizations to reach conjugation efficiencies and yields comparable to chemical modifications. Furthermore, the specific requirements of the coupling enzyme might limit the applicability of these techniques (e.g., by having constraints for the conjugation site, which for some enzymes must be located at the *C*-terminus, or by requiring particular recognition sequences).

One of the most popular conjugation techniques for protein modification is the reaction between a primary *N*-terminal amine and an *N*-hydroxy-succinimide (NHS)-ester (*see* Table 1). However, most proteins contain a number of lysine residues and thus a number of amine nucleophiles. As the reaction relies on the deprotonation of the reacting amine, and since the pK_a values of the *N*-terminal amine is only slightly lower than that of the ε-amino group of a fully exposed lysine (typically 8 vs. 10.5), specificity for the *N*-terminus is difficult to reach in practice. Thus, the reaction can only be optimized *either* for yield (higher pH) *or* for specificity for *N*-terminal conjugation (lower pH) [3]. Hence, amine conjugation is rather impractical when *both* the quantitative and the site-specific modifications at the *N*-terminus are crucial, and when the protein has multiple lysines. The engineering of proteins that are free of lysines except where needed is possible [4], but not many proteins tolerate this strategy.

Alternatively, the maleimide–thiol Michael addition offers a very fast reaction to quantitatively conjugate proteins under mild conditions and independent of particular buffers (*see* Table 1) [5, 6]. Proteins free of cysteine or with no surface-exposed cysteine are suited for this method, but also the intermolecular disulfide bridges of IgG molecules can be exploited to generate drug conjugates upon mild reduction [7–10]. Site-specificity can be assured by the incorporation of a single, surface-exposed cysteine in the protein of interest carrying a unique thiol residue. The maleimide group is then provided by the chemical payload (e.g., a maleimide-dye or maleimide–toxin). In contrast to NHS-chemistry, the maleimide–thiol reaction shows very little to no side reactions when reagent excess and reaction time are kept reasonable, while at long times and/or high reagent excess a slower reaction with primary amines is seen [11]. Because of the rather robust nature of the maleimide–thiol reaction, a comprehensive catalogue of protein-

Table 1
Conjugation chemistries

Reaction chemistry	Reaction scheme	Reaction rate	Conjugation site in protein	Incorporation method
Amine-NHS	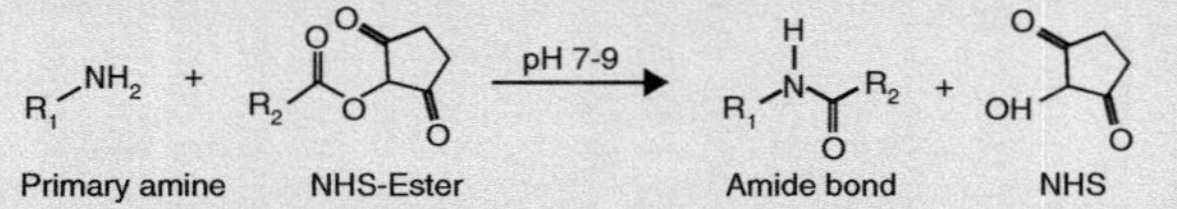	Fast	Lysine, terminal amine	Natural
Maleimide–thiol Michael addition	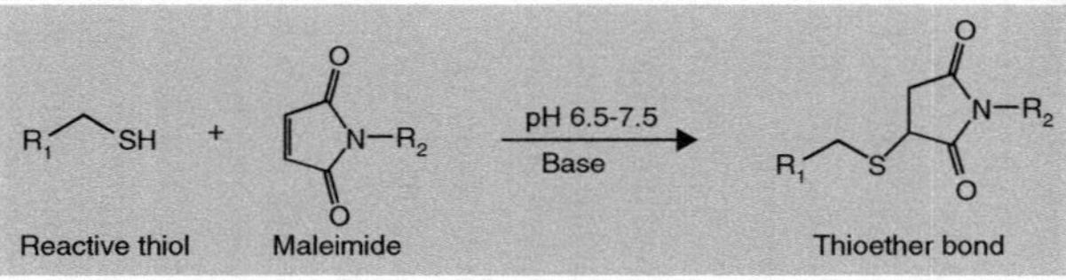	Fast	Cysteine	Natural
Cycloaddition CuAAC	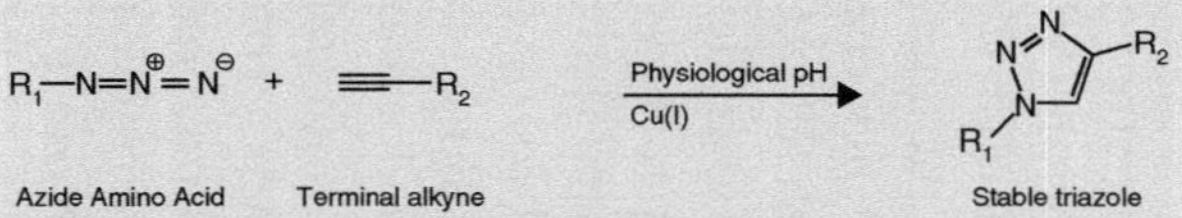	Slow	Azidohomoalanine	Methionine-auxotrophic *E. coli*
Cycloaddition SPAAC	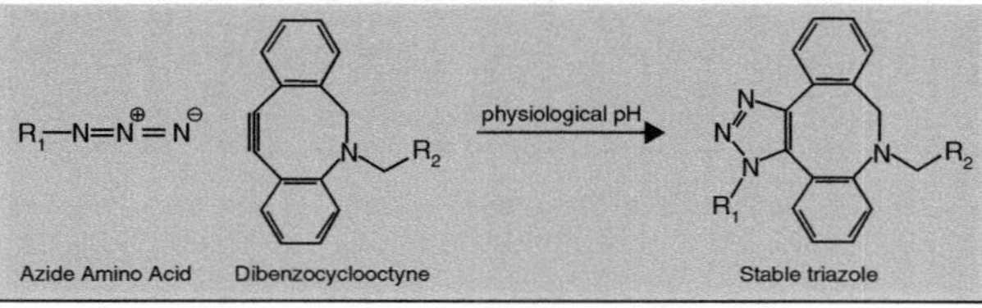	Slow	Azidohomoalanine	Methionine-auxotrophic *E. coli*

R_1: protein (e.g., DARPin), R_2: payload (e.g., PEG, dye, or cytotoxin). Table adapted from [40]. For reaction details see [16]

modifying agents (e.g., polyethylene glycol (PEG) and a plethora of dyes and chelators) are commercially available. However, a drawback of this reaction is that its efficiency depends on the complete reduction of the exposed cysteine and on avoiding its reoxidation during conjugation. Whereas this is easily achievable for originally cysteine-free proteins where an artificial cysteine is introduced, it is more challenging for proteins carrying internal disulfide bridges because prevention of oxidation of the exposed cysteine must not lead to reduction of the native disulfide bonds.

The principal reversibility of the thiol–maleimide reaction has been much discussed [12], but for most applications the back reaction is slow enough to be irrelevant, and the product can also be further stabilized by ring opening of the maleimide.

A number of bioorthogonal chemistries have been developed [13–16], all with their pros and cons. A very robust bioorthogonal conjugation approach exploits the cycloaddition reaction of an azide and an alkyne forming a stable triazole. Good reaction rates and stability of the reactants have made this type of reaction widely used. The uncatalyzed reaction is very slow, and has thus become useful only when accelerated, either with a terminal alkyne as Cu (I)-catalyzed azide–alkyne cycloaddition (CuAAC) [17, 18] or by using ring-strained cyclooctynes as strain-promoted azide–alkyne cycloaddition (SPAAC) [14], the latter usually being denoted as "click chemistry" (*see* Table 1) [19]. (It might be noted that in some of the literature, all biorthogonal reactions are collectively termed "click chemistry," but we will use this term here synonymous with SPAAC). To improve the slow reaction rates of cyclooctyne, various derivatives have been synthesized, of which the cyclopropanated variant bicyclo[6.1.0]nonyne (BCN) and the benzoannulated derivative dibenzocyclooctyne (DBCO), also denoted as aza-dibenzocyclooctyne (DIBAC), are the most important ones [20–22]. Both reactants improve reaction rates ~100-fold over the unmodified cyclooctyne, but in our hands, DBCO could best combine high stability and solubility with useful reaction kinetics. The CuAAC, conversely, has sometimes led to undesired protein precipitation, protein oxidation and comparably low reaction yields and furthermore required optimization for every protein individually (*see* **Note 1**). Therefore, we discuss here only the use of click chemistry with DBCO to modify proteins. Please note that, despite the use of rate-optimized cyclooctynes like DBCO and BCN, the described reactions are still relatively slow (0.1–0.3 M^{-1} s^{-1}) [16]. Moreover, unspecific coupling to irrelevant reaction partners (e.g., thiol coupling to the triple bond) cannot be excluded, and this limits the orthogonality of the reaction [21–24]. However, since the side reactions are slow and since proteins manufactured for bioconjugation are usually very pure, such side reactions are unlikely to be of concern under the recommended reaction conditions, and furthermore, the stability of the components used for click chemistry remains as an undisputed advantage. Nonetheless, DBCO has been reported to be unstable to strong acids and strong bases [21, 25], and the azide can in principle be converted to a primary amine under strong reducing conditions [26, 27].

For all protein conjugation reactions, including those using click chemistry, it is beneficial if the conjugation site in the protein can be freely chosen. This can be achieved with nonnatural amino acids containing azides to replace L-methionine. The methionine codon can be incorporated in the coding sequence of the protein at a desired position and the aminoacyl tRNA synthetase for methionine in *E. coli* accepts the methionine-surrogate L-azidohomoalanine (Aha). Thus, in methionine-auxotrophic *E. coli* strains like B834 (DE3), after simple methionine depletion of the culture, Aha can be introduced without any other changes in the *E. coli* strain [28]. However, this technique requires that the protein can be produced in *E. coli* (see detailed protocol below). The amber-suppression technology [29, 30] for the site-specific incorporation of bicyclo[6.1.0]nonyne (BCN)-lysine [31, 32] or other nonnatural amino acids carrying quinones [33, 34] or norbornenes with a nitrile-imine reagent [35, 36] offers an alternative approach. Nevertheless, even if the rate constants of click chemistry can be highly accelerated with these nonnatural amino acids [37], due to its rather low expression yield this technology is better suited to prepare imaging agents, for which only small amounts are needed, than for larger scale preparation of, for example, therapeutic proteins. Another disadvantage is that the efficiency of the amber suppression strongly depends on the surrounding sequences of the amber site and thus requires intensive engineering and optimization for each protein, which limits the freedom of protein design (unpublished data). Although these drawbacks can in principle be overcome using sequence optimization [38] and adapted *E. coli* strains [39], the problem remains that, compared to azide-containing nonnatural amino acids like Aha, nonnatural amino acids like BCN-lysine are expensive, and, to our knowledge, bacterial strains and vectors carrying tRNA–RNA synthase pairs for the incorporation of many relevant nonnatural amino acids have not been made commercially available. Therefore, this technology will not be discussed further herein.

Antibody–drug conjugates (ADCs) are amongst the most popular examples of FDA-approved protein conjugates currently applied in the clinic. However, for reasons described above, site-specific incorporation of bioorthogonal chemical moieties in full-length antibodies produced in eukaryotic systems with high expression yields and without interfering with the multiple functions of the IgG-molecule, is very difficult [40]. Hence, high-affinity binding proteins [41] that are not derived from IgGs and can be produced in bacterial hosts allow the use of these new bioorthogonal chemistries for fast and site-specific drug conjugation with high yield. They enable thus the screening of a wide range of pharmacological and molecular properties, and have become of interest and are thus currently under widespread preclinical investigation [42]. One such new scaffold are Designed Ankyrin Repeat Proteins (DARPins) [43]. Their robust nature allows the expression in large quantities in *E. coli*, they can be engineered to various formats from monovalent to bivalent to tetravalent and tolerate the introduction of reactive groups for site-specific and bioorthogonal conjugation [40, 44, 45]. DARPins can be easily selected to bind targets with high specificity and affinity and usually carry neither a free cysteine nor an essential methionine, thus providing a high freedom of engineering using the maleimide–thiol and click chemistry reactions described above [46]. Typically, a cysteine (e.g., at the *C*-terminus or anywhere else in the sequence) is introduced to allow the maleimide-based conjugation of effector functions, including the cytotoxic payloads known from classical ADCs (*see* Fig. 1). For production, the DARPin is expressed in minimal medium using a methionine-auxotrophic *E. coli* strain to introduce Aha at the *N*-terminus (methionine start codon, or if the initiator Met residue is cleaved off in front of a small residue, at a methionine introduced elsewhere). This results in a binding protein carrying two unique functional groups, at freely choosable positions, useful for various combinations of protein-modifying moieties, such as PEG, dyes or cytotoxins (*see* Fig. 1) [45, 47]. As mentioned, the *N*-terminus is not the only site allowing Aha incorporation. If the second amino acid is small (e.g., Ala, Gly, or Ser), the *N*-terminal Aha is cleaved off [48, 49], thereby enabling the introduction of a unique Aha at various desired positions elsewhere in the protein.

Here, we provide a detailed protocol how to express DARPins in minimal medium to incorporate a unique azide (Aha) and thiol (cysteine) residue, and describe how to use both functional groups sequentially for site-specific, bioorthogonal and stoichiometrically defined conjugation reactions. Furthermore, methods for the biochemical and functional analysis of the engineered DARPins and the conjugates will be described.

2 Materials

Prepare all solutions using ultrapure water and analytical grade reagents, if not stated otherwise. Autoclave all media, but make sure to not autoclave the nutrient mix used for the minimal expression medium. Filter all solutions and reagents using a 0.22 μm filter. Handling of bacteria including plating, cultivation, transformation, and cell lysis is conducted according to standard protocols, if not stated otherwise. We only list reagents and media, which are required to employ methods described here in detail without mentioning standard reagents.

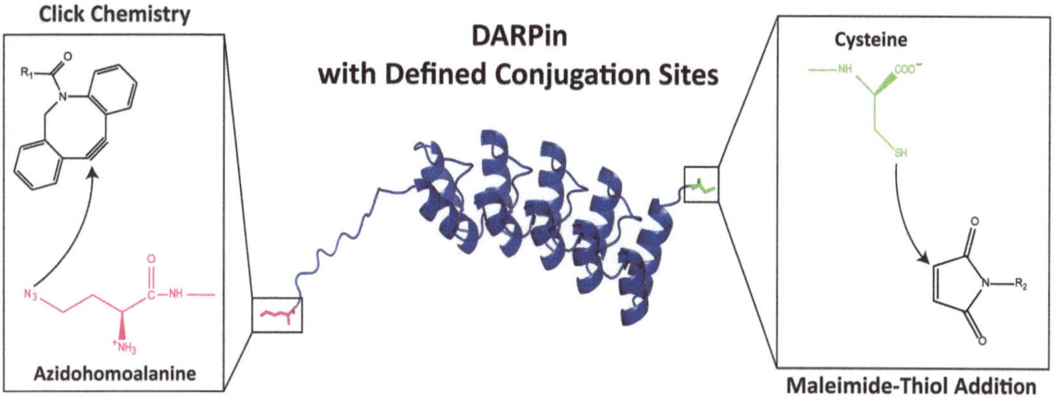

Fig. 1 DARPin protein (blue) equipped with two unique chemical moieties for the bioconjugation of different payloads (R_1 and R_2) at defined and freely choosable positions in the DARPin-scaffold. The terminal positions are merely drawn here for the clarity of illustration. A unique azidohomoalanine (pink) (for illustration shown at the *N*-terminus) and cysteine (green) (for illustration shown at the *C*-terminus) can be used for click chemistry or maleimide thiol coupling, resulting in stoichiometrically defined bioconjugates

2.1 Expression of an Azide–DARPin–-Cysteine Construct in Minimal Medium

1. Chemically competent cells of the *E. coli* B-strain B834 (DE3) (*F⁻ ompT gal hsdS*$_B$ *(r*$_B$$^-$ *m*$_B$$^-$*) met dcm lon λ(lacI, lacUV5-T7 gene 1, ind1, sam7, nin5)*).
2. Vector: Use a plasmid containing the modified construct with the respective cysteine and/or methionine codon and a His_6-tag for purification. It should allow IPTG-inducible expression and contain the *lacI* gene under the strong *lacI*^{q} promoter (e.g., a modified pQE30 (Qiagen) or pET-28b(+) (Novagen)).
3. M9 pre-expression medium: Mix the medium base (SelenoMet™ Base Medium & Nutrients Glucose Free, Molecular Dimensions) with the nutrient mix according to the manufacturer's instructions and add 100 mg/L ampicillin, 1% (w/v) glucose, and 40 mg/L L-methionine. This medium is used for the pre-pre-culture, the o/n culture, and the pre-expression culture.
4. M9 expression medium: Mix the medium base (SelenoMet™ Base Medium & Nutrients Glucose Free, Molecular Dimensions) with the nutrient mix according to the manufacturer's instructions and add 100 mg/L ampicillin, 0.4% (v/v) glycerol, and 40 mg/L of the nonnatural amino acid L-azidohomoalanine (H-Dab(N3)·HCl, Bapeks, Riga, Latvia). This medium is used for the expression-culture only.
5. 1 M isopropyl-β-D-thio-galactopyranoside (IPTG) solution: Dissolve IPTG in water and filter in a sterile vessel using a 0.22 μm filter. Aliquot and store at −20 °C.

2.2 Dual Conjugation Using Maleimide–Thiol and Click Chemistry

1. Appropriate degassing device (e.g., vacuum pump and adapter for flasks or a helium supply).
2. Inert gas supply (e.g., argon or N_2) to flush reaction vessels and buffer flasks.
3. Dimethyl sulfoxide (DMSO) to dissolve chemical reagents like short PEG linkers. Use DMSO of very high quality (e.g., for molecular biology, LC/MS grade) to not interfere with the desired reaction.
4. Desalting column or device allowing the fast and efficient depletion of reducing agent (e.g., a PD-10 or PD-Minitrap™ G25 column, GE Healthcare).
5. DBCO and/or maleimide component solutions either in PBS or in DMSO. Solutions in DMSO should be prepared at high concentrations (e.g., 5–10 mM) to keep DMSO concentrations in the reaction mix low.

2.3 Analyzing Reactivity and Conjugation Yields

1. 1–5 mM PEG solutions for analytical reactions: Make sure the respective PEG is soluble in PBS; therefore, use long PEG molecules (e.g., PEG_{10k} or PEG_{20k}) with high homogeneity and quality. We obtained good results with the manufacturers JenKem Technology, Laysan Bio Inc. and Click Chemistry Tools. For a 1 mM solution of a PEG_{20k} molecule, dissolve 20 mg in 1 mL PBS and mix thoroughly. Always prepare thiol or maleimide-containing PEG solutions freshly.
2. 0.1 M perchloric acid solution: Add 407 μL of a 70% perchloric acid solution (commercially available) to 50 mL water. Do not filter afterward.
3. 5% (w/v) $BaCl_2$ solution: Dissolve 5 g $BaCl_2$ in 100 mL 1 M HCl solution and mix thoroughly. Do not filter afterward.
4. 0.05 M iodine (I_2) solution (commercially available).
5. Appropriate SDS-PAGE gels. We tested a variety of suppliers and could not detect differences in PEG-staining efficiency.
6. 5× SDS-PAGE sample buffer to prepare protein samples: 175 mM Tris–HCl pH 6.8, 50% (v/v) glycerol, 10% (w/v) sodium dodecyl sulfate, 5% (v/v) 2-mercaptoethanol, 0.15% (w/v) bromophenol blue.

3 Methods

3.1 Expression of an Azide–DARPin–Cysteine Construct in Minimal Medium

The expression of a clickable DARPin requires the incorporation of the nonnatural amino acid L-azidohomoalanine (Aha) by cultivating the bacteria in minimal medium. Both the pre-expression and the expression medium are based on the glucose-free synthetic M9 minimal medium SelenoMethionine Medium Base supplemented with a nutrient mix. After reaching a particular cell density, the cells are transferred from the pre-expression medium to the expression medium.

1. Transform the methionine-auxotrophic *E. coli* B-strain B834 (DE3) with the expression plasmid coding for the azide–DARPin–cysteine construct and containing the *lacI* gene under the strong *lacI*^{q} promoter control (e.g., pQE30 (Qiagen) modified as in [45, 47, 50]).
2. Inoculate a 5 mL pre-pre-culture in M9 pre-expression medium from a single colony or glycerol stock. Incubate it for 8 h at 37 °C during the day and use this culture to inoculate a 100 mL o/n-culture for each 1 L of expression culture. Incubate this culture with agitation at 37 °C (*see* **Note 2**). Use a flask-to-medium volume ratio of at least 5:1 for all pre-, o/n, and expression cultures for sufficient aeration.
3. The next morning, inoculate 1 L M9 pre-expression medium to an OD_{600} of 0.1 (approximate dilution 1:25). The cells are cultivated at 37 °C with agitation.
4. Once an OD_{600} of 1.0–1.2 is reached (usually after 3–4 h), centrifuge the expression culture at 4000 × *g*, 10 min, 4 °C. Wash the pellet thoroughly three times by resuspending in ice-cold PBS (30 mL for a 1 L culture pellet) to deplete all extracellular L-methionine. Cells should be constantly cooled during the wash procedure.
5. Resuspend the pellet in 30 mL M9 expression medium and inoculate 1 L M9 expression medium with the solution.
6. The expression culture is agitated for 15–20 min at 37 °C in a shaker to additionally deplete all intracellular L-methionine pools of *E. coli*.
7. Induce the expression of the clickable DARPin using 1 mM isopropyl-β-D-thio-galactopyranoside (IPTG) from a sterile 1 M stock solution and incubate for 4–5 h at 37 °C with agitation.
8. Harvest the expression culture by centrifugation for 10 min at 4000 × *g*, 4 °C to pellet the cells. Wash the pellet three times in ice-cold PBS as described above and subsequently snap-freeze it in liquid nitrogen. Store the pellet at −20 °C for short-term storage or at −80 °C for longer storage times.

As the expressed DARPins contain a hexa-His-tag, their purification can be performed by immobilized metal ion affinity chromatography (IMAC) as a single-step purification (*see* **Note 3**). The reader is referred to [3, 51] for a detailed description.

Most DARPin constructs used for bioconjugation contain a unique cysteine residue at the *C*-terminus, hence, it is recommended to add 5 mM 2-mercaptoethanol to all washing buffers to reduce disulfide bonds that might have formed between individual DARPins and with other endogenous cellular components. In addition, it is advantageous to store the purified protein in a reducing buffer by dialyzing it after the elution step to, for example, PBS supplemented with 5 mM 2-mercaptoethanol (*see* **Note 4**).

After this IMAC purification, a yield of 30–50 mg/L can be expected, which depends strongly on the individual construct and the employed expression conditions. If low expression yields are encountered, expression optimizations should be performed by varying first the expression temperature (e.g., to 30 °C [50]) and then the time before harvest.

3.2 Dual Conjugation Using Maleimide–Thiol and Click Chemistry

The expressed and purified proteins (here an azide–DARPin–cysteine construct) carry either a thiol group (introduced by cysteine) or an azide group (introduced by Aha) or both, which can subsequently be used for the conjugation of various payloads. The payload can, for example, be a dye for visualization and imaging, a cytotoxin for tumor cell killing or a half-life extension module like PEG or serum albumin [45, 47]. In addition, the use of bifunctional chemical linkers carrying a DBCO and a maleimide group to, for example, link an azide–DARPin to a thiol–DARPin to generate bispecific binders or other protein-protein conjugates, have also been used. Here we only describe in detail the conjugation of a DARPin carrying an *N*-terminal azide and a *C*-terminal thiol to two reactants with the respective countergroup (DBCO or maleimide) as examples illustrating the procedure. All provided protocols can be adapted to either protein–protein conjugations using a small DBCO–maleimide linker or single conjugations when proteins carrying only either an azide or a thiol group are expressed, since all described workflows and considerations are applicable.

Theoretically, click chemistry offers an orthogonal coupling reaction that does not interfere with other functionalities present in proteins such as thiols. It is, hence, possible to perform the maleimide–thiol and click chemistry conjugation simultaneously in a one-step approach. However, as recently shown [23, 24] side reactions cannot be completely avoided and the simultaneous approach complicates the troubleshooting if a reaction step does not work or is incomplete. We thus recommend performing the two reactions separately to maximize the yield by minimizing side-reactions. As the maleimide–thiol reaction exhibits the much faster kinetics, this reaction should be performed first (*see* **Note 5**). The conjugated protein is then purified before performing the azide-alkyne reaction. For the latter, we will only explain the click chemistry (SPAAC) as we found CuAAC to result in much lower reaction product yields and large amounts of side products (*see* **Note 1**).

1. Reduce the protein to provide a reactive thiol group not blocked by disulfide bridges (e.g., between proteins or between

the protein and media components). Omit this step if the protein was stored in reducing buffer after purification (*see* Subheading 3.1). Incubate the protein in an appropriate buffer (e.g., PBS) supplemented with 20 mM 2-mercaptoethanol for 1 h at 25–30 °C. Keep in mind that for the dual-conjugation approach neither DTT nor TCEP can be used as both reduce the azide group fast, making click chemistry impossible (*see* **Note 4**). In addition, always degas all buffers used and flush reaction tubes with argon or another inert gas to avoid reoxidation of the protein.

2. Remove the reducing agent by loading the reduced protein to an appropriate desalting column flushed with degassed buffer, collect fractions and identify the protein-containing fraction by absorbance measurements at 280 nm. It is crucial to establish a desalting method that depletes the reducing agent completely as it interferes with the subsequent maleimide–thiol conjugation. Perform an analytical PEGylation (*see* Subheading 3.3) or a mass spectrometry analysis to confirm complete reactivity of the desalted protein and the absence of remaining reducing agent. If the reduction is not complete and not all thiols are reactive, consider **Note 6**.
3. Immediately mix the desired protein fraction with a 2–4 × molar excess of the maleimide component (e.g., a dye) and incubate it for 1–2 h at 25–30 °C or 4 °C for 24 h with agitation (*see* **Note 7**). Optionally, quench the reaction by adding 2-mercaptoethanol at a 10–30 × molar excess over the maleimide component and incubate for 10 min at 25–30 °C with agitation.

   Note that if the maleimide component is expensive or tedious to synthesize, the protein can also be applied in 2–4 × molar excess, and the derivatized protein must then be separated from unreacted proteins (e.g., by ion exchange chromatography). Make sure the protein concentration is at least in the micromolar range or increase the concentration of the maleimide component to maintain the fast reaction kinetics. If the maleimide component is dissolved in organic solvents (e.g., DMSO) make sure to have a highly concentrated (millimolar range) stock solution to keep the DMSO concentrations in the final reaction low (5–10% maximum).
4. Investigate the yield of your reaction by either performing mass spectrometry, or an analytical PEGylation on unreacted starting material or other analytical methods (*see* Subheading 3.3). If the yield is not satisfactory, increase reaction time, concentrations and temperature. Note that some thiols might be unreactive since they are hidden in a cavity thus decreasing the pK_A (making it less reactive at lower pH) or resulting in sterically blocking the thiol group. This can be tested using

smaller maleimide components and performing subsequent analytical reactions (*see* Subheading 3.3).

5. Purify the conjugated protein from remaining educts. For many dyes or other small maleimide components, this can be done using appropriate desalting columns or dialysis membranes. Further purifications options are ion exchange chromatography if a charge was introduced (e.g., with a charged dye) or size exclusion chromatography if the hydrodynamic radius was significantly increased (e.g., by PEG conjugation). These separation steps are described in detail elsewhere [45, 47, 50].
6. Perform the click chemistry reaction by mixing the purified protein with a 2–4 × molar excess of the DBCO component (dye, PEG, etc.) and incubate for 24–48 h at 4 °C or at least 4 h at 25 °C with agitation. Note that click chemistry has slow reaction rates; hence, increasing reactants concentrations (high micromolar range) as well as the temperature and reaction time has a large influence on the reaction yield. Thereafter, use suitable purification methods if needed to deplete the remaining DBCO component [45, 47, 50]. Again, the protein can also be applied in molar excess over the DBCO component, if the latter is expensive or a complex synthesis is needed, and then the conjugated protein must be separated from the unconjugated one (*see* **Note 7**).

3.3 Analyzing Reactivity and Conjugation Yields

Proteins that were expressed and purified as described above carry a unique azide group, a unique thiol group, or a combination of both. Prior to the following conjugation of the desired molecule, which might be costly or laborious to produce, the reactivity of the respective reactive group incorporated into the protein should be tested. A convenient option is the conjugation of a PEG molecule carrying the respective countergroup, followed by a gel-shift assay. If the SDS-PAGE gel band of the protein is shifted to higher molecular weights, then the respective azide or thiol group incorporated in the protein is reactive. By subsequent PEG staining of the same gel, the PEGylation of the protein can be confirmed, thereby excluding that the band shift was caused by, for example, the formation of covalent protein multimers. Hence, this facile assay shows whether the incorporation of the reactive group into the protein was successful and whether its reactivity was maintained after the various protein expression and purification steps.

Several functionalized PEG molecules are commercially available which greatly facilitates the testing of all possible types of reactions with the respective countergroup. The following PEG variants are particularly useful to test the modified proteins for their reactivity: maleimide–PEG (to test the reactivity of the thiol group in the protein), DBCO–PEG (to test the reactivity of the azide group in the protein), thiol–PEG (to test the reactivity of a maleimide group incorporated in the protein by a second functionalization reaction), or azide–PEG (to test the reactivity of a DBCO incorporated in the protein by a secondary functionalization reaction). Proceed as follows:

1. Prepare a 1 mM solution of the respective polyethylene glycol (PEG) in PBS and, if the maleimide-thiol should be investigated, degas the buffer thoroughly (*see* **Note 8**).
2. To test the reactivity of the protein thiol group, prepare a freshly reduced protein aliquot using 2-mercaptoethanol (*see* **Notes 4** and **6**) as described previously. Use degassed buffers during the reduction and desalting process and flush reaction vessels with argon or a similar inert gas. Ensure that the reducing agent is completely removed as thiol components interfere with the analytical maleimide-PEG conjugation, for example, by using a desalting column with very good separation capability (*see* Subheading 2.2) in a very conservative manner collecting only the main protein peak fractions. This reduces recovery but allows more efficient protein conjugations.
3. Add maleimide-PEG in 4–8 × molar excess (*see* **Note 8**) and mildly shake the reaction at 25 °C for 2 h or at 4 °C for 24 h. Degas all buffers and flush the reaction vessels with argon or another inert gas to avoid reoxidation of thiols during the test reaction. Note that this reaction can also be inverted, that is, a thiol–PEG solution can be added and the reactivity of a maleimide incorporated into the protein can be examined.

   To test the reactivity of an azide group incorporated in the protein, add a 4–8 × molar excess of DBCO–PEG and mildly agitate at 4 °C for 24 h. Again, this reaction can be inverted, that is, the incorporation of a DBCO group in a protein can be tested by adding azide–PEG. As the click chemistry reaction is not oxidation sensitive, no degassing or use of inert gas is required. If two reactive groups in one protein should be tested (e.g., the bifunctional DARPin carrying an azide and a thiol group), we recommend to perform two separate test reactions for each reactive group.
4. Analyze the complete reaction mix by SDS-PAGE. Load higher amounts of protein (~5 μg) than usual, as otherwise remaining unreactive educts might not be visible. Inspect and scan the gel after Coomassie staining. A gel shift will show a successful reaction (*see* Fig. 2) and the gel can be PEG-stained as described below.

The staining of PEG in a previously Coomassie stained gel is performed at 25 °C and was adapted from [52, 53]:

1. Rinse the gel 3× with ~50 mL water.
2. Cover the gel with ~20 mL 0.1 M perchloric acid and shake mildly for 15 min.

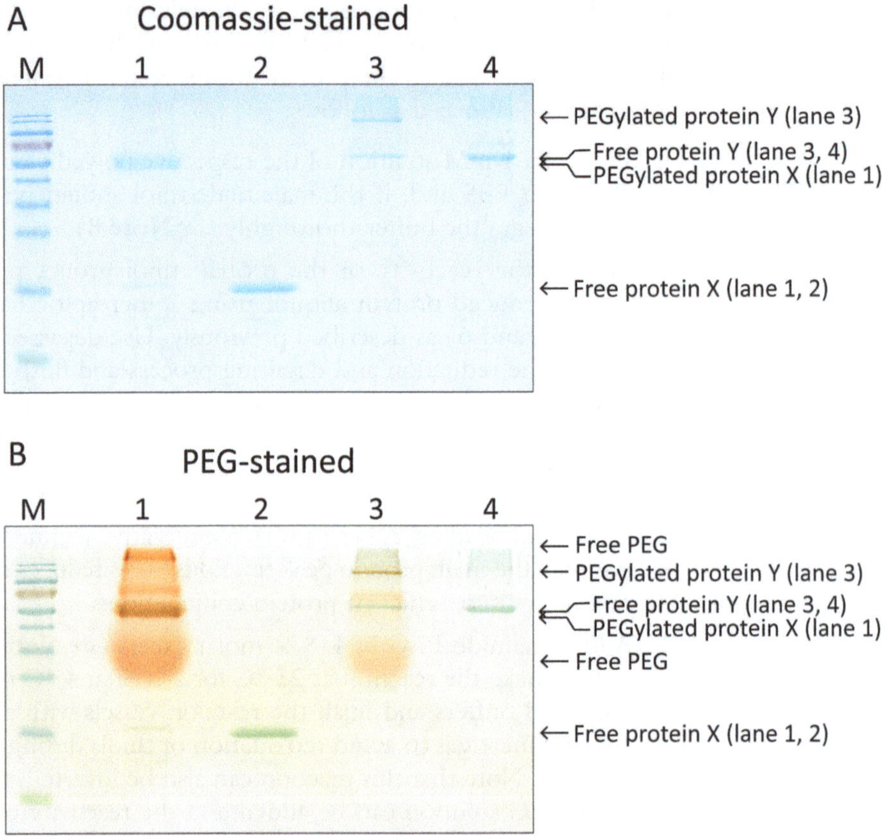

Fig. 2 15% SDS-PAGE gel for testing the quality of model proteins, ("X," a cys-containing model protein, and "Y," a maleimide-containing model protein) for their efficiency in conjugation, by coupling maleimide–PEG_{20k} as an easily detectable test compound. For protein X (lanes 1 and 2), maleimide–PEG_{20k} was used, while for Y, thiol–PEG_{20k} was used. 6 μg of each model protein was loaded and stained first with Coomassie (**a**) and subsequently for PEG (**b**) using the protocol described in the text. 1: Protein X reacted almost quantitatively with maleimide–PEG_{20k}, 2: Protein X w/o maleimide–PEG_{20k}, 3: Protein Y reacted partially with thiol–PEG_{20k}, 4: Protein Y w/o thiol–PEG_{20k}. A substantial portion of protein Y did not react with the thiol–PEG, presumably due to undesired side reactions in previous modification steps, compromising the integrity of the maleimide group

3. Add 5 mL 5% $BaCl_2$ in 1 M HCl to the solution and mix well.
4. Add 2 mL 0.05 M iodine solution and mix well.
5. Incubate on a shaker for ~10–15 min until the gel shows a dark brown color.
6. Destain the gel by discarding the solution and cover it with ~25 mL water, frequently check the destaining process and exchange the water. It takes approx. 10–15 min to destain the gel. As the destaining cannot be reverted, it is important to permanently check for a satisfying stain. Replace the water again before scanning the gel. *See* Fig. 2 for a PEG-stained example. Note that PEG molecules can form multimers and

be heterogeneous in length depending on the available PEG quality. Therefore, the free PEG can be visible as multiple bands at different heights in SDS-PAGE. Consequently, it is important to add the respective PEG in excess to the reaction and observe the free protein band.

Analytical PEGylation is a fast and convenient method to investigate reaction yields and reactive groups. However, the assessment of a gel band intensity and its comparison are not quantitative and SDS-PAGE exhibits a limited sensitivity. Hence, more precise, alternative methods are also listed in the following sections. These also enable the analysis of protein expression and integrity and the investigation of its conjugation to different chemical moieties and provide a broader dataset for final assessments.

The quantitative and exclusive incorporation of Aha in the protein can be investigated by *N*-terminal Edman analysis [45, 50]. Consider that blotting the protein to a PVDF membrane prior to Edman analysis might add interfering components to the analysis reaction. Try to use the protein in solution directly and apply only buffer components of very high-grade purity. The non-natural amino acid Aha usually elutes 1–2 min earlier than L-methionine [3]. Furthermore, the amino acid composition of the expressed protein can be analyzed with amino acid hydrolysis using commercial kits. Subsequently, the resulting amino acid mix is analyzed on an UHLPC system, where no L-methionine should be detectable [3, 45]. Finally, mass spectrometry analysis offers a facile method to determine the incorporation of Aha, as its mass is −5 Da different from L-methionine. For example, ESI-MS yields good results with DARPins as these proteins are comparably small. However, keep in mind that the analysis of large proteins in mass spectrometry might be complicated and such small mass shifts might not be detectable. Furthermore, it should not be used to compare conjugation rates quantitatively as peak heights might vary with the biochemical properties of the conjugate and the nonconjugated protein, and thus the peak heights cannot be used to compare molar amounts. Nevertheless, mass spectrometry is well suited for the analysis of any remaining educts in the reaction mix.

The conjugation of a chemical moiety (e.g., a dye or a small linker like maleimide–biotin or maleimide–DBCO for the cross-linking of proteins) can be quantified with various methods. Keep in mind that before using these methods, it is important to determine the quality of the respective conjugation partner itself, for example, by mass spectrometry, NMR, or RP-HPLC. An inhomogeneous reaction partner (e.g., dye, maleimide–DBCO linker) will otherwise result in ambiguous results. For the investigation of dyes and DBCO-containing molecules, it is possible to use the absorption of the respective moiety at a given wavelength for conjugation quantification. For example, for a maleimide–DBCO linker, we calculated the molar absorption coefficient in DMSO to be 12,388 L·mol^{-1} cm^{-1} at 280 nm and 13,375 L·mol^{-1} cm^{-1} at 309 nm. Next to its use in conjugation quantification, these high absorption coefficients can also impede protein quantification and must thus be taken into account here. Again, for homogeneous small molecules, mass spectrometry offers a viable method to determine mass shifts and quantify conjugation (see above). The conjugation of a small moiety often also changes the hydrophilicity of the modified protein, which can thus be analyzed by reverse phase HPLC, for example, using a C18-column as described in [47] (*see* **Note 9**). If the conjugated molecule contains a biotin residue, for example, when a maleimide–biotin linker is conjugated, a streptavidin gel shift assay as described in [54] can be performed. Finally, many chemical moieties like dyes and also, for example, a maleimide-PEG4–DBCO linker add detectable mass to the protein, large enough to be detectable in a gel-shift assay using high-quality commercial SDS-PAGE gels run at low voltage (e.g., 4–12% Bis/-Tris with MOPS running buffer or 10% Bis/Tris with MES buffer). Mass differences down to 0.5–1 kDa can be detected on such gels. This method offers a cheap screening method and furthermore, fluorescent moieties can additionally be quantified by a gel documentation device equipped with a fluorescence detector [45].

4 Notes

1. The azide-alkyne Huisgen cycloaddition reaction has an inherently very low reaction rate [16] which can be increased to a reasonable level by using either strained alkynes (SPAAC) like DBCO or copper catalysis (CuAAC). Using DARPins, we also performed the latter according to Presolski et al. [55], but found that the proteins contained multiple oxidation species in mass spectrometry analysis, high amounts of covalently linked protein by side-reactions and precipitates. Furthermore, we observed reaction yields of a maximum of 50%. Please note that the amount of copper and other components must be optimized for every protein individually when using CuAAC, whereas SPAAC offers a broadly applicable technology with yields of almost 100% in our hands and no detectable side reactions or undesired protein modifications, when following the recommended protocols. Currently, more strained alkynes are available such as BCN-based reagents. These reagents show decreased lipophilicity while maintaining high reaction rates. Furthermore, these compounds provide very high reaction rates also with, for example, aromatic azides. However, the incorporation of such azides into proteins is unfortunately much more laborious and the reaction with Aha does not show faster kinetics, whereas side-reaction rates might be

facilitated. Finally, a number of DBCO components such as dyes are now commercially available in good qualities and at affordable prices. We therefore recommend using the SPAAC reaction (click chemistry) with DBCO and an azide group provided by Aha.

2. In theory, 2YT or other rich growth media can also be used for the o/n and pre-expression cultures when thorough washing steps are applied. However, the pre-expression minimal medium offers the advantage that the metabolism of the bacteria has already been adapted to the growth conditions in minimal medium, resulting in significantly higher expression yields in our experience.

3. It is important to prevent the copurification of endogenous *E. coli* proteins which might bind to the Ni-NTA resin due to a native metal-binding function or the presence of surface clusters of histidine residues [56]. Therefore, next to avoiding the use of an oversized IMAC column, we highly recommended to include both a low salt washing step (20 mM NaCl, counteracting hydrophobic interactions) and a high salt washing step (at least 1 M NaCl, counteracting ionic interactions).

4. In principle, the cysteine residue can be reduced by using either dithiothreitol (DTT) or tris(2-carboxyethyl)phosphine (TCEP). However, for the dual-conjugation approach described here, DTT and TCEP must not be used as they react with the azide-group and inactivate it for click chemistry. Both molecules efficiently reduce azides to amino groups [26, 55] and TCEP furthermore undergoes a Staudinger ligation by providing a phosphine group. However, it seems that 2-mercaptoethanol can be used as a reducing agent for disulfide bridges keeping the azide reactive as the rate constants of the undesired reduction are much slower [27]. Furthermore, 2-mercaptoethanol is more stable in solution and, hence, also in the storage buffers used after IMAC purification [57].

5. The maleimide-thiol reaction exhibits much faster reaction kinetics compared to the strain-promoted click chemistry [16]. We found that when performing protein-protein conjugation using a small maleimide-DBCO linker with one protein species ("X") carrying an exposed thiol and another protein species ("Y") carrying an exposed azide group, performing the maleimide-thiol reaction first and then conjugating protein X-DBCO to protein Y-azide, greatly increased the reaction yields (unpublished data). Presumably, this is due to maleimide hydrolysis and/or side-reactions taking place during the lengthy DBCO–azide reaction, which is performed at 4 °C for at least 24 h.

6. If the reduction is incomplete, we recommend optimizing the reduction conditions by either increasing the reaction temperature, time, concentration of reducing agent or the pH of the used buffer. At higher pH the amount of thiolate ions is increased, which is the reactive species in the reduction of disulfides, and thus higher pH strongly increases the reaction rate (e.g., instead of PBS pH 7.4 use an appropriate buffer at pH 8.0–8.5 if your protein is stable under these conditions). It is also crucial to establish a fast and efficient desalting method. This must be fast (within 10–20 min) to avoid reoxidation of the previously reduced thiols and efficient so that the reducing agent 2-mercaptoethanol is *completely* depleted as it competes with the desired maleimide–thiol conjugation.
7. If the maleimide component is expensive or laborious to synthesize, the protein can also be applied in a 2–4 × molar excess, provided that an effective method to separate the conjugated from the nonmodified protein is available (e.g., by preparative anion exchange chromatography). Also make sure that the protein concentration is at least in the micromolar range or increase the concentration of the maleimide component to maintain the fast reaction kinetics. If the maleimide component is dissolved in organic solvents (e.g., DMSO), we recommend to prepare a highly concentrated (millimolar range) stock solution to keep the DMSO concentrations in the final reaction low (5–10% maximum, depending on the stability of the target protein).
8. Highly concentrated PEG solutions are viscous and difficult to pipet and mix. They reduce reaction rates and, consequently, yield. We therefore do not recommend to prepare PEG solutions above 5 mM PEG concentration. Consider using a high molecular weight PEG of at least 10–20 kDa to ensure solubility in PBS and to obtain a clearly visible gel shift also for larger proteins. Especially when maleimide–PEG or thiol–PEG is used, solutions should be prepared freshly before each experiment as the hydrolysis of the maleimide or disulfide formation of thiols can occur. Degas the PBS prior to dissolving thiol–PEG to avoid undesired oxidation reactions. If the protein is highly concentrated (>100 μM) the molar excess of PEG can be reduced or the protein diluted. However, we do not recommend using less than 4 × molar excess of PEG as, depending on the quality of the PEG and its storage conditions, not all molecules carry the desired reactive group. This can possibly lead to a false interpretation of the conjugation results, if the effective concentrations of the reactive group provided by the PEG molecule are too low.

9. Many modifying molecules like dyes or small linkers are dissolved in DMSO and then mixed with the respective protein for conjugation. If this reaction mix is loaded to RP-HPLC for analysis, high DMSO concentrations will influence the running behavior on the column. Therefore, the sample must be diluted (<1% final DMSO concentration) before loading it on the column to minimize this effect.

References

1. van Berkel SS, van Delft FL (2013) Metal-free bioconjugation reactions. Drug Discov Today Technol 10:45–51
2. Rashidian M, Dozier JK, Distefano MD (2013) Enzymatic labeling of proteins: techniques and approaches. Bioconjug Chem 24:1277–1294
3. Tamaskovic R, Simon M, Stefan N et al (2012) Designed ankyrin repeat proteins (DARPins) from research to therapy. Methods Enzymol 503:101–134
4. Hansen S, Stüber JC, Ernst P et al (2017) Design and applications of a clamp for Green fluorescent protein with picomolar affinity. Sci Rep 7:16292
5. Gorin G, Martic PA, Doughty G (1966) Kinetics of the reaction of N-ethylmaleimide with cysteine and some congeners. Arch Biochem Biophys 115:593–597
6. Makmura L, Hamann M, Areopagita A et al (2001) Development of a sensitive assay to detect reversibly oxidized protein cysteine sulfhydryl groups. Antioxid Redox Signal 3:1105–1118
7. Panowksi S, Bhakta S, Raab H et al (2014) Site-specific antibody drug conjugates for cancer therapy. MAbs 6:34–45
8. Hamblett KJ, Senter PD, Chace DF et al (2004) Effects of drug loading on the antitumor activity of a monoclonal antibody drug conjugate. Clin Cancer Res 10:7063–7070
9. Pillow TH, Tien J, Parsons-Reponte KL et al (2014) Site-specific trastuzumab maytansinoid antibody-drug conjugates with improved therapeutic activity through linker and antibody engineering. J Med Chem 57:7890–7899
10. Junutula JR, Raab H, Clark S et al (2008) Site-specific conjugation of a cytotoxic drug to an antibody improves the therapeutic index. Nat Biotechnol 26:925–932
11. Sharpless NE, Flavin M (1966) The reactions of amines and amino acids with maleimides. Structure of the reaction products deduced from infrared and nuclear magnetic resonance spectroscopy. Biochemistry 5:2963–2971
12. Fontaine SD, Reid R, Robinson L et al (2015) Long-term stabilization of maleimide-thiol conjugates. Bioconjug Chem 26:145–152
13. McKay CS, Finn MG (2014) Click chemistry in complex mixtures: bioorthogonal bioconjugation. Chem Biol 21:1075–1101
14. Dommerholt J, Rutjes FP, van Delft FL (2016) Strain-promoted 1,3-dipolar cycloaddition of cycloalkynes and organic azides. Top Curr Chem 374:16
15. Hu QY, Berti F, Adamo R (2016) Towards the next generation of biomedicines by site-selective conjugation. Chem Soc Rev 45:1691–1719
16. Lang K, Chin JW (2014) Bioorthogonal reactions for labeling proteins. ACS Chem Biol 9:16–20
17. Rostovtsev VV, Green LG, Fokin VV et al (2002) A stepwise Huisgen cycloaddition process: copper(I)-catalyzed regioselective "ligation" of azides and terminal alkynes. Angew Chem Int Ed Engl 41:2596–2599
18. Tornøe CW, Christensen C, Meldal M (2002) Peptidotriazoles on solid phase: [1,2,3]-triazoles by regiospecific copper(I)-catalyzed 1,3-dipolar cycloadditions of terminal alkynes to azides. J Org Chem 67:3057–3064
19. Agard NJ, Prescher JA, Bertozzi CR (2004) A strain-promoted [3 + 2] azide-alkyne cycloaddition for covalent modification of biomolecules in living systems. J Am Chem Soc 126:15046–15047
20. Ning X, Guo J, Wolfert MA et al (2008) Visualizing metabolically labeled glycoconjugates of living cells by copper-free and fast Huisgen cycloadditions. Angew Chem Int Ed Engl 47:2253–2255
21. Debets MF, van Berkel SS, Schoffelen S et al (2010) Aza-dibenzocyclooctynes for fast and efficient enzyme PEGylation via copper-free (3+2) cycloaddition. Chem Commun 46:97–99
22. Dommerholt J, Schmidt S, Temming R et al (2010) Readily accessible bicyclononynes for bioorthogonal labeling and three-dimensional

imaging of living cells. Angew Chem Int Ed Engl 49:9422–9425

23. van der Linden WA, Li N, Hoogendoorn S et al (2012) Two-step bioorthogonal activity-based proteasome profiling using copper-free click reagents: a comparative study. Bioorg Med Chem 20:662–666
24. van Geel R, Pruijn GJ, van Delft FL et al (2012) Preventing thiol-yne addition improves the specificity of strain-promoted azide-alkyne cycloaddition. Bioconjug Chem 23:392–398
25. Witte MD, Theile CS, Wu T et al (2013) Production of unnaturally linked chimeric proteins using a combination of sortase-catalyzed transpeptidation and click chemistry. Nat Protoc 8:1808–1819
26. Faucher A, Grand-Maître C (2003) Tris(2-carboxyethyl)phosphine (TCEP) for the reduction of sulfoxides, sulfonylchlorides, N-oxides, and azides. Synth Commun 33:3503–3511
27. Handlon AL, Oppenheimer NJ (1988) Thiol reduction of 3′-azidothymidine to 3′-aminothymidine: kinetics and biomedical implications. Pharm Res 5:297–299
28. Kiick KL, Saxon E, Tirrell DA et al (2002) Incorporation of azides into recombinant proteins for chemoselective modification by the Staudinger ligation. Proc Natl Acad Sci U S A 99:19–24
29. Plass T, Milles S, Koehler C et al (2011) Genetically encoded copper-free click chemistry. Angew Chem Int Ed Engl 50:3878–3881
30. Wang L, Brock A, Herberich B et al (2001) Expanding the genetic code of Escherichia coli. Science 292:498–500
31. Borrmann A, Milles S, Plass T et al (2012) Genetic encoding of a bicyclo[6.1.0]nonyne-charged amino acid enables fast cellular protein imaging by metal-free ligation. ChemBioChem 13:2094–2099
32. Lang K, Davis L, Wallace S et al (2012) Genetic encoding of bicyclononynes and trans-cyclooctenes for site-specific protein labeling in vitro and in live mammalian cells via rapid fluorogenic Diels-Alder reactions. J Am Chem Soc 134:10317–10320
33. Alfonta L, Zhang Z, Uryu S et al (2003) Site-specific incorporation of a redox-active amino acid into proteins. J Am Chem Soc 125:14662–14663
34. Borrmann A, Fatunsin O, Dommerholt J et al (2014) Strain-promoted oxidation-controlled cyclooctyne-1,2-quinone cycloaddition (SPOCQ) for fast and activatable protein conjugation. Bioconjug Chem 26:257–261
35. Kaya E, Vrabel M, Deiml C et al (2012) A genetically encoded norbornene amino acid for the mild and selective modification of proteins in a copper-free click reaction. Angew Chem Int Ed Engl 51:4466–4469
36. Lang K, Davis L, Torres-Kolbus J et al (2012) Genetically encoded norbornene directs site-specific cellular protein labelling via a rapid bioorthogonal reaction. Nat Chem 4:298–304
37. Dommerholt J, van Rooijen O, Borrmann A et al (2014) Highly accelerated inverse electron-demand cycloaddition of electron-deficient azides with aliphatic cyclooctynes. Nat Commun 5:5378
38. Pott M, Schmidt MJ, Summerer D (2014) Evolved sequence contexts for highly efficient amber suppression with noncanonical amino acids. ACS Chem Biol 9:2815–2822
39. Johnson DB, Xu J, Shen Z et al (2011) RF1 knockout allows ribosomal incorporation of unnatural amino acids at multiple sites. Nat Chem Biol 7:779–786
40. Merten H, Brandl F, Plückthun A et al (2015) Antibody-drug conjugates for tumor targeting-novel conjugation chemistries and the promise of non-IgG binding proteins. Bioconjug Chem 26:2176–2185
41. Owens B (2017) Faster, deeper, smaller-the rise of antibody-like scaffolds. Nat Biotechnol 35:602–603
42. Deonarain MP, Yahioglu G, Stamati I et al (2015) Emerging formats for next-generation antibody drug conjugates. Expert Opin Drug Discovery 10:1–19
43. Plückthun A (2015) Designed ankyrin repeat proteins (DARPins): binding proteins for research, diagnostics, and therapy. Annu Rev Pharmacol Toxicol 55:489–511
44. Boersma YL, Plückthun A (2011) DARPins and other repeat protein scaffolds: advances in engineering and applications. Curr Opin Biotechnol 22:849–857
45. Simon M, Zangemeister-Wittke U, Plückthun A (2012) Facile double-functionalization of designed ankyrin repeat proteins using click and thiol chemistries. Bioconjug Chem 23:279–286
46. Binz HK, Stumpp MT, Forrer P et al (2003) Designing repeat proteins: well-expressed, soluble and stable proteins from combinatorial libraries of consensus ankyrin repeat proteins. J Mol Biol 332:489–503
47. Simon M, Frey R, Zangemeister-Wittke U et al (2013) Orthogonal assembly of a designed ankyrin repeat protein-cytotoxin conjugate with a clickable serum albumin module for half-life extension. Bioconjug Chem 24:1955–1966

48. Hirel PH, Schmitter MJ, Dessen P et al (1989) Extent of N-terminal methionine excision from Escherichia coli proteins is governed by the side-chain length of the penultimate amino acid. Proc Natl Acad Sci U S A 86:8247–8251
49. Wang A, Winblade Nairn N, Johnson RS et al (2008) Processing of N-terminal unnatural amino acids in recombinant human interferon-beta in Escherichia coli. Chembiochem 9:324–330
50. Stefan N, Zimmermann M, Simon M et al (2014) Novel prodrug-like fusion toxin with protease-sensitive bioorthogonal PEGylation for tumor targeting. Bioconjug Chem 25:2144–2156
51. Block H, Maertens B, Spriestersbach A et al (2009) Immobilized-metal affinity chromatography (IMAC): a review. Methods Enzymol 463:439–473
52. Kurfürst MM (1992) Detection and molecular weight determination of polyethylene glycol-modified hirudin by staining after sodium dodecyl sulfate-polyacrylamide gel electrophoresis. Anal Biochem 200:244–248
53. Skoog B (1979) Determination of polyethylene glycols 4000 and 6000 in plasma protein preparations. Vox Sang 37:345–349
54. Petris G, Vecchi L, Bestagno M et al (2011) Efficient detection of proteins retro-translocated from the ER to the cytosol by in vivo biotinylation. PLoS One 6:23712
55. Presolski SI, Hong VP, Finn MG (2011) Copper-catalyzed azide-alkyne click chemistry for bioconjugation. Curr Protoc Chem Biol 3:153–162
56. Bolanos-Garcia VM, Davies OR (2006) Structural analysis and classification of native proteins from E. coli commonly co-purified by immobilised metal affinity chromatography. Biochim Biophys Acta 1760:1304–1313
57. Hansen RE, Winther JR (2009) An introduction to methods for analyzing thiols and disulfides: reactions, reagents, and practical considerations. Anal Biochem 394:147–158

Chapter 18

Site-Specific Photocrosslinking to Immunoglobulin G Using Photoreactive Antibody-Binding Domains

Fabiana Zappala and Andrew Tsourkas

Abstract

The high specificity and strong binding affinity of antibodies, most commonly immunoglobulin G (IgG), have led to their use in a wide range of research, diagnostic and therapeutic applications. Many of these applications require the antibody to be labeled with additional chemical or biological moieties. Here, we describe a method for the rapid and site-specific labeling of nearly any "off-the-shelf" IgG. Our method utilizes small photoreactive antibody-binding domains (pAbBDs) that are produced by modifying the IgG-binding domains of Protein A and Protein G with the unnatural amino acid benzoylphenylalanine (BPA). The pAbBDs are covalently linked to IgG heavy chains upon exposure to ultraviolet light. Fusion of pAbBDs to a given protein of interest or conjugation of pAbBDs with drugs, fluorophores, and/or other chemical moieties, enables the facile production of a diverse range of antibody conjugates.

Key words Bioconjugation, Antibody conjugates, Photocrosslinking, Antibody labeling, Immunoglobulin G, Site-specific modification, Unnatural amino acids

1 Introduction

Immunoglobulin G (IgG) antibodies provide a powerful platform for therapeutic, diagnostic and research applications due to their high binding specificity and affinity, as well as established methods for their discovery and production [1–3]. These features result in key advantages associated with the use of IgG, such as reduced side effects of biologic therapies versus small molecules and high sensitivity in antibody-based assays, like enzyme-linked immunosorbent assay (ELISA), immunofluorescent imaging, and flow cytometry. Many of the technologies that exploit IgG's binding capability require the antibody to be covalently modified with chemical or biological moieties in a controlled manner [4–6].

Despite the expansive uses of chemically and biologically modified antibodies, challenges associated with the modification of IgG, such as low bioconjugation efficiencies and heterogeneity of the final conjugates, continue to limit their potential [7, 8]. For

Sam Massa and Nick Devoogdt (eds.), *Bioconjugation: Methods and Protocols*, Methods in Molecular Biology, vol. 2033, https://doi.org/10.1007/978-1-4939-9654-4_18,

example, chemical modifications with dyes, drugs, or other small molecules are often completed via nonspecific labeling of lysine or cysteine residues. These methods require excess materials and optimization of chemical-to-antibody ratio and still suffer from heterogeneity in the number and location of the labels. Antibody conjugate uniformity is key to control pharmacokinetics and to achieve optimal immunoassay sensitivity and specificity [9–12].

The production of antibody–protein conjugates, using chemical conjugation methods, can be even more problematic, resulting in aggregates, precipitation, and loss of functionality [13, 14]. The most common approach to eliminating the heterogeneity of chemical conjugations is to produce antibody–protein fusions. However, this requires each construct to be optimized individually, which can be a lengthy and complex process, hindering research and development cycles. Reoptimizing every new antibody–protein conjugate is particularly detrimental to discovery phases that sort through and compare various candidate targets and pairings. In some cases, after extensive optimization, antibody binding and function are found to be compromised. A facile strategy to screen through antibody–protein conjugates would speed up the identification of novel targets as well as the development of optimal pairings that avoid unexpected interference with binding and functionality.

The following sections describe a method for rapid and site-specific labeling of nearly any "off-the-shelf" IgG with biological or chemical moieties. This technology alleviates the previously summarized problems by taking a modular approach that directs conjugation of chemicals and proteins to the IgG constant region site, thus avoiding interference with binding. Specifically, we have engineered two small photoreactive antibody-binding domains (pAbBDs) by modifying the IgG-binding domains of Protein A and Protein G with the unnatural amino acid benzoylphenylalanine (BPA) [15]. Under ultraviolet light, the BPA within the IgG-binding domain cross-links covalently to IgG at the Fc CH2-CH3 junction (Fig. 1). Efficient cross-linking, up to 100%, can be achieved with antibodies from a broad range of hosts in under 2 h. Any chemical or biological modification that is included on the pAbBD is site-specifically attached to the IgG heavy chain. An additional key aspect of this technology is the high bacterial expression yield of the pAbBDs. Lastly, using this technology, modified IgGs retain their original binding affinities post-cross-linking with pAbBDs.

pAbBDs can be expressed in fusion with any protein of interest to enable biological modification of the IgG heavy chains with an antibody fragment, peptide or enzyme to name a few examples. Additionally, we describe sortase-tagged expressed protein ligation (STEPL), a complementary technology that can be used to incorporate small molecules (e.g., fluorophore, click chemistry handle, or drug) onto the pAbBD during affinity purification (Fig. 2)

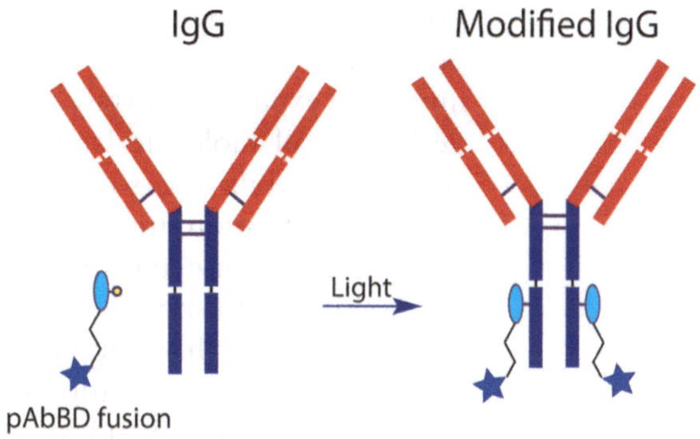

Fig. 1 IgG chemical or biological modification photocrosslinking. Under UV exposure, pAbBD constructs cross-link covalently to the IgG heavy chains at the constant region. In 2 h, nearly 100% IgGs are site-specifically modified with a 2:1 ratio. The 6.5 kDa pAbBDs can be expressed in fusion with any protein and/or labeled with a small molecule as desired

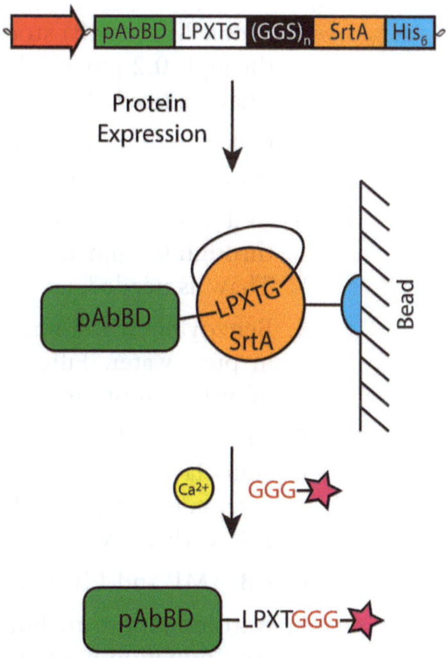

Fig. 2 pAbBD fusion protein N-to-C Sequence and sortase-tagged expressed protein ligation (STEPL). STEPL combines protein purification and labeling into one step. When flanked by a given protein of interest and an affinity tag, the sortase A (SrtA) transpeptidase drives elution from an affinity resin. It releases the captured recombinant protein through ligation of a triglycine peptide with an optional small molecule label (fluorophore, click chemistry handle, drugs, additional affinity tags) onto the sortase recognition motif, LPXTG. This technology, in combination with pAbBDs, facilitates the chemical modification of IgG. C-terminal LPXTG and Sortase can be removed if STEPL labeling is not needed. $(GGS)_n$ linkers or other linkers of choice between each domain can be optimized for sufficient length and flexibility. Reproduced with minor modifications from Warden-Rothman 2013 with Permission from ACS Publications [16]

[16]. This strategy involves fusing the pAbBD (or pAbBD-fusion protein) with Sortase, capturing this construct on an affinity column, and exploiting sortase to ligate a short peptide containing the desired small molecule label onto the c-terminus of the pAbBD construct, while the sortase itself is excised. STEPL combines affinity purification and labeling into one step and only requires cloning the sortase and motif sequences, without adding steps to the overall procedure.

In summary, this chapter reviews a quick route for the covalent and site-specific modification of IgG. The strategy detailed in the following sections includes: cloning, expression, purification of pAbBD constructs with chemical and/or biological modifications and IgG cross-linking. These steps facilitate the generation of various types of antibody conjugates.

2 Materials

1. Ampicillin (AMP) stock: 100 mg/mL in pure water. Filtered through 0.2 μm and kept frozen in 2 mL aliquots at −20 °C. Thaw as needed.
2. Chloramphenicol (CAM) stock: 25 mg/mL in 95% (v/v) ethanol. Filtered through 0.2 μm and stored at −20 °C.
3. Arabinose stock: 10% (w/v) L-arabinose in pure water. Filtered through 0.2 μm and kept frozen in 50 mL aliquots at −20 °C. Thaw as needed.
4. Benzoylphenylalanine (BPA) stock: 0.05 M H-p-Bz-Phe-OH in pure water. Filtered through 0.2 μm and kept frozen in 50 mL aliquots at −20 °C. Thaw as needed.
5. LB-AMP and LB-AMP-CAM: 25 g/L LB broth in pure water. Autoclave 15 min at 121 °C. After cooling to room temperature, add 1 mL of AMP stock, and 1 mL of CAM stock for LB-AMP-CAM.
6. LB-AMP and LB-AMP-CAM bacterial plates.
7. Autoinduction medium: dissolve 34.85 g of LB broth-based autoinduction medium (Formedium, Norfolk, UK) including trace elements in 1 L pure water and autoclave 15 min at 121 °C. After cooling to room temperature, add 1 mL AMP stock, 1 mL CAM stock, 6 mL sterile glycerol, 10 mL arabinose stock and 6 mL BPA stock (*see* **Note 1**).
8. Bacterial expression plasmid of choice to drive the production of the pAbBD-fusion protein: In the current protocol, we assume Lac operon induction control and AMP selection (*see* **Note 2**).

9. *E. coli* cloning strain and media.
10. pEVOL-pBpF plasmid (Addgene plasmid #31190) [17] encoding the *E. coli* tRNA synthetase–tRNA pair for BPA incorporation in response to the amber stop codon.
11. T7 Express (New England Biolabs) or *E. coli* expression cell line of choice (*see* **Note 3**).
12. GGG-X peptides: 1 mM stock. 2 mL aliquots stored frozen at −20 °C. Examples: GGG-biotin, GGG-azide, GGG-dibenzocyclooctyl (DBCO), and GGG-5-TAMRA. Triglycine can also be used when a chemical modification is not needed—for example, when conjugating a pAbBD fusion protein onto the IgG. The choice of label depends on what is required for subsequent assays (*see* **Note 4**).
13. TALON cobalt resin or other affinity resin of choice and purification columns, such as 10 mL disposable gravity chromatography columns.
14. Phosphate buffered saline (PBS): 137 mM NaCl, 2.7 mM KCl, and 10 mM phosphate buffer.
15. Calcium stock: 10 mM in pure water. Stored at room temperature.
16. DNAse stock: 1 mg/mL DNAse in pure water. 10 mL aliquots stored frozen at −20 °C.
17. Lysozyme stock: 100 mg/mL lysozyme in pure water. 10 mL aliquots stored frozen at −20 °C.
18. Lysis buffer: 50 mL of 1% octylthioglucoside (OTG) in PBS, 200 μL DNAse stock, 200 μL lysozyme stock, 1 cOmplete EDTA-free protease inhibitor cocktail tablet (Roche).
19. UV lamp: 365 nm (*see* **Note 5**).
20. 4–12% Bis-Tris protein gel, loading dye, and ladder.
21. Spin filter tubes with required molecular weight cutoff for protein of interest. We suggest using 0.5, 2, or 4 mL Amicon Ultra Centrifugal Filters depending on working volume.
22. BCA Protein Assay kit or alternative protein quantification method.

3 Methods

3.1 Cloning pAbBD Fusion Construct: pAbBD-Protein of Interest-Affinity Tag

1. Clone pAbBD into the bacterial expression plasmid, if it is only desirable to label IgG with a small molecule. If an IgG-protein conjugate is desired, clone pAbBD at the N-terminus and in series with the given protein of interest, separated by a $(GGS)_3$ linker (*see* **Note 6**). *See* Table 1 for choice and sequence of pAbBD [2].

Table 1
Summary of IgG subtypes found to bind to pAbBD versions based on protein G (PGA24) and protein A (PZQ32)

pAbBD Version	Mouse IgG subtypes	Human IgG subtypes	Others
PGA24: MTFKLIINGKTLKGEITIEAVDA*EAEKIF KQYANDYGIDGEWTYDDATKTFTVTE	2a, 2b, 2c, 3	1, 2, 3	Rat 2c, rabbit polyclonal
PZQ32: MVDNKFNKEQQNAFYEILHLPNLNEE QRNAFI*SLKDDPSQSANLLAEAKKLNDAQAPK	1	–	–

2. If opting to use STEPL one-step affinity purification and labeling, clone Sortase at the C-terminus (amino acid sequence: QAKPQIPKDKSKVAGYIEIPDADIKEPVYPGPATPEQLNR GVSFAEENESLDDQNISIAGHTFIDRPNYQFTNLKAAKK GSMVYFKVGNETRKYKMTSIRDVKPTDVEVLDEQKGKD KQLTLITCDDYNEKTGVWEKRKIFVATEVK). The pAbBD or pAbBD fusion protein should be separated from sortase by the LPXTG sortase motif, flanked by $(GGS)_5$ linkers. Add a 6His affinity handle for capture (Fig. 2). In the current protocol, it is assumed that the backbone contains the ampicillin resistance gene and pAbBD expression is under the control of a lac operon.

3.2 Bacterial Expression in T7 Express

1. Cotransform sequence-confirmed plasmid containing pAbBD with the pEVOL plasmid containing the necessary unnatural amino acid machinery for BPA incorporation into the T7 Express expression *E. coli* cell line. The pEVOL plasmid utilizes CAM selection and expression is controlled by the L-arabinose operon. This step should therefore be completed using LB-AMP-CAM plates.
2. Incubate plates at 25 °C for 2–3 days. Colonies usually take longer to arise in comparison to one-plasmid transformations.
3. Pick a single colony and grow a 1 mL starter culture in LB-AMP-CAM media at 37 °C, shaking at 200 rpm until OD reaches 0.6.
4. Dilute the 0.6 OD starter culture 1:1000 in autoinduction medium using bacterial culture flasks. Grow these 12–16 h at 37 °C, shaking at 155 rpm in autoinduction medium (*see* **Note 7**). The final OD can be optimized to maximize protein yield. For T7 Express, optimal final OD generally lies within 1.5–2, but this number can vary vastly depending on the protein of interest. For insoluble and difficult-to-express proteins, final ODs ranging 0.5–1 can be beneficial.

3.3 Cell Centrifugation and Lysis

1. Centrifuge cultures for 8 min at 3000 × *g* or until complete separation is reached.
2. Prepare lysis buffer immediately before use by diluting one protease inhibitor cocktail tablet in 50 mL of 1% OTG buffer. Add thawed DNase and lysozyme stock to obtain a 1:250 dilution.
3. Resuspend pellets fully using 4 mL of lysis buffer per gram of cells. There should be no visible pieces of pellet. Rotate 30 min at room temperature (*see* **Note 8**).
4. If lysate remains viscous after **step 3**, further cell breakdown can be achieved with 30 s sonication cycles. Protein samples should be kept on ice during sonication and between cycles. After two to five 30 s sonication cycles the lysate should be visibly less viscous. If further breakdown is required, the lysate can be frozen in a −80 °C freezer or using liquid nitrogen until completely solid and thawed in a 37 °C water bath or at room temperature. For lysates greater than 10 mL vortexing can speed up thawing.
5. The lysate can be kept in −80 °C until ready to start purification.

3.4 One-Step Affinity Purification and Labeling Using STEPL (See Note 9)

1. Clarify thawed lysate by centrifuging 10 min at 3000 × *g* or until complete separation is reached.
2. Use 300 μL of TALON cobalt resin per 10 mL of clarified lysate supernatant. Load resin onto column and wash three times with 10 mL PBS (*see* **Note 10**).
3. Cap bottom of column and add clarified lysate supernatant to the resin column. Cap the top of the column and rotate for 30 min at room temperature.
4. Uncap the column and wash resin twice with 10 mL PBS, then return cap to the bottom of the column and add 350 μL elution buffer per column. The elution buffer should be prepared as needed, but should contain 200 μM of GGG-X peptide and 0.05 mM Calcium.
5. Allow 4 h for elution/sortase reaction at room temperature before collecting three consecutive elution fractions by readding 350 μL PBS into the column. Keep the column vertical and do not rotate; collecting the 350 μL elution will be difficult if it spreads over the walls of the column.
6. Run elution fractions on a 4–12% Bis-Tris protein gel to select those containing protein.
7. Buffer exchange into PBS to get rid of excess GGG-X peptides and calcium (*see* **Note 11**).
8. Quantify final protein concentration using Pierce's BCA kit, or any other method of choice, and calculate molarity.

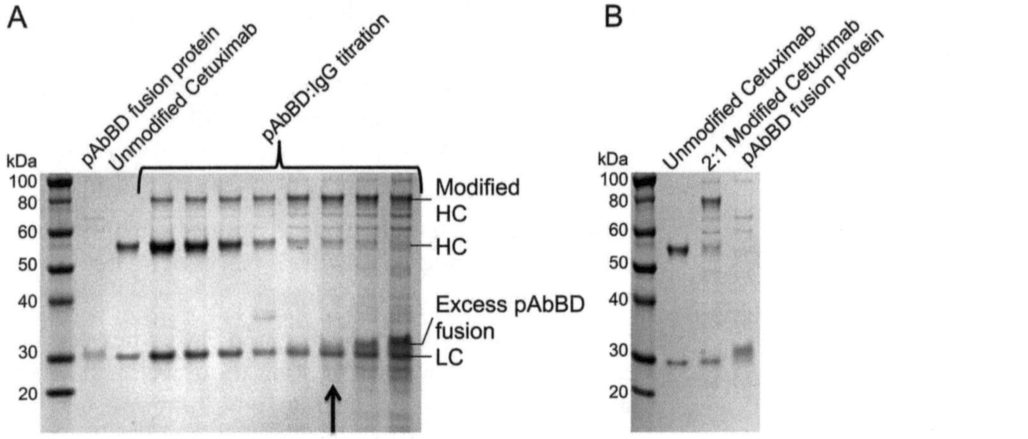

Fig. 3 pAbBD to IgG molar ratio titration to maximize cross-linking extent. (**a**) IgG heavy and light chains are marked as HC and LC, respectively. pAbBD to IgG molar ratios used in this titration are as follows (lanes 3–10): 1:4, 1:3, 1:2, 1:1, 2:1, 2.5:1, 4.5:1, and 9:1. 20 μL reactions were exposed to UV for 2 h. (**b**) Four parallel 200 μL cross-linking reactions were exposed to UV for 2 h using 2.5 pAbBD moles per mole of IgG (the condition indicated by the arrow in panel **a**). The combined 800 μL of modified product was centrifuged 5 min using 3000 × *g* and a 100 kDa molecular weight cutoff spin filter tube

3.5 Small-Scale Cross-Linking Titration

1. Set up various 20 μL small-scale cross-linkings in clear PCR tubes using increasing molar ratios of the pAbBD construct to IgG. The molar ratio of the pAbBD construct to IgG should range between 1:1 and 20:1. A minimum final concentration of 0.1 μM IgG is recommended to easily visualize the heavy chain on a protein gel.
2. Allow 2 h of 365 nm light exposure. The UV cross-linker should be set up at 4 °C and the sample should be placed in a secondary container filled with water (*see* **Note 12**).
3. Run a protein gel with all ratios and an unmodified IgG control side-by-side (*see* **Notes 13–15**). *See* Fig. 3a for an example titration gel.

3.6 Scaling Up Cross-Linking

1. Use findings from Subheading 3.5 to choose the molar ratio that maximizes cross-linking of the pAbBD construct to IgG heavy chains. To ensure maximum IgG modification, the ratio of the pAbBD construct to IgG usually exceeds 2:1 slightly.
2. Preferably, parallel 200 μL reactions should be set up in translucent PCR tubes and 2 h should be allowed under the UV source. The final IgG concentration should range between 0.1 and 1 μM. If larger volumes are required, translucent 1.5 or 2 mL Eppendorf tubes can be used. In this case, the reaction volume should not exceed 80% of the tube volume and 4 h of UV exposure are suggested to maximize modification extent (*see* **Note 16**).

3. Run final sample on a 4–12% Bis-Tris protein gel to confirm complete cross-linking.
4. Remove excess pAbBD from the final cross-linked sample. Often, IgGs can be separated from excess pAbBD constructs using 50 or 100 kDa cutoff spin filter tubes (*see* **Note 17**). *See* Fig. 3b for an example final product.

4 Notes

1. Autoinduction media provides the advantage of transitioning from bacterial growth to protein purification without the need to monitor culture OD and manual IPTG induction. It also reduces the number of factors that need to be optimized for each protein since the induction OD is bypassed. However, pAbBD fusion proteins can be expressed using a standard IPTG protocol as well, and the yield is not significantly impacted.
2. This protocol uses bacterial expression, but it can also be transferred to yeast if needed. In this case, the pAbBD fusion would need to be cloned into a yeast expression plasmid, and this would be cotransformed with the yeast pEVOL equivalent [18].
3. Both pAbBD versions express with high yield and purity in T7 Express with the manufacturer's recommended protocol as described above. For certain constructs this cell line choice will not yield high amounts of protein and an alternative should be chosen to better suit the protein being expressed in fusion with pAbBD. For example, for pAbBD fused to a single-chain variable fragment we have found Origami 2(DE3) to yield maximum protein expression, in most cases.
4. Another alternative is to use GGG-cysteine for STEPL elution. Additionally, simply cloning a C-term cysteine at or near the N- or C-terminus is also an option. Since there are no cysteines in either of the pAbBD versions, these cysteines can be used for additional labeling.
5. Our recommended ultraviolet cross-linker (UVP CL-1000) operates at a maximum intensity of 999,900 μJ per cm^2. With these settings and the previously described volumes, the distance between UV source and sample does not impact cross-linking extent significantly.
6. The linker type and length can be further optimized for each pAbBD-protein fusion, as well as the N versus C-terminal orientation. If preferred, pAbBD can be fused C-terminal.
7. Expression conditions as described above work as go-to settings for a variety of proteins, but the optimized culture

temperature and final OD will vary depending on the protein fused to pAbBD. One of the main suggestions to improve protein yield is to slow down growth and induction by reducing culture temperature. For T7 Express, 25 °C and 18 °C are reasonable alternatives.

8. Alternatively, lysis can be completed at 4 °C for 2 h. This reduces the background proteins in final product postpurification, and can be a key adjustment for difficult-to-express proteins.
9. We present a complementary strategy that combines labeling and purification into one step (STEPL), but if no chemical modification is desired, this technique is not necessary. In the cases where only a biological modification of the IgG is intended, a common 6His tag or any other method for purification suffices.
10. The specified ratio of affinity resin to lysate is a starting point suggestion, but depending on the yield of each pAbBD fusion it should be adjusted to maximize capture and purity.
11. Depending on the protein of interest, buffer exchanging into different solvents might be advantageous in terms of stability and storage. Purified pAbBDs (not in fusion) can be stored at 4 °C in PBS for at least 2 weeks without compromising cross-linking capability. Possible methods for this step include spin filters, dialysis or desalting columns. While spin filters allow fastest buffer exchange, it is sometimes best to avoid this method since some proteins can get adsorbed onto the filter membrane. To develop a protein-specific protocol, small samples can be tested with spin filters made of different materials (cellulose, hydrophilic polyethersulfone, etc.) to determine which minimizes sample loss during centrifugation. Dialysis is the slowest of the three alternatives, but it provides a gentler gradient to buffer exchange, which can help with protein aggregation. To dialyze small sample volumes with minimum loss, cassettes or tubes can be used (such as Sigma Slide-A-Lyzer and Pur-A-Lyser kits). If protein aggregation occurs, these buffer exchange methods can be completed at 4 °C as well as in a 10% glycerol solution.
12. If the UV cross-linker cannot be set up at 4 °C, the samples can instead be placed on ice.
13. To obtain full modification with a 2:1 ratio a slight excess above 2:1 is typically needed. It is worth mentioning that a 1:1 ratio can also be achieved by modifying the molar ratio used for cross-linking and capturing IgGs that have one unmodified heavy chain on a common Protein A or G resin. The small portion of IgGs that are modified with two pAbBD fusions will

not be captured and can be collected in the flow-through post binding.

14. In many cases, a positive control cross-linking experiment with an antibody that has previously been validated for efficient cross-linking with pAbBD is helpful [15]. Cetuximab and Rituximab are two such examples.
15. For small pAbBD fusion proteins, such as pAbBD-biotin for example, it will be hard to see size shift of the heavy chain on a protein gel by Coomassie staining. In these cases, tagging the pAbBD fusion with fluorophore is suggested.
16. The cross-linking upper limit of IgG concentration is defined by the intensity of UV the source used to cross-link. The higher the final IgG concentration and the larger the reaction volume, the longer cross-linking time will be needed to maximize conjugation extent.

    Although not recommended, if further scale-up is needed, translucent 15 or 50 mL conical tubes can be used, and additional time beyond 4 h should be allowed for sufficient UV sample penetration.
17. The optimal purification method could vary for each IgG. While with some pAbBD-IgG conjugates the amount of product lost to spin filtering is minimal, other conjugates might require alternative protein separation strategies. Possible options include size exclusion chromatography or dialysis. One key suggestion is to choose the most exact molar ratio of pAbBD to IgG that maximizes the desired 2:1 cross-linking and minimizes the amount of excess uncrosslinked pAbBD construct.

References

1. Rodgers KR, Chou RC (2016) Therapeutic monoclonal antibodies and derivatives: historical perspectives and future directions. Biotechnol Adv 34:1149–1158
2. Chan CEZ, Lim APC, MacAry PA, Hanson BJ (2014) The role of phage display in therapeutic antibody discovery. Int Immunol 26:649–657
3. Diamantis N, Banerji U (2016) Antibody-drug conjugates—an emerging class of cancer treatment. Br J Cancer 114:362
4. Byrne H, Conroy PJ, Whisstock JC, O'Kennedy RJ (2013) A tale of two specificities: bispecific antibodies for therapeutic and diagnostic applications. Trends Biotechnol 31:621–632
5. McCombs JR, Owen SC (2015) Antibody drug conjugates: design and selection of linker, payload and conjugation chemistry. AAPS J 17:339–351
6. Goodall S, Jones ML, Mahler S (2015) Monoclonal antibody-targeted polymeric nanoparticles for cancer therapy—future prospects. J Chem Technol Biotechnol 90:1169–1176
7. Sochaj AM, Świderska KW, Otlewski J (2015) Current methods for the synthesis of homogeneous antibody–drug conjugates. Biotechnol Adv 33:775–784
8. Agarwal P, Bertozzi CR (2015) Site-specific antibody–drug conjugates: the nexus of bioorthogonal chemistry, protein engineering, and drug development. Bioconjug Chem 26:176–192
9. Junutula JR, Raab H, Clark S et al (2008) Site-specific conjugation of a cytotoxic drug to an

antibody improves the therapeutic index. Nat Biotechnol 26:925–932

10. Shen B-Q, Xu K, Liu L et al (2012) Conjugation site modulates the in vivo stability and therapeutic activity of antibody-drug conjugates. Nat Biotechnol 30:184–189

11. Beck A, Goetsch L, Dumontet C, Corvaïa N (2017) Strategies and challenges for the next generation of antibody–drug conjugates. Nat Rev Drug Discov 16:315–337

12. Lyon RP, Bovee TD, Doronina SO et al (2015) Reducing hydrophobicity of homogeneous antibody-drug conjugates improves pharmacokinetics and therapeutic index. Nat Biotechnol 33:733–735

13. Beck A, Wurch T, Bailly C, Corvaia N (2010) Strategies and challenges for the next generation of therapeutic antibodies. Nat Rev Immunol 10:345–352

14. Spiess C, Zhai Q, Carter PJ (2015) Alternative molecular formats and therapeutic applications for bispecific antibodies. Mol Immunol 67:95–106

15. Hui JZ, Tamsen S, Song Y, Tsourkas A (2015) LASIC: light activated site-specific conjugation of native IgGs. Bioconjug Chem 26:1456–1460

16. Warden-Rothman R, Caturegli I, Popik V, Tsourkas A (2013) Sortase-tag expressed protein ligation: combining protein purification and site-specific bioconjugation into a single step. Anal Chem 85:11090–11097

17. Young TS, Ahmad I, Yin JA, Schultz PG (2010) An enhanced system for unnatural amino acid mutagenesis in E. coli. J Mol Biol 395:361–374

18. Mohibullah N, Hahn S (2008) Site-specific cross-linking of TBP in vivo and in vitro reveals a direct functional interaction with the SAGA subunit Spt3. Genes Dev 22:2994–3006

Chapter 19

SpyTag–SpyCatcher Chemistry for Protein Bioconjugation In Vitro and Protein Topology Engineering In Vivo

Xiao-Wei Wang and Wen-Bin Zhang

Abstract

The emergence of "molecular superglue," such as SpyTag–SpyCatcher chemistry, has tremendously expanded our capability in manipulating protein shape and architecture via conjugation. Telechelic proteins bearing the SpyTag and SpyCatcher reactive sequences can be expressed and purified for bioconjugation in vitro, giving protein conjugates, branched proteins, and circular proteins. By encoding both reactive sequences in the same construct for expression in vivo, the nascent protein undergoes programmed posttranslational modification guided by protein folding and reaction, leading to diverse nonlinear topologies in situ. In this chapter, we present the SpyTag–SpyCatcher chemistry as a versatile platform for protein bioconjugation and topology engineering.

Key words SpyTag, SpyCatcher, Protein bioconjugation, Topology Engineering

1 Introduction

Protein ligation techniques are important for assembling protein building blocks into sophisticated structures with nontrivial topologies and unique functions. Traditionally, they are mostly based on (a) the intrinsic reactivity of functional groups in protein/peptides such as NHS amidation and native chemical ligation [1], (b) introducing noncanonical amino acids with spring-loaded reactivity [2], (c) native splicing protein segments such as inteins, and (d) enzyme-mediated covalent bond formation such as sortase and butelase. The first strategy is also the simplest, but the specificity of reaction is limited and the number of reaction sites is difficult to control. For native chemical ligation, terminal cysteine and thioester moiety are needed and thus it is more often used in ligating synthesized peptides. Introducing noncanonical amino acids can bring bioorthogonal "click" functionalities, such as alkynes and azides. However, it is nontrivial to do so. It often leads to much lower yield and sometimes disrupts native structure and function. The intein technology is very powerful and can be genetically

Sam Massa and Nick Devoogdt (eds.), *Bioconjugation: Methods and Protocols*, Methods in Molecular Biology, vol. 2033, https://doi.org/10.1007/978-1-4939-9654-4_19,

encoded. However, the chemistry relies on a folded structure and is highly directional. It can hardly be used to generate branched architectures. Enzyme-mediated bioconjugation techniques are much more common. Subtiligase is among the first enzymes to be used in peptide ligation [3]. But, it requires the substrates to possess *C*-terminal glycolate phenylalanylamide for efficient ligation, which is difficult for native proteins. Sortase A is a more powerful transpeptidase in this regard. It was discovered and optimized to mediate ligation of both peptides and proteins based on native peptide sequences [4, 5]. The enzyme recognizes a specific amino acid sequence LPXTG (X can be any amino acid), cleaves the amide bond between T and G, and further fuses LPXT-pendent segment with the other segment containing *N*-terminal oligoglycines. Various cases, such as linearly conjugated proteins [6], cyclized peptides/proteins [7–11] and hybrid conjugates [12–15] have been reported. Recently, one novel asparaginyl endopeptidase, butelase 1, was extracted and isolated from the medicinal plant *Clitoria ternatea* and found to possess extremely high ligation efficiency and fast coupling kinetics [16]. This Asx-specific ligase facilitates the rapid cyclization of peptides/proteins or ligation of two peptide segments with short recognition sequences [16–21]. However, this powerful enzyme still could not be expressed heterologously in *E. coli* or in yeast, which severely limits their usage. In addition, most enzymes mentioned above have limited activity for ligation in cellular environments. To overcome the limits, genetically encodable reaction pairs are highly desired. Protein *trans*-splicing (PTS) approaches based on split inteins have been widely applied for this purpose [22, 23]. Nonetheless, the protein partners could only support the head-to-tail cyclization and/or linear ligation [24, 25]. It calls for more versatile and powerful protein ligation tools that are based entirely on natural amino acids and that work independent of their locations in the protein sequence.

Nature is always a source of inspiration. To enhance the mechanical properties of the proteins, bacteria can accomplish amazing chemistry for interunit and/or intraunit crosslinking guided by folding and autocatalysis. Howarth et al. selected one fibronectin binding domain CnaB2 from the fibronectin binding protein FbaB from *Streptococcus pyogenes* and split it into two parts, the peptide segment (i.e., SpyTag) and the protein partner (i.e., SpyCatcher), containing reactive aspartic acid and lysine, respectively [26]. Reconstitution in situ brings two residues in close proximity to facilitate the autocatalytic formation of an isopeptide bond (Fig. 1) [27]. SpyTag–SpyCatcher chemistry has proven to be highly efficient, very fast, independent of its location on the protein chain and robust under various environments [26]. Moreover, it is completely genetically encoded and well expressed in a variety of host cells [28]. Its remarkable features are reminiscent of the

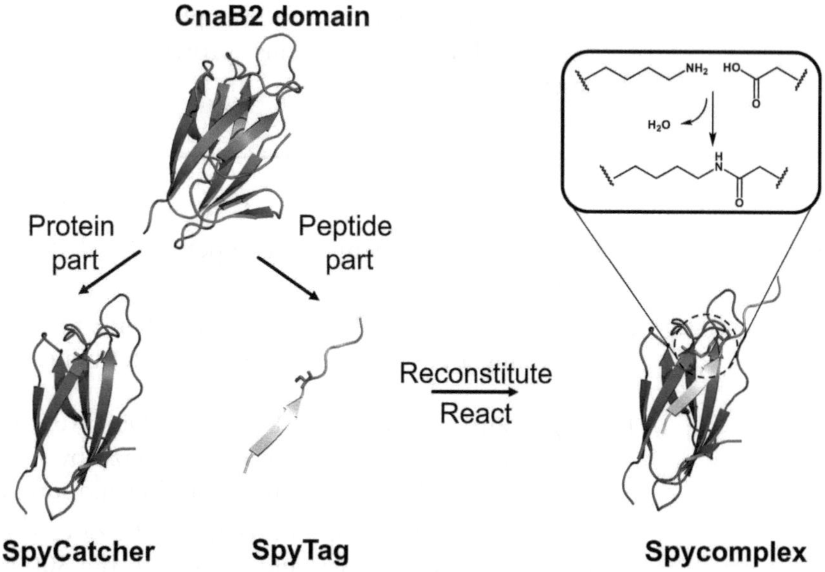

Fig. 1 Structures of CnaB2 domain and the SpyTag–SpyCatcher complex. Both SpyTag and SpyCatcher are split from CnaB2 domain in *Streptococcus pyogenes* (PDB entry: 2×5P). They reconstitute in situ and react to form the SpyTag–SpyCatcher complex (PDB entry: 4MLI) with an isopeptide chemical bonding

"click" chemistry in materials science [29, 30]. The potential benefits of this technology have further promoted the development of orthogonal pairs such as the SnoopCatcher–SnoopTag reaction [31] and a faster version of the SpyCatcher–SpyTag reaction [32]. We envision that there will be an evolution toward a family of genetically encoded peptide–protein reactive pairs with diverse features. Nowadays, this tool has already gained in popularity in protein topology engineering [33], for making protein-based biomaterials [34, 35] and in synthetic biology [36]. Specifically, as a powerful tool for bioconjugation, the SpyTag–SpyCatcher pair has successfully been used to couple proteins in vitro [37–40], to cyclize structural proteins in vivo [41–43] and to facilitate the cellular synthesis of various topological proteins [33, 44, 45].

In this chapter, we describe the general procedure of protein ligation in vitro and topology engineering in vivo by SpyTag–SpyCatcher chemistry, as it is applied on elastin-like polypeptides (ELPs). To simplify molecular design, ELPs featuring repeated sequences and disordered structures were chosen as a model biopolymer [46]. It shall help provide insight on the topological effects in the self-assembly and material properties of proteins [37]. For in vitro coupling, telechelic proteins containing either SpyTag (s) or SpyCatcher(s) are each individually expressed and purified, and then directly mixed in equimolar stoichiometry. For in vivo topology engineering, fusion protein containing both SpyTag and SpyCatcher can be expressed in cells with programmed posttranslational modification in situ as guided by protein folding (e.g., the intertwining dimerization of p53dim) to give topological proteins including cyclic proteins, protein tadpoles, and protein catenanes.

2 Materials

2.1 Strains and Vectors

1. Bacterial strains: Chemically competent *E. coli* BL21(DE3) (*see* **Note 1**).
2. Vectors: pQE80L (Qiagen) with resistance to ampicillin and containing a His-tag at the protein's *N*-terminus (*see* **Note 2**).

2.2 Antibiotics and Selective Solid Media

1. 100 mg/mL ampicillin stock solution: Dissolve 5 g of ampicillin in 50 mL of ultrapure water. Sterilize by filtration with 0.22 μm filter. Aliquot the stock solution and store at −20 °C.
2. LB agar-plates: Weigh 4 g of tryptone, 2 g of yeast extract, 4 g of NaCl, and 6 g of agar and transfer to the 500 mL reagent bottle. Add deionized water to 400 mL and mix thoroughly. Sterilize by autoclaving at 121 °C for 20 min. Cool to about 50 °C, and sterilely add 400 μL ampicillin stock solution (Final concentration is 100 μg/mL). Transfer the medium to plates. Once the agar turns to a solid gel, transfer the plates to 4 °C and use them within 1 month.

2.3 Protein Expression

1. 2×YT Medium: Weigh 16 g of tryptone, 10 g of yeast extract, and 5 g of NaCl, transfer to a 3 L baffled flask. Add deionized water to 1 L. Shake until all solutes have dissolved. Sterilize by autoclaving at 121 °C for 20 min (*see* **Note 3**).
2. 1 M IPTG stock solution: Dissolve 9.5 g of isopropyl-*β*-D-thiogalactopyranoside in 40 mL ultrapure water. Sterilize by filtration with a 0.22 μm filter. Aliquot the solution and store at −20 °C.

2.4 Protein Purification

1. Native lysis buffer: 50 mM NaH_2PO_4, 300 mM NaCl, 10 mM imidazole, pH 8.0. Dissolve 6.9 g of $NaH_2PO_4{\cdot}H_2O$, 17.5 g of NaCl, and 0.68 g of imidazole in 900 mL of ultrapure water. Shake until all solutes have dissolved. Adjust pH to 8.0 with 5 M NaOH. Adjust the volume of solution to 1 L with ultrapure water. Sterilize by filtration with 0.22 μm filter. Store at 4 °C (*see* **Note 4**).
2. Native wash buffer: 50 mM NaH_2PO_4, 300 mM NaCl, 25 mM imidazole, pH 8.0. Dissolve 6.9 g of $NaH_2PO_4{\cdot}H_2O$, 17.5 g of NaCl, and 1.36 g of imidazole in 900 mL of ultrapure water. Shake until all solutes have dissolved. Adjust pH to 8.0 with 5 M NaOH. Adjust the volume of solution to 1 L with ultrapure water. Sterilize by filtration with 0.22 μm filter. Store at 4 °C.

3. Native elution buffer: 50 mM NaH_2PO_4, 300 mM NaCl, 250 mM imidazole, pH 8.0. Dissolve 6.9 g of $NaH_2PO_4 \cdot H_2O$, 17.5 g of NaCl, and 17.0 g of imidazole in 900 mL of ultrapure water. Shake until all solutes have dissolved. Adjust pH to 8.0 with 5 M NaOH. Adjust the volume of solution to 1 L with ultrapure water. Sterilize by filtration with 0.22 μm filter. Store at 4 °C.
4. Denature lysis buffer: 100 mM NaH_2PO_4, 10 mM Tris–HCl, 8 M urea, pH 8.0. Dissolve 13.8 g of $NaH_2PO_4 \cdot H_2O$, 1.2 g of Tris, and 480.5 g of urea in 900 mL of ultrapure water. Shake until all solutes have dissolved. Adjust pH to 8.0 with 6 M HCl. Adjust the volume of solution to 1 L with ultrapure water. Sterilize by filtration with 0.22 μm filter. Store at room temperature. Use in 1 week.
5. Denature wash buffer: 100 mM NaH_2PO_4, 10 mM Tris–HCl, 8 M urea, pH 6.3. Dissolve 13.8 g of $NaH_2PO_4 \cdot H_2O$, 1.2 g of Tris, and 480.5 g of urea in 900 mL of ultrapure water. Shake until all solutes have dissolved. Adjust pH to 6.3 with 6 M HCl. Adjust the volume of solution to 1 L with ultrapure water. Sterilize by filtration with 0.22 μm filter. Store at room temperature. Use in 1 week.
6. Denature elution buffer: 100 mM NaH_2PO_4, 10 mM Tris–HCl, 8 M urea, 10 mM imidazole, pH 4.5. Dissolve 13.8 g of $NaH_2PO_4 \cdot H_2O$, 1.2 g of Tris, and 480.5 g of urea in 900 mL of ultrapure water. Shake until all solutes have dissolved. Adjust pH to 4.5 with 6 M HCl. Adjust the volume of solution to 1 L with ultrapure water. Sterilize by filtration with 0.22 μm filter. Store at room temperature. Use in 1 week.
7. Denature buffer A: 20 mM NaH_2PO_4, 500 mM NaCl, 25 mM imidazole, 8 M urea, pH 8.0. Dissolve 2.4 g of anhydrous NaH_2PO_4, 29 g of NaCl, 1.7 g of imidazole and 480.5 g of urea in 900 mL ultrapure water. Shake until all solutes have dissolved. Adjust pH to 8.0 with 5 M NaOH. Adjust the volume of solution to 1 L with ultrapure water. Sterilize by filtration with 0.22 μm filter. Use in 1 week.
8. Denature buffer B: 20 mM NaH_2PO_4, 500 mM NaCl, 250 mM imidazole, 8 M urea, pH 8.0. Dissolve 2.4 g of anhydrous NaH_2PO_4, 29 g of NaCl, 17 g of imidazole, and 480.5 g of urea in 900 mL of ultrapure water. Shake until all solutes have dissolved. Adjust pH to 8.0 with 5 M NaOH. Adjust the volume of solution to 1 L with ultrapure water. Sterilize by filtration with 0.22 μm filter. Use in 1 week.
9. 10× PBS stock solution: 1.37 M NaCl, 27 mM KCl, 100 mM Na_2HPO_4, 20 mM KH_2PO_4, pH 7.5. Dissolve 32 g NaCl, 0.8 g of KCl, 5.8 g of Na_2HPO_4, 1.0 g of KH_2PO_4 in 350 mL of ultrapure water. Adjust pH to 7.5 with 6 M HCl. Adjust the

volume of solution to 400 mL with ultrapure water. Sterilize by filtration with 0.22 μm filter. Store at room temperature.

10. 1× PBS: 137 mM NaCl, 2.7 mM KCl, 10 mM Na_2HPO_4, 2 mM KH_2PO_4, pH 7.5. Dilute 100 mL of 10× PBS stock solution to 1 L with ultrapure water.
11. Ni Sepharose™ 6 Fast Flow resin (GE Healthcare Life Sciences).
12. Econo-Pac® Chromatography Columns (Bio-Rad).
13. Superdex 200 Increase 10/300GL (GE Healthcare Life Sciences) (*see* **Note 5**).
14. ÄKTA pure (GE Healthcare Life Sciences).

3 Methods

3.1 In Vitro Ligation

3.1.1 Transformation and Protein Expression

1. Transform the chemical competent *E. coli* BL21 (DE3) with pQE80L plasmids containing the genes coding for constructs EA, EB, EAE, EBE, AA, and BB vectors (*see* Fig. 2 and **Note 6**).
2. Plate the cultures on LB agar plates containing 100 μg/mL ampicillin. Incubate plates at 37 °C for 12–16 h.
3. Select a single colony and inoculate it into 5 mL of 2×YT medium containing 100 μg/mL ampicillin. Incubate seed cultures overnight at 37 °C in a shaker at 250 rpm.

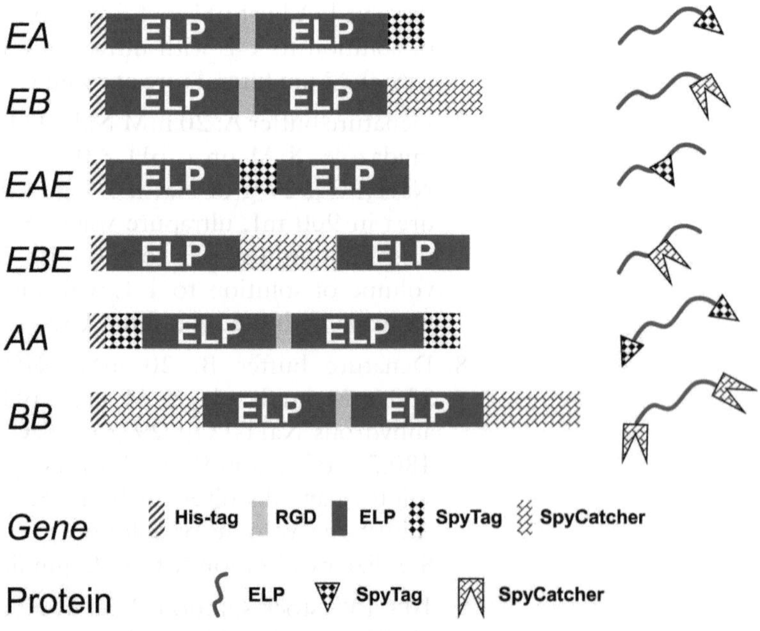

Fig. 2 Schematic representation of fusion proteins for in vitro coupling. Genes contain sequences of His-tag, elastin-like peptides (ELP), RGD-peptide, SpyTag, and SpyCatcher

4. Transfer seed cultures to 1 L of fresh 2×YT containing 100 μg/mL ampicillin. Incubate at 37 °C in a shaker at 250 rpm until OD_{600} reaches 0.5–0.7.
5. Add 1 mL of 1 M of IPTG (1 mM final) to induce target protein expression. Incubate 8 h at 37 °C/220 rpm (*see* **Note** 7).
6. Collect cells via centrifugation (20 min, 5000 × *g*, 4 °C). Discard supernatant. Resuspend the pellet in either native or denature lysis buffer (1 g of cells per 5–8 mL of lysis buffer), depending on the nature of the construct (*see* **Note 8**). At this stage, dry or resuspended cell pellets can be stored at −80 °C for later purification if needed.
7. Disrupt cells in ice-water bath using ultrasonication (amplitude 40%, 5 s on and 5 s off, total 20 min).
8. Clarify the mixture via centrifugation (30 min × 16,000 × *g* at 4 °C). Transfer the supernatant to a clean tube and keep the sample on ice till purification.

3.1.2 Purification of Reactive Proteins by Immobilized Metal Ion Affinity Chromatography (IMAC)

1. Mix supernatant with Ni Sepharose™ 6 Fast Flow resin (1 L of medium usually uses 5 mL of resin). Agitate at 4 °C for 1 h (*see* **Note 9**).
2. Load the lysate–resin mixture into an empty Econo-Pac® Chromatography Column with the bottom cap attached. Collect the flow-through.
3. Wash the resin with 6 column volumes (CV) of corresponding (native or denature) wash buffer. Collect the flow-through.
4. Elute proteins with 3 CV of corresponding (native or denature) elution buffer. Collect the flow-through.
5. Analyze all eluates by SDS-PAGE.
6. Dialyze target fractions against deionized water for 4 days. Change water every day.
7. Filter the solution with a 0.22 μm filter. Lyophilize the filtered supernatant for 3 days (*see* **Note 10**).
8. Store lyophilized powder at −80 °C for further usage.

3.1.3 In Vitro Coupling

1. Measure the mass of lyophilized protein powder and dissolve samples in 1× PBS (20 μM final).
2. Mix solutions containing SpyTag-encoding components A (*see* Fig. 2) and solutions containing SpyCatcher-encoding components B (*see* Fig. 2) with same stoichiometric ratios (*see* Fig. 3 and **Note 11**).
3. Incubate at room temperature for at least 2 h (or overnight at 4 °C). The mixture can be directly used for further characterization or be stored at −20 °C.

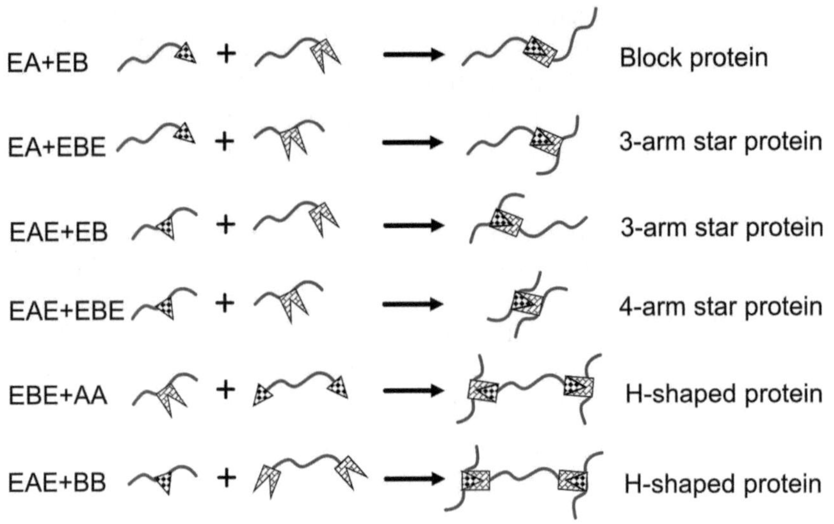

Fig. 3 Schematic representation of in vitro coupling products

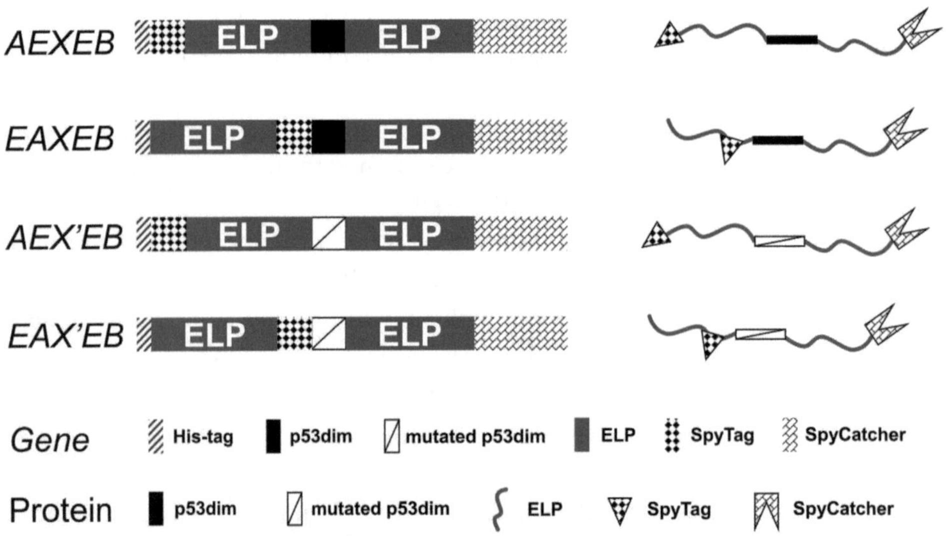

Fig. 4 Schematic representation of fusion proteins for in vivo coupling. Genes contain His-tag, SpyTag, ELP, p53dim (or mutated p53dim), and SpyCatcher

3.2 In Vivo Coupling

3.2.1 Transformation and Biosynthesis of Topological Proteins

1. Transform the chemically competent *E. coli* BL21 (DE3) with pQE80L plasmids encoding for constructs AEXEB, EAXEB, AEX'EB, and EAX'EB (*see* Fig. 4 and **Note 12**).
2. Plate the cultures on LB agar containing 100 μg/mL ampicillin. Incubate plates at 37 °C for 12–16 h.
3. Select a single colony and inoculate it into 5 mL of 2×YT medium containing 100 μg/mL ampicillin. Incubate seed cultures overnight at 37 °C in a shaker at 250 rpm.

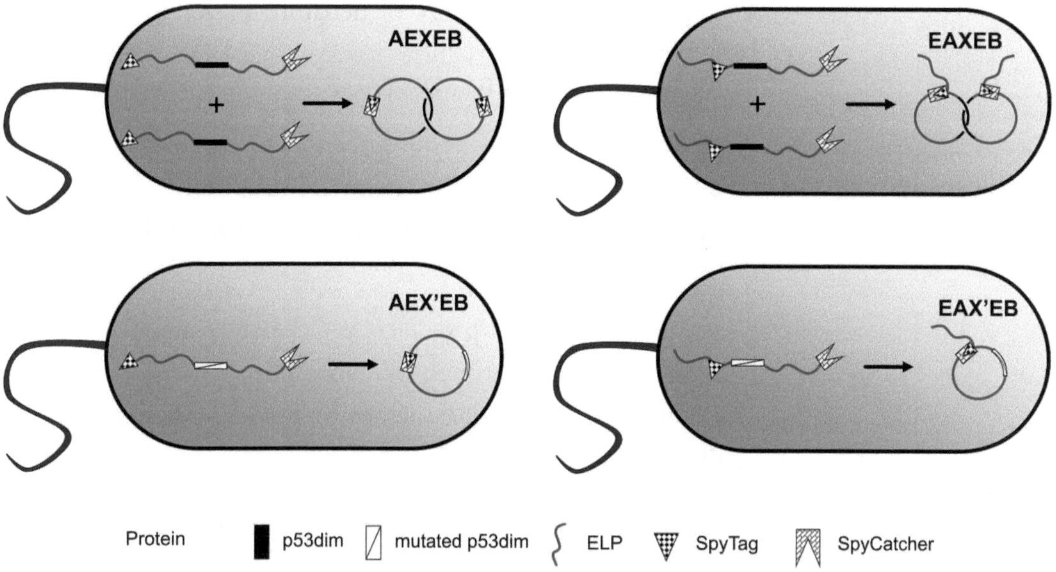

Fig. 5 Schematic representation of in vivo coupling and assembly, leading to diverse topological proteins

4. Transfer seed cultures to 1 L of fresh 2×YT containing 100 μg/mL ampicillin. Incubate at 37 °C in a shaker at 250 rpm until OD_{600} reaches 0.6–0.8.
5. Add 1 mL of 1 M of IPTG (1 mM final) to induce target protein expression (*see* Fig. 5). Incubate overnight at 16 °C/220 rpm (*see* **Note 13**).
6. Collect cells via centrifugation (20 min × 5000 × *g* at 4 °C). Discard supernatant. Resuspend the pellet in denature buffer A (1 g of cells per 5–8 mL of lysis buffer) (*see* **Note 14**). At this stage, dry or resuspended cell pellets can be stored at −80 °C for later purification.
7. Disrupt cells in an ice-water bath using ultrasonication (amplitude 40%, 5 s on and 5 s off, total 20 min).
8. Clarify the mixture via centrifugation (30 min × 16,000 × *g* at 4 °C). Transfer the supernatant to a clean tube and keep the sample on ice till purification.

3.2.2 Purification of Topological Proteins by Immobilized Metal Ion Affinity Chromatography (IMAC)

1. Mix supernatant with Ni Sepharose™ 6 Fast Flow resin (1 L of medium usually uses 5 mL of resins). Agitate at 4 °C for 1 h.
2. Load the lysate–resin mixture into an empty Econo-Pac® Chromatography Column with the bottom cap attached. Collect the flow-through.
3. Wash the resin with 6 column volumes (CV) of denature buffer A. Collect the flow-through.

4. Elute proteins with 3 CV of denature buffer B. Collect the flow-through.
5. Analyze all samples by SDS-PAGE.

3.2.3 Further Purification by Size Exclusion Chromatography (SEC)

1. Equilibrate a Superdex 200 increase 10/300GL column with 2 CV of 1× PBS at a flow rate of 0.5 mL/min.
2. Load 1 mL sample (total protein concentration should not be higher than 10 mg/mL) on the column at a flow rate of 0.5 mL/min.
3. Elute column with 1.25 CV of 1× PBS. Collect the eluted peaks in the flow-through (*see* **Note 15**).
4. Combine fractions and analyze by SDS-PAGE to determine purity.
5. Determine protein concentration by the molar extinction coefficient or using dye-based protein assays such as Bradford or BCA assays.
6. Concentrate target product using Amicon® Ultra-15 centrifugal filter if concentration is lower than 0.2 mg/mL.
7. Aliquot samples and store at −80 °C for further characterization.

4 Notes

1. BL21 (DE3) competent cells are commonly used for recombinant protein expression. Other strains, such as the Origami B series for proteins containing disulfide bonds, or the Rosetta series for proteins coming from eukaryotic cells can also be used for certain cases.
2. pQE80L (Qiagen) is used to carry genes of target proteins. The vector contains a T5 promoter transcription-translation system and a 6× His-tag, which are helpful for high-level expression and purification. Other vectors with different resistance genes and fusion tags can also be used, depending on target protein of interest.
3. 2×YT medium has simple formulation and can support the growth of cells, resulting in acceptable yields of target proteins (typically ~40 mg/L of culture). Other medium can also be used if desired.
4. Purification under native condition is necessary to maintain the folding and biofunctions of proteins. However, for intrinsically disordered proteins or polypeptides with few structural features, purification under denaturing conditions works as well.

5. Superdex 200 increase 10/300GL has high resolution and can separate proteins with M_r ~10 kDa to 600 kDa. For smaller proteins or larger-scale purification, other columns such as Superdex 75 10/300GL, HiLoad Superdex series, or HiPrep Sephacryl series can also be used depending on the size of the target protein.
6. Fusion proteins for in vitro coupling are cloned in the pQE80L plasmids and have sequences represented in Fig. 2. Segments of elastin-like polypeptides (ELP), SpyTag, and SpyCatcher are represented as E, A, and B. For the design of the construct and the cloning, it is noted that when SpyTag is positioned right next to the initiation codon at the *N*-terminus, poor induction in bacteria may occur with certain codon usage owing to secondary structure formation with vector-derived sequences in the mRNA. To address this, it is recommended that the codon of gcacacatagtaatggtagacgcctacaagccgacgaag to be used in the T7 vector and the codon of gctcatatcgtcatggttgacgcgtataaaccgaccaaa to be used in the T5 vector for SpyTag (AHIVMVDAYKPTK) [47]. It is advised that the construct should be checked using an online Ribosome Binding Site calculator to ensure satisfactory binding [48, 49].
7. Expression conditions are dependent on the target proteins and shall preserve the solubility and reactivity of the reactive components. Mild expression conditions (i.e., lower temperatures) are recommended for fusion proteins containing folded structural domains. In most cases, fusion proteins containing SpyCatcher are found to have enhanced solubility.
8. Proteins like EA, EAE, and AA only contain disordered ELP and peptidyl SpyTag, which allows purification under denaturing condition. EB, EBE, and BB can only be purified under native condition to maintain the reactivity of SpyCatcher.
9. Mixing and agitating lysates with resin usually captures more His-tagged proteins. Dynamic capture (directly loading the lysate into a column with equilibrated resins) can also be used and is more time-saving.
10. Lyophilization is not required for all proteins. In cases that lyophilization would impair protein activity and make them insoluble, one can perform buffer exchange to adjust the concentration of the telechelic protein and store them under proper conditions.
11. SpyTag and SpyCatcher tend to react quantitatively. Final coupling products are shown in Fig. 3.
12. Fusion proteins for in vivo coupling are the following constructs (Fig. 4): Segments of elastin-like polypeptides, SpyTag, SpyCatcher, p53dim, and mutated p53dim are represented as E, A, B, X, and X′.

13. Low temperature of expression will decrease intermolecular reactions. The scheme for coupling and assembly in a cellular environment is shown in Fig. 5.
14. Purification under denaturing condition is supposed to quench the reaction to prevent possible future coupling after cell lysis. Purification under native conditions can also be performed.
15. After affinity chromatography, it is recommended to further purify the samples by SEC to separate the desired topological protein from other side products such as chain-extension products. The procedure also allows the removal of denaturants and facilitates the refolding of the target protein. The SpyTag--SpyCatcher complex is well folded after the SEC purification. In most cases, its presence is beneficial for preventing the aggregation and promoting the efficient refolding of the target proteins upon the removal of denaturants.

Acknowledgments

We thank Prof. Mark Howarth at Oxford University for helpful discussions and suggestions. We are grateful for the financial support from the National Natural Science Foundation of China (Grants 21474003, 91427304), 863 Program (2015AA020941), and "1000 Plan (Youth)."

References

1. Dawson PE, Muir TW, Clark-Lewis I et al (1994) Synthesis of proteins by native chemical ligation. Science 266:776–779
2. Beatty KE, Tirrell DA (2009) Noncanonical amino acids in protein science and engineering. In: Köhrer C, RajBhandary UL (eds) Protein engineering. Springer, Berlin/Heidelberg, pp 127–153
3. Jackson DY, Burnier JP, Wells JA (1995) Enzymatic cyclization of linear peptide esters using subtiligase. J Am Chem Soc 117:819–820
4. Mazmanian SK, Liu G, Ton-That H et al (1999) *Staphylococcus aureus* Sortase, an enzyme that anchors surface proteins to the cell wall. Science 285:760–763
5. Chen I, Dorr BM, Liu DR (2011) A general strategy for the evolution of bond-forming enzymes using yeast display. Proc Natl Acad Sci U S A 108:11399–11404
6. Mao H, Hart SA, Schink A et al (2004) Sortase-mediated protein ligtion: a new method for protein engineering. J Am Chem Soc 126:2670–2671
7. Wu Z, Guo X, Guo Z (2011) Sortase A-catalyzed peptide cyclization for the synthesis of macrocyclic peptides and glycopeptides. Chem Commun 47:9218–9220
8. Stanger K, Maurer T, Kaluarachchi H et al (2014) Backbone cyclization of a recombinant cystine-knot peptide by engineered Sortase A. FEBS Lett 588:4487–4496
9. Antos JM, Popp MW, Ernst R et al (2009) A straight path to circular proteins. J Biol Chem 284:16028–16036
10. Popp MW, Dougan SK, Chuang TY et al (2011) Sortase-catalyzed transformations that improve the properties of cytokines. Proc Natl Acad Sci U S A 108:3169–3174
11. Rasche N, Tonillo J, Rieker M et al (2016) Prolink-single step circularization and purification procedure for the generation of an improved variant of human growth hormone. Bioconjug Chem 27:1341–1347
12. Samantaray S, Marathe U, Dasgupta S et al (2008) Peptide-sugar ligation catalyzed by transpeptidase sortase: a facile approach to

neoglycoconjugate synthesis. J Am Chem Soc 130:2132–2133

13. Hou Y, Yuan J, Zhou Y et al (2016) A concise approach to site-specific topological protein-poly(amino acid) conjugates enabled by in situ-generated functionalities. J Am Chem Soc 138:10995–11000
14. Pritz S, Wolf Y, Kraetke O et al (2007) Synthesis of biologically active peptide nucleic acid-peptide conjugates by sortase-mediated ligation. J Org Chem 72:3909–3912
15. Guo X, Wang Q, Swarts BM et al (2009) Sortase-catalyzed peptide-glycosylphosphatidylinositol analogue ligation. J Am Chem Soc 131:9878–9879
16. Nguyen GK, Wang S, Qiu Y et al (2014) Butelase 1 is an Asx-specific ligase enabling peptide macrocyclization and synthesis. Nat Chem Biol 10:732–738
17. Nguyen GK, Kam A, Loo S et al (2015) Butelase 1: a versatile ligase for peptide and protein macrocyclization. J Am Chem Soc 137:15398–15401
18. Cao Y, Nguyen GK, Tam JP et al (2015) Butelase-mediated synthesis of protein thioesters and its application for tandem chemoenzymatic ligation. Chem Commun 51:17289–17292
19. Nguyen GKT, Cao Y, Wang W et al (2015) Site-specific N-terminal labeling of peptides and proteins using butelase 1 and thiodepsipeptide. Angew Chem Int Ed 54:15694–15698
20. Nguyen GK, Hemu X, Quek JP et al (2016) Butelase-mediated macrocyclization of d-amino-acid-containing peptides. Angew Chem Int Ed 55:12802–12806
21. Hemu X, Qiu Y, Nguyen GK et al (2016) Total synthesis of circular bacteriocins by butelase 1. J Am Chem Soc 138:6968–6971
22. Muralidharan V, Muir TW (2006) Protein ligation: an enabling technology for biophysical analysis of proteins. Nat Methods 3:429–438
23. Elleuche S, Poggeler S (2010) Inteins, valuable genetic elements in molecular biology and biotechnology. Appl Microbiol Biotechnol 87:479–489
24. Iwai H, Lingel A, Pluckthun A (2001) Cyclic green fluorescent protein produced *in vivo* using an artificially split PI-*PfuI* intein from *Pyrococcus furiosus.* J Biol Chem 276:16548–16554
25. Iwai H, Zuger S, Jin J et al (2006) Highly efficient protein trans-splicing by a naturally split DnaE intein from nostoc punctiforme. FEBS Lett 580:1853–1858
26. Zakeri B, Fierer JO, Celik E et al (2012) Peptide tag forming a rapid covalent bond to a protein, through engineering a bacterial adhesin. Proc Natl Acad Sci U S A 109:E690–E697
27. Li L, Fierer JO, Rapoport TA et al (2014) Structural analysis and optimization of the covalent association between SpyCatcher and a peptide tag. J Mol Biol 426:309–317
28. Veggiani G, Zakeri B, Howarth M (2014) Superglue from bacteria: Unbreakable bridges for protein nanotechnology. Trends Biotechnol 32:506–512
29. Reddington SC, Howarth M (2015) Secrets of a covalent interaction for biomaterials and biotechnology: SpyTag and SpyCatcher. Curr Opin Chem Biol 29:94–99
30. Sun F, Zhang WB (2017) Unleashing chemical power from protein sequence space toward genetically encoded click chemistry. Chin Chem Lett 28:2078–2084
31. Veggiani G, Nakamura T, Brenner MD et al (2016) Programmable polyproteams built using twin peptide superglues. Proc Natl Acad Sci U S A 113:1202–1207
32. Keeble AH, Banerjee A, Ferla MP et al (2017) Evolving accelerated amidation by SpyTag/SpyCatcher to analyze membrane dynamics. Angew Chem Int Ed 56:16521–16525
33. Liu D, Wu WH, Liu YJ et al (2017) Topology engineering of proteins *in vivo* using genetically encoded, mechanically interlocking SpyX modules for enhanced stability. ACS Cent Sci 3:473–481
34. Sun F, Zhang WB, Mahdavi A et al (2014) Synthesis of bioactive protein hydrogels by genetically encoded SpyTag-SpyCatcher chemistry. Proc Natl Acad Sci U S A 111:11269–11274
35. Gao X, Fang J, Xue B et al (2016) Engineering protein hydrogels using SpyCatcher-SpyTag chemistry. Biomacromolecules 17:2812–2819
36. Banerjee A, Howarth M (2017) Nanoteamwork: covalent protein assembly beyond duets towards protein ensembles and orchestras. Curr Opin Biotechnol 51:16–23
37. Zhang WB, Sun F, Tirrell DA et al (2013) Controlling macromolecular topology with genetically encoded SpyTag-SpyCatcher chemistry. J Am Chem Soc 135:13988–13997
38. Gilbert C, Howarth M, Harwood CR et al (2017) Extracellular self-assembly of functional and tunable protein conjugates from bacillus subtilis. ACS Synth Biol 6:957–967
39. Stein V, Nabi M, Alexandrov K (2017) Ultrasensitive scaffold-dependent protease sensors with large dynamic range. ACS Synth Biol 6:1337–1342

40. Brune KD, Buldun CM, Li Y et al (2017) Dual plug-and-display synthetic assembly using orthogonal reactive proteins for twin antigen immunization. Bioconjug Chem 28:1544–1551
41. Schoene C, Bennett SP, Howarth M (2016) SpyRing interrogation: analyzing how enzyme resilience can be achieved with phytase and distinct cyclization chemistries. Sci Rep 6:21151
42. Schoene C, Fierer JO, Bennett SP et al (2014) SpyTag/SpyCatcher cyclization confers resilience to boiling on a mesophilic enzyme. Angew Chem Int Ed 53:6101–6104
43. Si M, Xu Q, Jiang L et al (2016) SpyTag/SpyCatcher cyclization enhances the thermostability of firefly luciferase. PLoS One 11: e0162318
44. Wang X-W, Zhang WB (2016) Cellular synthesis of protein catenanes. Angew Chem Int Ed 55:3442–3446
45. Wang X-W, Zhang WB (2017) Protein catenation enhances both stability and activity of folded structural domains. Angew Chem Int Ed 56:13985–13989
46. MacEwan SR, Chilkoti A (2010) Elastin-like polypeptides: biomedical applications of tunable biopolymers. Biopolymers 94:60–77
47. Howarth M. http://www.bioch.ox.ac.uk/howarth/reagents.htm. Accessed 20 Jan 2018
48. Salis HM. https://salislab.net/software/. Accessed 20 Jan 2018
49. Salis HM (2011) The ribosome binding site calculator. Methods Enzymol 498:19–42

Chapter 20

FN3 Protein Conjugates for Cancer Diagnosis and Imaging Studies

Arutselvan Natarajan and Lotfi Abou-Elkacem

Abstract

Bioconjugation of biologically useful proteins is in great demand (e.g., conjugation to biotins, metal chelators, and drug carriers to target specific tissues for both in vitro and in vivo use). These conjugates provide widespread opportunities for various biological and biomedical applications. Evolving state-of-the-art protein conjugation strategies have led to the development of many affinity ligands, including for cancer imaging and diagnosis. However, to achieve the desirable protein conjugates, there are many challenges that remain to be addressed in order to obtain a reproducible procedure for all proteins and ligands. These include a control over the protein modification and the efficiency of the conjugation while retaining the original biological protein affinity postmodification. Here we present detailed conjugation methods for the human fibronectin tenth type III domain (FN3) protein scaffold for use in preclinical PET imaging. More specifically, this chapter provides detailed methods to produce a FN3 and a FN3-chelator-conjugate, its labeling with the radionuclide 64-Cu, and its use for noninvasive PET imaging in mice.

Key words FN3 protein, 64-Cu, DOTA, hCD20, PET/CT

1 Introduction

Several years ago, Paul Ehrlich proposed the existence of a biological molecule (an antibody) that conjugates with a toxin to achieve targeted selective delivery and killing of a disease-causing organism [1]. This was the first ever example of protein bio-conjugation to generate an antibody cytotoxic drug conjugate (ADC). Since then, many humanized monoclonal antibodies (mAb) and engineered protein binders bearing long circulating half-lives with minimal immunogenicity, have been developed for specifically targeted tumor-associated antigens. Thus, mAbs and protein ligands provide an ideal delivery platform for selective targeting of tumor cells for therapy and imaging [2, 3].

With the discovery and validation of new receptors and a deeper understanding of cancer biology and disease progression, there is a growing need for a robust platform to allow rapid production of

Sam Massa and Nick Devoogdt (eds.), *Bioconjugation: Methods and Protocols*, Methods in Molecular Biology, vol. 2033, https://doi.org/10.1007/978-1-4939-9654-4_20,

the next generation of diagnostic and therapeutic molecules [4]. Over the last decade, antibodies have been frequently used as the binding reagents of choice because of their great binding capacity and ability to be generated naturally by immunization and hybridoma technology [5]. Novel specific engineered antibodies capable of high affinity binding to a molecular target are in high demand and vital to immunotherapy and in vivo and in vitro diagnostics. However, the use of antibodies as binding ligands can be problematic because of potential immune reactions as well as the time-consuming and high cost of antibody production. Full immunoglobulins are large (~150 kDa), glycosylated molecules connected by hydrophobic interactions and disulfide bonds [6]. Its multidomain structural complexity impedes its production, which is generally conducted by expensive and time-consuming mammalian cell culture. To overcome these limitations, evolving the engineered protein scaffold technique paves the way for developing novel protein binders to a target of interest. Thus, one of the potential alternatives to antibodies is the engineered fibronectin scaffold (FN3) that can be produced more readily in recombinant bacterial culture with the advantage of generating small, high-affinity ligands in a cost-efficient way [7–9].

1.1 FN3 Proteins

The human FN3 is an attractive protein platform composed of homologous repeats of three prototypical types of domains known as types I, II, and III [10]. The human tenth type III (FN3) is a small protein (~10 kDa) domain that maintains its native folding without disulfide bonds [6]. This FN3 protein has already been utilized to target many different biomarkers (Table 1) with minimal differences in major framework residues, except for the three binding loop regions (Fig. 1). These loop regions are BC, DE, and FG, and are each composed of ~15 amino acids. These varying loop regions of FN3 provide an opportunity to select a unique binder against any target of interest from a vast library. A library of FN3 proteins was previously generated (with diversity of 2.5×10^8 transformants and 60% of them being full-length FN3) that had varying loop lengths compared to that of wild-type FN3, with an amino acid repertoire similar to that of antibody CDR-H3 [8]. For example, the recently developed FN3 binders $FN3_{CD20}$, $FN3_{EGFR}$, $FN3_{VEGFR2}$, and $FN3_{hPD\text{-}L1}$ have unique amino acid configurations at the loop regions [9, 14, 19, 22]. There are many advantages of using these small, stable domains that maintain binding capabilities similar to that of antibodies [23]. They have a small size and high thermostability and solubility; they lack post-translational modifications and disulfide bonds. Also, FN3-scaffolds can be recombinantly produced at high levels in bacteria and are easy to purify using simple column chromatography [24]. In the last decade, strong inducible expression systems have been developed such as the T7-promoter-based pET vectors

Table 1
Examples of targeted receptors and applications with FN3 protein conjugates

FN3 binder conjugates used in various biomedical applications			
Targeted molecules	**Scope/application**	**Year**	**Reference**
SH3 domain of human c-Src	Binder selection technology	2004	[11]
Lysozyme	Binder selection technology	2008	[12]
VEGFR2	Phase I clinical trial with a cancer therapeutic	2011	[13]
EGFR	PET imaging	2012	[14]
Human CD20	PET imaging	2013	[15]
Various human proteins	Binder selection technology	2015	[16]
Human EphA2	In vivo optical tumor imaging	2015	[17]
EGFR	pH responsive antigen binding	2015	[18]
VEGFR2	Ultrasound imaging of tumor vasculature	2016	[19]
VEGFR2	CAR-T and -NK cells for targeted cellular immune therapy	2018	[20]
Mesothelin	Binder selection technology	2018	[21]

providing high yields of stable and soluble FN3-based scaffolds. In the current protocol we have used a FN3 scaffold that targets human CD20 ($FN3_{CD20}$) as a model protein to demonstrate a conjugation strategy to develop a clinically translatable immuno-PET tracer.

1.2 Conjugation Chemistry

Protein modifications can be performed with a variety of chemistry approaches, depending on the nature of the application. In general, protein chemistry can be performed with three different chemical methods, which involve the amine (NH2–), carboxylic acid (COOH–), and thiol (SH–) groups. Most chemical reactions are achieved by using many different organic solvents, varying pH, and increased temperature. On the contrary, proteins are extremely sensitive to organic solvents and need to be processed at physiological pH and temperature. Further, the ratio of the functional molecules conjugated to the proteins should be carefully designed in such a way that the scaffold proteins are able to retain their original binding capacity and conformation.

1.2.1 Carbodiimide Coupling Reaction

Carbodiimide cross-linking reactions have been one of the more frequently applied bioconjugation techniques for coupling of functional proteins, peptides, nanoparticles, and medical devices to aid in the diagnosis of various diseases. In general, the carbodiimide

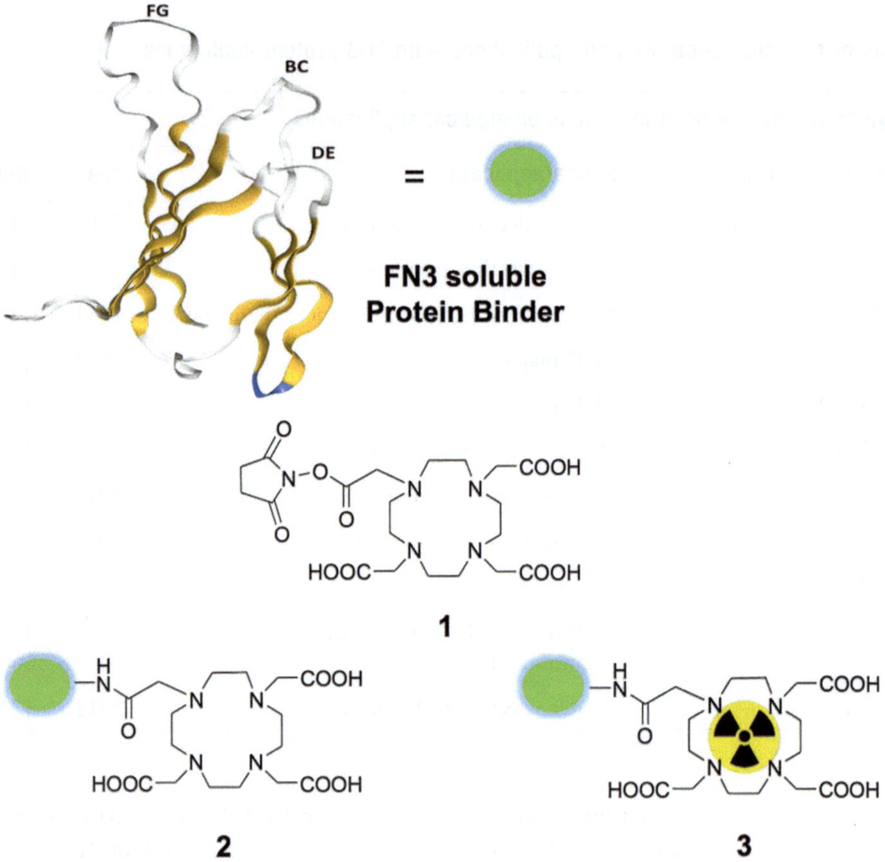

Fig. 1 Structure of the $FN3_{CD20}$ protein (1), its conjugation to a chelator (2), and labeling with a radionuclide (3) to generate an anti-hCD20 PET-tracer

containing molecules/linkers are converted to *N*-hydroxyl-succinimide (NHS) or sulfo-NHS esters. NHS-esters are more stable than carboxyl esters and improve the efficiency of the overall reaction. Although this chemistry is popular and widely practiced for many protein conjugations, the stability of the carbodiimide and NHS cross-linking agents in aqueous medium is poor.

1.2.2 Copper-Catalyzed Click Chemistry

Sharpless and colleagues introduced a new strategy in organic chemistry, denoted "click chemistry" [25]. In brief, "click chemistry" reactions are modular, wide in scope and high-yield, leaving no toxic by-products (even in the absence of chromatography), are stereospecific, simple to perform and require benign or easily removed solvent. Since their introduction, click chemistries have been adopted in many chemical conjugation fields, more recently with applications in the fields of bioconjugation, materials science, and drug discovery [25–27]. The copper (I)-catalyzed click reaction (CuAAC) results in a 1,4 regioisomer from the original alkynated-R2 and azido-R1 species. Several publications have shown yields for the CuAAC reaction to be above 90%, with reactions running to completion in a relatively short time period (12–24 h) [28, 29]. While Cu-catalyzed cycloadditions have proven to be a robust and versatile chemistry, the cytotoxicity of copper limits this chemistry application in living subjects [30]. Further, the use of copper is limited in conjugation reactions with metal chelators like DOTA, since copper can bind tightly the chelator and block subsequent radiolabeling.

1.2.3 Copper-Free Click Chemistry

A novel "clickable" cycloaddition conjugation with cyclooctynes has recently been reported [31] to overcome the problems related to copper toxicity, reducing agents, and accelerating ligands, in "click chemistry" reactions. These cyclooctynes have yielded even more accelerated reaction kinetics, even in mild reaction conditions [32]. Two of the most popularly employed cyclooctyne derivates are difluorocyclooctyne (DIFO) and dibenzocyclooctyne (DIBO). In spite of these advantages, still this cyclooctyne based clickable conjugation approach has some hindrances: (a) region specificity of the conjugated product, (b) the formation of a mixture of 1, 4-substituted conjugates; (c) the aqueous solubility of the cyclooctyne reagents [33].

1.3 DOTA Conjugation to FN3$_{CD20}$

Proteins, peptides, and biomedical compounds that are covalently conjugated to chelate complexes are known to bind radiometals (lanthanides as the metal core). These radiometal complexes are well established in clinical utilization as imaging agents. For example, the 1,4,7,10-tetraazacyclododecane-1,4,7,10-tetraacetic acid mono-*N*-hydroxysuccinimide ester (DOTA-NHS) is an important metal chelator used in radiolabeling of many proteins for biomarker imaging (64-Cu, 111-In, 68-Ga) and cancer radioimmunotherapy (90-Y, 177-Lu) [34]. DOTA-NHS is one of the most stable metal chelators and shows high water solubility. It is a nontoxic small molecule approved by the FDA [34, 35]. The DOTA-NHS can easily be conjugated to amine groups on many proteins, polymers, enzymes, and antibodies without any significant change in their biological activities [36]. The FN3$_{CD20}$ protein of interest needs to be conjugated with DOTA prior to radiolabeling with the radioisotope (64-Cu) for PET imaging [15]. This can be achieved by covalent chemistry using the DOTA-NHS ester (Fig. 1).

In this protocol, summarized in Fig. 1, we provide procedures for FN3$_{CD20}$ protein production from bacteria, conjugation of DOTA-NHS to FN3$_{CD20}$, radiolabeling of the DOTA-FN3$_{CD20}$ precursor with 64-Cu, and imaging of a human CD20 transgenic (hCD20T) mouse using a small animal PET/CT system.

2 Materials

2.1 Production and Purification of FN3 Protein

1. Expression of pET plasmid (Genscript) coding for the $FN3_{CD20}$ protein and a hexahistidine-tag at the C-terminus (*see* **Note 1**).
2. 1 L Lysogeny broth (LB) media (Sigma-Aldrich) for bacterial growth, with appropriate antibiotics.
3. LB agar (Thermo Fisher Scientific) with appropriate antibiotics.
4. BL21 (DE3) chemocompetent *E. coli* cells.
5. Super Optimal broth with Catabolite repression (SOC) medium: 2% Tryptone, 0.5% Yeast Extract, 10 mM NaCl, 2.5 mM KCl, 10 mM $MgCl_2$, 10 mM $MgSO_4$, 20 mM glucose (all from Sigma-Aldrich).
6. 0.5 M isopropyl β-D-thiogalactopyranoside (IPTG) in water (Sigma-Aldrich).
7. Lysis buffer: 50 mM sodium phosphate, pH 8.0, 5% glycerol, 500 mM sodium chloride, 5 mM 3-[(3-cholamidopropyl) dimethylammonio]-1-propanesulfonate (CHAPS), and 25 mM imidazole (Sigma-Aldrich). Prepare in sterile double distilled water. Freshly add 1× complete EDTA-free protease inhibitor cocktail (Roche, Indianapolis, IN) prior to use.
8. HisPur™ cobalt resin gravity-flow column (Thermo Fisher Scientific).
9. Immobilized metal affinity chromatography (IMAC) wash buffer: 20 mM Imidazole.
10. IMAC elution buffer: 200 mM Imidazole.
11. Acetonitrile (Sigma-Aldrich).
12. 0.1% Trifluoroacetic acid (Sigma-Aldrich).
13. Reversed-phase high-performance liquid chromatography (RP-HPLC) connected in line with ultraviolet/visible (UV) detector and radioactivity detector (HPLC-Ultimate 3000, Phenomenex).
14. C4 HPLC column (10 nm, 250 × 10 mm, Higgins Analytical Inc.).
15. Centrifuge (Sorvall RC 6+, Thermo Scientific).
16. Vortex Mixer (Thermomixer R, Eppendorf).

2.2 Synthesis of DOTA-FN3 (Fig. 1, Compound 2)

1. RP-HPLC C4 column (250 × 10 mm, Higgins Analytical Inc.).
2. *N*-succinimidyl-DOTA (NHS-DOTA, Macrocyclics, Cat # B280) (Fig. 1, compound **1**).

3. pH testing papers (pH paper 3–8, 0.5 pH resolution, Thermo Fisher Scientific).
4. Phosphate buffered saline (PBS; 0.1 M phosphate, 0.15 M sodium chloride; pH 7.2).
5. 1% v/v dimethylformamide (Sigma-Aldrich, Cat # 227056) in PBS.
6. 2% triethylamine (Sigma-Aldrich, Cat # 471283).
7. 1 M Tris–HCl pH 8.0
8. Sinapinic acid (Sigma-Aldrich, Cat # 85429).
9. 0.1 M ammonium acetate (pH 5.5) (*see* **Note 2**).
10. Zeba™ Spin Desalting Columns, (7 K MWCO, 0.5 mL; Thermo Scientific Cat # 89882).
11. Matrix-assisted laser desorption ionization mass spectrometry (MALDI) connected with a TOF/TOFTM 5800 detector (AB SCIEX, Concord, Canada).

2.3 Radiolabeling of DOTA-FN3$_{CD20}$ (Fig. 1, Compound 2)

1. 64-$CuCl_2$ in 0.1 N HCL with ~400 mCi/μmol specific activity (University of Wisconsin, Madison, WI),
2. 10 mg/mL DOTA-FN3 in 0.1 ammonium acetate buffer, from Subheading 2.2.
3. 0.1 M ammonium acetate (pH 5.5) (*see* **Note 2**).
4. pH testing papers (0.5 pH resolution, Thermo Fisher Scientific).
5. 0.1 M Ethylenediaminetetraacetic acid (EDTA; pH = 8).
6. Size exclusion column (SEC 2000, 300 × 7.8 mm with 5 μm hydrophilic bonded silica support and 400 Å pore size, Phenomenex, CA).
7. Nalgene Rapid-Flow Syringe filters (0.2 μm, Thermo Fisher Scientific).
8. Reversed-phase high-performance liquid chromatography (RP-HPLC) connected in line with ultraviolet/visible (UV) detector and radioactivity detector (HPLC-Ultimate 3000, Phenomenex).

2.4 PET/CT Imaging with the 64-Cu-DOTA-FN3$_{CD20}$ Tracer in hCD20 Transgenic Mice (Fig. 2)

1. Human CD20 transgenic (hCD20T) mice (Stanford animal facility, Genentech, South San Francisco, CA).
2. Isoflurane USP (Stanford Vet Pharmacy, CA).
3. BD U-100 Insulin Syringes with Ultra-Fine Needle (Thermo Fisher Scientific).
4. Small animal anesthesia delivery system (Patterson Scientific, WI).

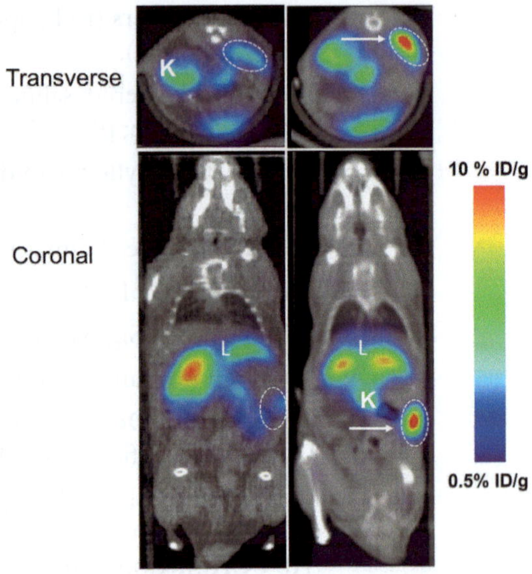

Fig. 2 PET/CT images of hCD20 transgenic mice injected with an anti-hCD20 PET tracer. Top and bottom rows show transverse and coronal views of the images, respectively. Mice were injected with the ^{64}Cu-DOTA-FN3$_{CD20}$ tracer (3.7 MBq). PET images were obtained 24 h after tracer injection. Preblocked mice (blocking group) received 50-fold mass excess of unmodified FN3$_{CD20}$ 2 h before tracer injection. The spleen (containing hCD20-expressing B cells that are targeted by the PET tracer) is indicated by the white arrow. The major clearance organs are marked in white fonts (*L* liver, *K* kidney). The scale bar shows tracer %ID/g

5. Sterile 64-Cu-DOTA-FN3$_{CD20}$ PET tracer prepared from Subheading 2.3.
6. PET/CT Inveon small animal multimodality system (Preclinical Solutions; Siemens Healthcare Molecular Imaging, Knoxville, TN).
7. Inveon Research Work-place (IRW) software (Preclinical Solutions; Siemens Healthcare Molecular Imaging, Knoxville, TN).

3 Methods

3.1 Production and Purification of FN3 Proteins

1. Thaw a tube of BL21 competent *E. coli* cells on ice for 10 min.
2. Add 100 ng of pET plasmid-FN3 to the cell mixture. Do not vortex. Incubate on ice for 5 min.
3. Heat-shock at exactly 42 °C for exactly 30 s. Place on ice for 5 min. Do not mix.
4. Pipet 950 μL of room temperature SOC into the mixture for outgrowth. Place at 37 °C for 1 h at 250 rpm.

5. Spread 100 μL of transformed *E. coli* onto a LB agar selection plate and incubate overnight at 37 °C.
6. Select a single bacterial colony and inoculate into 5 mL of LB media supplemented with antibiotics and culture overnight in a shaker at 250 rpm at 37 °C.
7. Transfer bacteria to 1 L of LB media and incubate at 37 °C with shaking until OD_{600} reaches 0.4–0.8.
8. Induce $FN3_{CD20}$ scaffolds expression by adding IPTG to a final concentration of 400 μM in 1 L LB media and express protein for 3–4 h at 37 °C (*see* **Note 3**).
9. After induction, harvest the cells by centrifugation at 10,000 × *g* for 10 min.
10. Resuspend the cell pellet of $FN3_{CD20}$ protein expressing bacteria in 3 mL of lysis buffer. Freeze overnight at −20 ° C.
11. Thaw the cell pellet and centrifuge at 12,000 × *g* for 10 min to remove the insoluble fraction. Collect supernatant.
12. Purify the $FN3_{CD20}$ protein from the supernatant by metal affinity chromatography using a HisPur™ cobalt resin gravity-flow column. Wash the column twice in IMAC wash buffer. Elute the $FN3_{CD20}$ protein with IMAC elution buffer, at room temperature.
13. Purify eluted $FN3_{CD20}$ protein by reversed-phase high-performance liquid chromatography on a C4 column using a 9–81% gradient of acetonitrile in water with 0.1% trifluoroacetic acid.
14. Lyophilize The HPLC fractions containing the $FN3_{CD20}$ protein. Resuspended in sterile DMSO or PBS (*see* **Note 4**).

3.2 DOTA-$FN3_{CD20}$ Precursor Synthesis

1. Dissolve the lyophilized $FN3_{CD20}$ protein in 1% dimethylformamide in metal free PBS (*see* **Note 5**).
2. Add a fivefold molar excess of NHS-DOTA to $FN3_{CD20}$ (*see* **Note 6**).
3. Add the appropriate volume of the 2% triethylamine to the $FN3_{CD20}$ solution to adjust the pH = 8 ± 0.5 (*see* **Note 7**).
4. Incubate this reaction mixture at room temperature for 60–90 min.
5. Quench the reaction by adding Tris–HCl (pH = 8) to a final concentration of 0.1 M.
6. Remove excess of NHS-DOTA by performing small-scale size-exclusion chromatography on a 7 kDa cutoff spin desalting column (*see* **Note 8**).
7. Determine the DOTA chelates per FN3 using MALDI-TOF mass spectroscopy operated in linear mode with sinapinic

acid as matrix. The number of DOTA chelates per FN3 should be <2.

8. The final DOTA-FN3$_{CD20}$ (Fig. 1, compound **2**) conjugate should be buffer exchanged with 0.1 M ammonium acetate (pH = 5.5). These DOTA-FN3$_{CD20}$ precursor conjugates should be placed in multiple vials of 1 mg/vial as a ready kit for radiolabeling.
9. Store these vials at −20 °C. These conjugates are stable for 12 at least months.

3.3 Preparation of 64-Cu-DOTA-FN3$_{CD20}$ Tracer

1. In a metal free microfuge mix 5 mCi of 64-CuCl$_2$ solution (pH = 5.5) with 50 μL of 0.1 ammonium acetate buffer, pH 5.5 and ~0.2–0.5 mg of DOTA-FN3$_{CD20}$ (Fig. 1, compound **2**) in 0.1 M ammonium acetate (pH 5.5).
2. Incubate the above reaction solution at 37 ° C for 1 h in a shaker.
3. After 1 h remove the solution and keep at room temperature for about 15 min.
4. Add 10–20 μL of 0.1 M EDTA to scavenge unchelated 64-CuCl$_2$ in the reaction vial and keep for 15–30 min at room temperature.
5. Purify the 64-Cu-DOTA-FN3$_{CD20}$ from the reaction mixture by again using the 7 kDa cutoff spin column (*see* Subheading 3.2, **step 6**) or using HPLC (SEC-2000 column) with a flow rate of 1.0 mL/min in PBS (pH 7.4).
6. Filter the final purified 64-Cu-DOTA-FN3 (Fig. 1, compound **3**) tracer through a 0.45 μm filter prior to injection in animals.

3.4 PET/CT Imaging Using hCD20 Mice with 64-Cu-DOTA-FN3$_{CD20}$ Tracer

1. Anesthetize hCD20T mice in an isoflurane chamber.
2. Intravenously inject into the "blocking mice" group a blocking dose (50-fold mass excess over tracer mass of 64-Cu-FN3$_{CD20}$) of unconjugated FN3$_{CD20}$ 2 h before radiotracer injection.
3. Administer into each mouse from two groups ($n = 3$; blocking and nonblocking) the 64-Cu-FN3 radiotracer (3.7 MBq in 200 μL of PBS) through tail vein injection.
4. Move the tracer-administered mice to the scanner 24 h post-tracer injection and perform PET/CT imaging during a 10 min scan.
5. Analyze the PET/CT scan data using Inveon Research Workplace (IRW) software.
6. The average radioactivity concentration was estimated for each organ from the small animal PET scan by drawing three-dimensional regions of interest (ROI).

7. ROI data was computed from mean pixel values within the ROI volume and converted to counts per milliliter per minute by using a predetermined conversion factor.
8. The percentage injected dose per gram of tissue (%ID/g) was computed from the injected dose and ROI of the image data.

4 Notes

1. The pET plasmid ligated with the gene encoding the binder of interest needs to be placed under an appropriate promotor and with a flanking amino-acid tag for proper detection and/or purification, such as a C-terminal 6xHis-tag.
2. 0.1 M ammonium acetate should be prepared in metal free water and adjusted to pH = 5.5.
3. Induction of protein can be varied with temperature and time. For example, 5 h at 30 °C or overnight at 16 °C or 23 °C.
4. Lyophilized powder of $FN3_{CD20}$ protein should be dissolved in DMSO when used in further conjugation reactions in non-aqueous medium. For cell binding and preclinical studies $FN3_{CD20}$ protein should be dissolved in PBS. In this formulation the content of DMSO should not be >10% v/v in PBS.
5. PBS used in the conjugation of DOTA-NHS and the protein of interest ($FN3_{CD20}$) should not contains amine compounds nor trace metals. It should be highly pure and free of metals to achieve high specific activity of the tracer.
6. The vial containing the NHS-DOTA-ester molecule should be opened right before the conjugation reaction as it is quickly hydrolyzes from the moisture in the room. Therefore, it should be aliquoted in small quantities (0.5–1 mg) in multiple vials and packed in an inert gas atmosphere for one-time use.
7. The pH of the conjugation reaction is critical. The pH should be 8 ± 0.5. This ensures optimal conjugation of chelate per protein molecule. DOTA chelates per $FN3_{CD20}$ molecule was optimized by controlling DOTA vs $FN3_{CD20}$ (1:5) ratio, temperature at 37 °C and reaction time (45–60 min) to achieve less than two DOTA chelates per $FN3_{CD20}$ molecule. This is an important step since more chelates per $FN3_{CD20}$ could interfere with the binding affinity, charge, and mass of the $FN3_{CD20}$ protein. Caution, some protein may be precipitated at this pH; careful observation is required.
8. Remove spin column's (Zeba™ Spin Desalting Columns) bottom closure, loosen the cap and place it in a 1.5–2.0 mL collection tube. Place column and tube in the microcentrifuge. Centrifuge at 1500 × *g* for 1 min and remove storage solution.

Repeat the procedure after column loading with 0.1 M ammonium acetate buffer (pH = 5.5) and discard the buffer. Add the DOTA-FN3$_{CD20}$ conjugate and centrifuge again to collect the flow-through.

Acknowledgments

We would like to acknowledge the SCi3 Core imaging facilities and the Canary Center at Stanford for providing technical support.

References

1. Schwartz RS (2004) Paul Ehrlich's magic bullets. N Engl J Med 350:1079–1080
2. Blaise L, Wehnert A, Steukers MP et al (2004) Construction and diversification of yeast cell surface displayed libraries by yeast mating: application to the affinity maturation of Fab antibody fragments. Gene 342:211–218
3. Weaver-Feldhaus JM, Lou J, Coleman JR et al (2004) Yeast mating for combinatorial Fab library generation and surface display. FEBS Lett 564:24–34
4. Gebauer M, Skerra A (2009) Engineered protein scaffolds as next-generation antibody therapeutics. Curr Opin Chem Biol 13:245–255
5. Tami JA, Parr MD, Brown SA et al (1986) Monoclonal antibody technology. Am J Hosp Pharm 43:2816–2825
6. Shukla AA, Thommes J (2010) Recent advances in large-scale production of monoclonal antibodies and related proteins. Trends Biotechnol 28:253–261
7. Koide A, Gilbreth RN, Esaki K et al (2007) High-affinity single-domain binding proteins with a binary-code interface. Proc Natl Acad Sci U S A 104:6632–6637
8. Hackel BJ, Wittrup KD (2010) The full amino acid repertoire is superior to serine/tyrosine for selection of high affinity immunoglobulin G binders from the fibronectin scaffold. Protein Eng Des Sel 23:211–219
9. Koide S, Koide A, Lipovsek D (2012) Target-binding proteins based on the 10th human fibronectin type III domain ((1)(0)Fn3). Methods Enzymol 503:135–156
10. Petersen TE, Thogersen HC, Skorstengaard K et al (1983) Partial primary structure of bovine plasma fibronectin: three types of internal homology. Proc Natl Acad Sci U S A 80:137–141
11. Karatan E, Merguerian M, Han Z et al (2004) Molecular recognition properties of FN3 monobodies that bind the Src SH3 domain. Chem Biol 11:835–844
12. Hackel BJ, Kapila A, Wittrup KD (2008) Picomolar affinity fibronectin domains engineered utilizing loop length diversity, recursive mutagenesis, and loop shuffling. J Mol Biol 381:1238–1252
13. Tolcher AW, Sweeney CJ, Papadopoulos K et al (2011) Phase I and pharmacokinetic study of CT-322 (BMS-844203), a targeted Adnectin inhibitor of VEGFR-2 based on a domain of human fibronectin. Clin Cancer Res 17:363–371
14. Hackel BJ, Kimura RH, Gambhir SS (2012) Use of (64)cu-labeled fibronectin domain with EGFR-overexpressing tumor xenograft: molecular imaging. Radiology 263:179–188
15. Natarajan A, Hackel BJ, Gambhir SS (2013) A novel engineered anti-CD20 tracer enables early time PET imaging in a humanized transgenic mouse model of B-cell non-Hodgkins lymphoma. Clin Cancer Res 19:6820–6829
16. Huang R, Gorman KT, Vinci CR et al (2015) Streamlining the pipeline for generation of recombinant affinity reagents by integrating the affinity maturation step. Int J Mol Sci 16:23587–23603
17. Park SH, Park S, Kim DY et al (2015) Isolation and characterization of a monobody with a fibronectin domain III scaffold that specifically binds EphA2. PLoS One 10:e0132976
18. Heinzelman P, Krais J, Ruben E et al (2015) Engineering pH responsive fibronectin domains for biomedical applications. J Biol Eng 9:6
19. Abou-Elkacem L, Wilson KE, Johnson SM et al (2016) Ultrasound molecular imaging of the breast cancer neovasculature using engineered fibronectin scaffold ligands: a novel class of targeted contrast ultrasound agent. Theranostics 6:1740–1752

20. Kulemzin SV, Gorchakov AA, Chikaev AN et al (2018) VEGFR2-specific FnCAR effectively redirects the cytotoxic activity of T cells and YT NK cells. Oncotarget 9:9021–9029
21. Sirois AR, Deny DA, Baierl SR et al (2018) Fn3 proteins engineered to recognize tumor biomarker mesothelin internalize upon binding. PLoS One 13:e0197029
22. Natarajan A, Patel CB, Ramakrishnan S et al (2018) A novel engineered small protein for positron emission tomography imaging of human programmed death ligand-1: validation in mouse models and human cancer tissues. Clin Cancer Res 25:1774. https://doi.org/10.1158/1078-0432.CCR-18-1871
23. Hackel BJ, Ackerman ME, Howland SW et al (2010) Stability and CDR composition biases enrich binder functionality landscapes. J Mol Biol 401:84–96
24. Wurch T, Pierre A, Depil S (2012) Novel protein scaffolds as emerging therapeutic proteins: from discovery to clinical proof-of-concept. Trends Biotechnol 30:575–582
25. Kolb HC, Sharpless KB (2003) The growing impact of click chemistry on drug discovery. Drug Discov Today 8:1128–1137
26. Bock VD, Perciaccante R, Jansen TP et al (2006) Click chemistry as a route to cyclic tetrapeptide analogues: synthesis of cyclo-[Pro-Val-psi(triazole)-Pro-Tyr]. Org Lett 8:919–922
27. Breinbauer R, Kohn M (2003) Azide-alkyne coupling: a powerful reaction for bioconjugate chemistry. Chembiochem 4:1147–1149
28. Hein JE, Fokin VV (2010) Copper-catalyzed azide-alkyne cycloaddition (CuAAC) and beyond: new reactivity of copper(I) acetylides. Chem Soc Rev 39:1302–1315
29. Zhu L, Brassard CJ, Zhang X et al (2016) On the mechanism of copper(I)-catalyzed azide-alkyne cycloaddition. Chem Rec 16:1501–1517
30. Gaetke LM, Chow CK (2003) Copper toxicity, oxidative stress, and antioxidant nutrients. Toxicology 189:147–163
31. Jewett JC, Bertozzi CR (2010) Cu-free click cycloaddition reactions in chemical biology. Chem Soc Rev 39:1272–1279
32. Agard NJ, Prescher JA, Bertozzi CR (2004) A strain-promoted [3 + 2] azide-alkyne cycloaddition for covalent modification of biomolecules in living systems. J Am Chem Soc 126:15046–15047
33. McKay CS, Finn MG (2014) Click chemistry in complex mixtures: bioorthogonal bioconjugation. Chem Biol 21:1075–1101
34. Dai L, Jones CM, Chan WTK et al (2018) Chiral DOTA chelators as an improved platform for biomedical imaging and therapy applications. Nat Commun 9:857
35. Tei L, Baranyai Z, Gaino L et al (2015) Thermodynamic stability, kinetic inertness and relaxometric properties of monoamide derivatives of lanthanide(III) DOTA complexes. Dalton Trans 44:5467–5478
36. Schwarz G, Mueller L, Beck S et al (2014) DOTA based metal labels for protein quantification: a review. J Anal At Spectrom 29:221–233

INDEX

Sam Massa and Nick Devoogdt (eds.), *Bioconjugation: Methods and Protocols*, Methods in Molecular Biology, vol. 2033, https://doi.org/10.1007/978-1-4939-9654-4,

MIX
Papier aus verantwortungsvollen Quellen
Paper from responsible sources
FSC® C105338

If you have any concerns about our products,
you can contact us on
ProductSafety@springernature.com

In case Publisher is established outside the EU,
the EU authorized representative is:
Springer Nature Customer Service Center GmbH
Europaplatz 3, 69115 Heidelberg, Germany

Printed by Libri Plureos GmbH
in Hamburg, Germany